RATCHABURI

Climatic comparison between the Ratchaburi trial site near Bangkok and 2795 locations in a half degree grid across Australia. Dark green shaded areas are most similar and red areas are least similar (see Chapter 4 for more details).

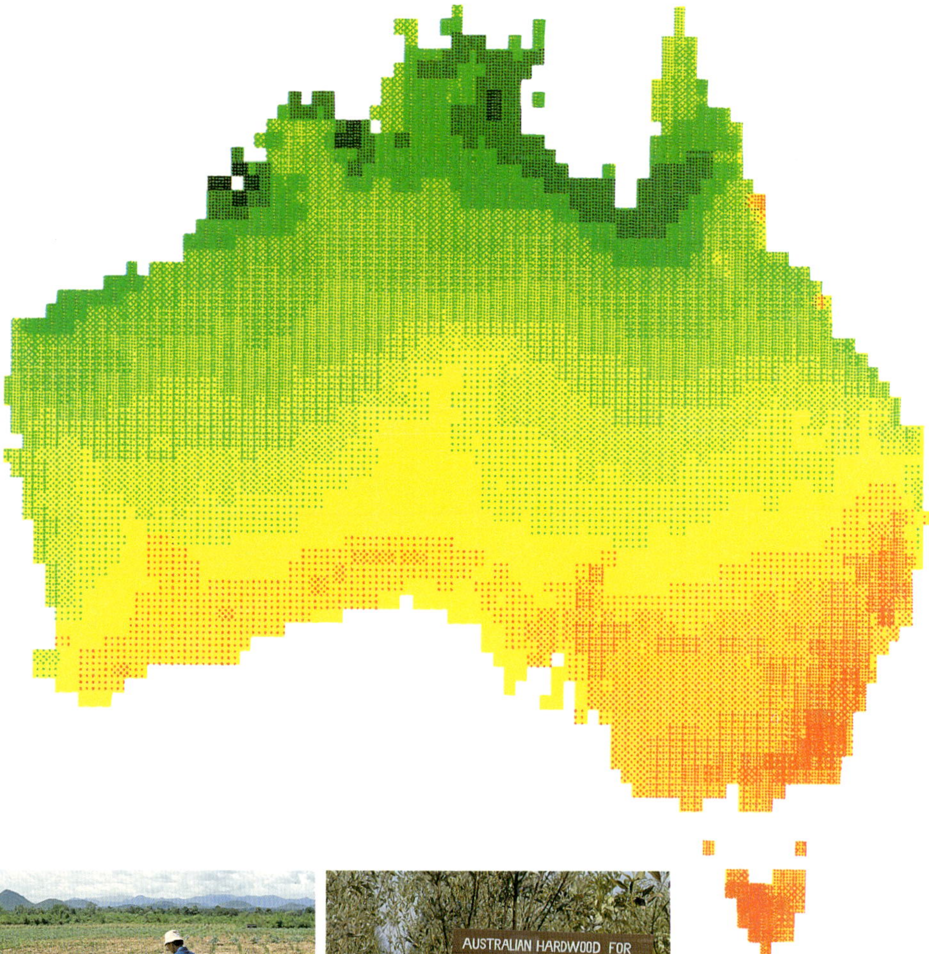

Development of the RFD/ACIAR species testing site at Ratchaburi, Thailand; the site (left) two months after planting in August 1985 (photo: G. Bowen), and (right) in April 1987 twenty-two months after planting.

TREES FOR THE TROPICS

Growing Australian Multipurpose Trees and Shrubs in Developing Countries

Editor: **D.J. Boland**

AUSTRALIAN CENTRE FOR INTERNATIONAL AGRICULTURAL RESEARCH
Canberra 1989

The Australian Centre for International Agricultural Research (ACIAR) was established in June 1982 by an Act of the Australian Parliament. Its mandate is to help identify agricultural problems in developing countries and to commission collaborative research between Australian and developing country researchers in fields where Australia has a special research competence.

ACIAR MONOGRAPH SERIES

This peer-reviewed series contains the results of original research supported by ACIAR, or material deemed relevant to ACIAR's research objectives. The series is distributed internationally, with an emphasis on the Third World.

Boland, D.J. 1989. Trees For The Tropics: growing Australian multipurpose trees and shrubs in developing countries. ACIAR Monograph No. 10, 247 p.

ISBN 0 949511 87 0

Computer typeset and laid out by Press Etching (Qld) Pty Ltd, Brisbane.
Printed by Watson Ferguson & Company, Brisbane.
Photos: D.J. Boland except where otherwise credited.
Cover design: John Best.
Technical Editing: Reginald MacIntyre.

Contents

Foreword, J.R. McWilliam 5
Acknowledgments 6
Editor's Preface 7
Contributors 9

Program Development 11

Chapter 1 Australian tree species for fuelwood and agroforestry in China, Kenya, Thailand and Zimbabwe
D.J. Boland and J.W. Turnbull 13

Chapter 2 *Acacia mearnsii*: its past and potential use with reference to the development of plantations in the People's Republic of China
W.E. Hillis 21

Chapter 3 Seed collections of lesser-known trees and shrubs in Queensland, Australia
S.D. Searle 27

Field Trials 35

Chapter 4 Climatic conditions at trial sites in China, Kenya, Thailand and Zimbabwe compared to similar regions in Australia
T.H. Booth 37

Chapter 5 Growth, coppicing and flowering of Australian tree species in trials in southeast Queensland, Australia
P.A. Ryan and R.E. Bell 49

Chapter 6 Temperate eucalypt trials in southwest People's Republic of China
Wang Huoran, Yan Hong and Zhang Rongqui 69

Chapter 7 Tropical eucalypt trials on Hainan Island, People's Republic of China
Zhou Wenlong and Bai Jiayu 79

Chapter 8 Tropical Australian acacia trials on Hainan Island, People's Republic of China
Yang Minquan, Bai Jiayu and Zeng Yutian 89

Chapter 9 *Acacia mearnsii* provenance trials in the People's Republic of China
Gao Chuanbi 97

Chapter 10 Growth and survival of Australian tree species in field trials in Kenya
P.B. Milimo 103

Chapter 11 Growth and survival of Australian tree species in field trials in Thailand
K. Pinyopusarerk **109**

Chapter 12 Growth and survival of Australian tree species in field trials in Zimbabwe
D.P. Gwaze **129**

Chapter 13 Response of Australian tree species to nitrogen and phosphorus in Thailand
R.N. Cromer **139**

Chapter 14 Statistical analysis of tree species trials and seedlot:site interaction in Thailand
E.R. Williams and V. Luangviriyasaeng **145**

Resource Evaluation **153**

Chapter 15 Vegetative propagation of *Casuarina* and *Acacia*: potential for success
L.D. Pryor **155**

Chapter 16 Fuelwood evaluation of four Australian-grown tree species
K.W. Groves and A.M. Chivuya **159**

Chapter 17 Fuelwood evaluation using a simple crib test
W.D. Gardner **171**

Chapter 18 Drying and burning properties of the wood of some Australian tree species
D.K. Gough, R.E. Bell, P.A. Ryan and C.T. Bragg **177**

Chapter 19 Fodder value of selected Australian tree and shrub species
T.K. Vercoe **187**

Chapter 20 Leaf essential oils of *Melaleuca* and *Leptospermum* species from tropical Australia
J.J. Brophy, D.J. Boland and E.V. Lassak **193**

Chapter 21 Leaf essential oil of *Eucalyptus bakeri*
J.J. Brophy and D.J. Boland **205**

Chapter 22 Managing nitrogen fixation in *Casuarina* species to increase productivity
P. Reddell, P.A. Rosbrook and P.A. Ryan **209**

Chapter 23 Susceptibility to termite attack of various tree species planted in Zimbabwe
M.R. Mitchell **215**

Future Perspectives **229**

Chapter 24 Realising the potential of Australia's lesser-known trees and shrubs: a summary and future perspectives
D.J. Boland **231**

Chapter 25 References **237**

Chapter 26 List of publications for ACIAR forestry projects **245**

Foreword

The Australian Centre for International Agricultural Research (ACIAR) was established to seek out agricultural and forestry problems in developing countries and then support collaborative research programs linking research institutions in Australia and overseas to help resolve these problems. ACIAR has a strong commitment to forestry research as the need for wood for fuel and shelter is basic to human needs. Trees also play an important role in beautifying, sustaining, and improving the environment.

The ACIAR forestry program is only a small part of ACIAR's activities, but is expanding. Central to all activities is the common theme that Australia contains a rich repository of unusual and little-known tree species of benefit to developing countries. Special attention has been focused on tropical trees suitable for growing on infertile soils that often suffer seasonal water stress. Most ACIAR forestry projects are aimed at exploiting this potential and assessing growth in trials under a range of climatic and soil conditions. The program has given special emphasis to nitrogen-fixing trees but is also examining ways of improving productivity economically through use of microorganisms such as bacteria (e.g. *Frankia*) or specialised mycorrhiza.

This book is a landmark in ACIAR's forestry program in that it consists of a series of papers summarising our attempts to exploit, evaluate and domesticate a wide range of lesser-known Australian tropical tree species. The book has been divided into four parts: Program Development, Field Trials, Resource Evaluation and Future Perspectives, reflecting very strongly the historical and philosophical development of the overall ACIAR forestry program. The monograph also reflects strongly the collaborative mode of ACIAR's research program in which overseas scientists have made a fine contribution, complementing the work of several major forest research centres in Australia. I would commend the book to all readers seeking to discover how Australia's lesser-known trees grow under cultivation, how we should evaluate them and to learn something of their potential. I believe, also, that the actual methods we have used to develop our own ACIAR forestry program should be of interest to other nations seeking to better understand the potential of their own lesser-known tree species. Collectively, such knowledge will benefit all people.

Australia still has much to learn about its own native forest resources and ways to best maximise the productivity of many lesser-known species. Despite this, Australian foresters have long had to contend with the vagaries of growing trees and forests in Australia on nutrient-poor soils in areas where droughts and bushfires are facts of life. Their skills and experience in coping with these difficulties in conditions similar to those experienced in developing countries make our foresters ideally suited for tackling similar problems overseas. This book also demonstrates their commitment and ability, and it is my belief that Australian foresters have much to offer and will in the future be increasingly sought internationally for their skills.

J.R. McWilliam
Director
ACIAR

Acknowledgments

The continued support of Dr J.W. Turnbull, ACIAR Forestry Program Coordinator, and Mr A.G. Brown, Deputy Chief, Division of Forestry and Forest Products CSIRO, is gratefully acknowledged. Mr Brown is, in addition, Australian project leader of ACIAR projects 8320, 8457 and 8458 as well as being leader of the 'Australian Tree Resources' program within the Division.

In preparing this publication numerous colleagues have aided in its preparation. I would like to pay special thanks to Mr Khongsak Pinyopusarerk for editorial assistance during the final stages of the work. In ACIAR I would like to thank Mr Reg MacIntyre for advice on editorial matters and for assistance in skillfully handling the manuscripts through to publication.

Each of the papers was reviewed by two or more scientists and to each of them I extend thanks. The reviewers were: Bryan Barlow, Alan Brown, Peter Burgess, Phil Cheney, Dick Date, John Doran, David Gardner, David Gough, Ken Groves, David Gwaze, Jamie Hartley, Chris Harwood, Ted Hillis, Jen McCombe, Nick Malajczuk, Nico Marcar, Colin Matheson, Dennis Minson, Mike Moncur, Jim Moriarty, Cliff Ohmart, Carolyn Raymond, Rod Roughley, Paul Ryan, Peter Snowdon, Ian Southwell, Hugh Stewart, Jens Svensson, Lex Thomson, John Turnbull, Tim Vercoe and Tony Watson.

The editor would especially like to thank Eva Morrow and Karin Munro (DFFP/CSIRO) who typed most of the manuscript of this book. Their cheerfulness and unflagging support, even when typing difficult drafts, was most commendable.

Editor's Preface

This book is an attempt to bring together a collection of forest research papers from five countries (Australia, China, Kenya, Thailand and Zimbabwe), highlighting the main achievements accomplished during the first 4 years of activities in ACIAR projects up until around the end of 1987. The projects are numbered 8320 (now 8808), 8331 (now 8809), 8457 and 8458. The monograph is also a concerted attempt to provide an historical record of the initiation and development of these projects.

The papers clearly demonstrate the wide range of activities in which ACIAR has been involved in forest research. The main thrust has been in seed collection of lesser-known species, species elimination and evaluation trials, climatic matching, fodder assessment, essential leaf oils, fuelwood studies, termite studies, and nutrition. In all studies the overall aim was to evaluate and assess Australian species for use in developing countries. There has been no diminution of our original belief that Australia contains a wealth of lesser-known trees and shrub species of value to humankind, and the results to date strengthen this supposition. Nevertheless we must continue our assessments for a few more years to accumulate the full benefits of our collaborative research efforts.

In some ways it may seem premature to release some of our results, especially preliminary results from field trials. Balancing this limitation is the recognition that an early publication draws together our collective ideas for sharing amongst ourselves, provides a focus for some of the better performing species and stimulates an extension of the research results. Some lines of research have been very useful. In particular, our research work on climatic matching, fuelwood, leaf oil and fodder studies and differential susceptibilities of tree species to termite damage deserves special mention and opens up new methods of research enquiry.

A special thank you is extended to the research leaders in China, Kenya, Thailand and Zimbabwe, firstly for believing in the aims of the project and secondly in helping to strengthen its development. Beyond all the materialistic accomplishments, the scientific development in research skills of individual scientists through collaboration and reciprocal visits has been a special feature of the program. In Australia, the Department of Forestry, Queensland, deserves special praise for developing the extensive field trials near Gympie which remain a cornerstone of our overall program. Mr P.A. Ryan of the Department was active in fostering collaboration with other research bodies and has spent considerable time in showing his trials to a wide range of local and overseas visitors.

The book has been divided into four parts roughly reflecting the phases of development of the work. The first part details the overall ACIAR forestry program development and indicates the reasons why China, Kenya, Thailand and Zimbabwe were initially chosen for field trial sites. A detailed account is given of the industrial history of a remarkable Australian species (*Acacia mearnsii*), at home and abroad, leading on to the development of an ACIAR program in the People's Republic of China to improve productivity and utility of the species. Details are also given of our early seed collection activities which, together with the activities of the Australian Tree Seed Centre, form the basis for the provision of seeds for trial. The second part documents the early results of ACIAR field trials overseas. The third part evaluates the resource in ways useful to people in developing countries such as fuelwood studies, fodder assessments, vegetative propagation and melaleuca leaf oils for potential development as a cottage industry. And finally, a paper is included to summarise the work to date and to discuss the future potential of the species and products covered in the book.

This book should be of value to government officials (mainly in forestry and agriculture) in other developing countries in helping to select new species for trials. The book will also be of use to other overseas aid agencies and research organisations. We hope that the articles presented will help stimulate other researchers to follow up some of our activities in more detail, or extend the results in other directions currently unforeseen by us. We also believe that the book will be of value to lecturers teaching forestry in tertiary institutions and to students.

Finally a list of published papers (or papers in an advanced state of publication) resulting from our work is included at the end of this book. This list demonstrates quite clearly the high level of activity that the four ACIAR projects have generated.

D.J. Boland

Contributors

Bai Jiayu	Research Institute of Tropical Forestry, Chinese Academy of Forestry, Longdong, Guangzhou, People's Republic of China, 510520.
Bell, R.E.	Queensland Department of Forestry, Forest Research Centre, MS483, Gympie, Qld, 4570, Australia.
Boland, D.J.	CSIRO Division of Forestry and Forest Products, PO Box 4008, Queen Victoria Terrace, Canberra, ACT, 2600, Australia.
Booth, T.H.	CSIRO Division of Forestry and Forest Products, PO Box 4008, Queen Victoria Terrace, Canberra, ACT, 2600, Australia.
Bragg, C.T.	Queensland Department of Forestry, Timber Research and Extension Branch, PO Box 631, Indooroopilly, Qld, 4068, Australia.
Brophy, J.J	Department of Organic Chemistry, University of New South Wales, PO Box 1, Kensington, NSW, 2033, Australia.
Chivuya, A.M.	Nyanga Estates, The Wattle Co. Ltd., P.B. J7107, Mutare, Zimbabwe.
Cromer, R.N.	CSIRO Division of Forestry and Forest Products, PO Box 4008, Queen Victoria Terrace, Canberra, ACT, 2600, Australia.
Gao Chuanbi	Research Institute of Subtropical Forestry, Chinese Academy of Forestry, Fuyang, Zheijang, People's Republic of China.
Gardner, W.D.	Wood Technology and Forest Research Division, Forestry Commission of New South Wales, PO Box 100, Beecroft, NSW, 2119, Australia.
Gough, D.K.	Queensland Department of Forestry, Timber Research and Extension Branch, PO Box 631, Indooroopilly, Qld, 4068, Australia.
Groves, K.W.	Department of Forestry, The Australian National University, PO Box 4, Canberra, ACT, 2600, Australia.
Gwaze, D.P.	Forest Research Centre, Zimbabwe Forestry Commission, PO Box HG595, Highlands, Harare, Zimbabwe.
Hillis, W.E.	CSIRO Division of Forestry and Forest Products, Private Bag 10, Clayton, Vic., 3168, Australia.
Lassak, E.V.	School of Technology, Kalgoorlie College, PMB22, Kalgoorlie, WA, 6430, Australia.
Luangviriyasaeng, V.	Silviculture Division, Royal Forest Department, Bangkok, 10900, Thailand.

Milimo, P.B. Kenya Forestry Research Institute, PO Box 20412, Nairobi, Kenya.

Mitchell, M.R. Forest Research Centre, Zimbabwe Forestry Commission, PO Box HG 595, Highlands, Harare, Zimbabwe.

Pinyopusarerk, K. CSIRO Division of Forestry and Forest Products, PO Box 4008, Queen Victoria Terrace, Canberra, ACT, 2600, Australia. Formerly Silviculture Division, Royal Forest Department, Bangkok 10900, Thailand.

Pryor, L.D. 69 Endeavour St., Red Hill, Canberra, ACT, 2603, Australia.

Reddell, P. CSIRO Division of Soils, Davies Laboratory, Private Mail Bag, PO Aitkenvale, Townsville, Qld, 4814, Australia.

Rosbrook, P.A. CSIRO Division of Soils, Davies Laboratory, Private Mail Bag, PO Aitkenvale, Townsville, Qld, 4814, Australia.

Ryan, P.A. Queensland Department of Forestry, Forest Research Centre, MS483, Gympie, Qld, 4570, Australia.

Searle, S.D. CSIRO Division of Forestry and Forest Products, PO Box 4008, Queen Victoria Terrace, Canberra, ACT, 2600, Australia.

Turnbull, J.W. Forestry Program Coordinator, ACIAR, GPO Box 1571, Canberra, ACT, 2601, Australia.

Vercoe, T.K. CSIRO Division of Forestry and Forest Products, PO Box 4008, Queen Victoria Terrace, Canberra, ACT, 2600, Australia.

Wang, Huoran Chinese Academy of Forestry, Wan Shou Shen, Beijing, 100091, People's Republic of China.

Williams, E.R. Biometrics Unit, Institute of Natural Resources and Environment, CSIRO, GPO Box 1666, Canberra, ACT, 2601, Australia.

Yan, Hong Chinese Academy of Forestry, Wan Shou Shan, Beijing, 100091, People's Republic of China.

Yang, Minquan Research Institute of Tropical Forestry, Chinese Academy of Forestry, Longdong, Guangzhou, People's Republic of China, 510520.

Zeng, Yutian Research Institute of Tropical Forestry, Chinese Academy of Forestry, Longdong, Guangzhou, People's Republic of China, 510520.

Zhang, Rongqui Forestry Research Institute, Kunming, Yunnan Province, People's Republic of China.

Zhou, Wenlong Research Institute of Tropical Forestry, Chinese Academy of Forestry, Longdong, Guangzhou, People's Republic of China, 510520.

Program Development

Grevillea robusta is perhaps Australia's most successful agroforestry tree species overseas. It is widely used in tree/crop mixtures in crops as diverse as tea, coffee, bananas and maize. Photograph above shows *Grevillea robusta* being used for high shade over coffee in the eastern highlands of Zimbabwe (1986). Photograph on right shows a pollarded tree of *Grevillea robusta* on a farm on the slopes of Mt Kenya, Kenya (1988).

Chapter 1

Australian Tree Species for Fuelwood and Agroforestry in China, Kenya, Thailand and Zimbabwe

D.J. Boland and J.W. Turnbull

Introduction

The Australian Centre for International Agricultural Research (ACIAR) was established in June 1982 with the specific aim of strengthening the agricultural and forestry research capacity of Australia's bilateral aid program. The Centre's brief is to mobilise Australia's research expertise to help solve problems limiting agricultural productivity. This is achieved by contracting scientific groups in Australia to set up collaborative research projects on problems of mutual interest with counterparts in other countries. ACIAR allocates funds to the Australian and developing country partners to complement the resources provided by the respective research institutions. The focus of the collaborative research program is Southeast Asia and the Pacific Islands but significant support is also provided for projects in South Asia, the People's Republic of China and Eastern Africa.

The world's natural forests and woodlands have been placed under heavy pressure by a rapidly increasing population requiring land for food crops and wood for domestic and industrial use. More than half the timber cut each year is used for fuelwood, and tree planting to meet fuelwood needs has emerged as a major development task facing many countries. The Food and Agriculture Organisation of the United Nations has estimated that the global demand for fuelwood will require the equivalent of 50 million ha of plantation before the year 2000 (Palmberg 1981). Such a task is beyond most governments in countries where an acute fuelwood deficit exists, and mobilisation of farmers and community groups to plant trees seems to be the only solution. The benefits of tree planting by rural communities can extend beyond fuelwood production. With the choice of appropriate species the same trees can have a multipurpose role providing animal forage, domestic building poles, tannins, honey and medicinal products. The same trees can also provide the shade, shelter and soil protection that contribute to sustainable agriculture.

The trees required for community forestry usually have very different characteristics to those used in the industrial plantations. ACIAR has recognised that the search for suitable trees and shrubs to include in the community tree-planting effort is a priority in many countries and has supported projects with this objective. The output from the research will be a technology package that enables useful trees to be established in a wide range of environmental and social conditions. In 1988 ACIAR allocated over $A1.3 million to its forestry program.

Australia has a unique flora adapted to nutrient-deficient sites in the tropics, and many trees and shrubs with characteristics useful for community forestry. The acacias and casuarinas are nitrogen-fixing species that can tolerate infertile sites and other unfavourable environmental conditions. The eucalypts show fast growth and have the ability to coppice, thus avoiding costly replanting. It is this vast genetic resource that gives Australian scientists a comparative advantage in the search for appropriate species for the reforestation effort in degraded tropical environments. The ACIAR Forestry Program has aimed to use its resources to exploit more fully the potential of Australian trees and shrubs for agroforestry and domestic fuelwood production in developing countries. It has not sought to promote Australian species to the exclusion of native trees or exotics from other countries, but rather to provide the villager, farmer or forester with a wider range of options in selecting an appropriate species to meet local requirements.

Commercial exploitation of Australia's tropical forests and woodlands has been limited mainly to selective logging of rainforests, sandalwood gathering, sawing of railway sleepers and an attempt to manage the Cypress pine *(Callitris)* forests of the Northern Territory. There has been little development or interest in the cultivation and utilisation of the tropical native woody flora, other than to rehabilitate land after mining operations and for ornamental purposes near habitation.

The financial support provided by ACIAR has enabled Australian scientists to explore the native trees and shrubs more thoroughly, to assess their growth performance over a wide range of environments and to determine their potential uses. The benefits of this exploration and testing will be diverse. Previously unrecognised species are now seen to have potential for pulp and paper production, rehabilitation of degraded lands, forage, chemical products and horticulture. In addition the program has contributed to the humanitarian, economic and trade objectives of Australia's foreign aid program.

The first ACIAR forestry project and the core activity for all subsequent projects is entitled 'Australian Hardwoods for Fuelwood and Agroforestry.' This project was implemented by the CSIRO Division of Forestry and Forest Products, and the Queensland Department of Forestry in collaboration with the Kenya Forestry Research Institute, the Royal Forest Department of Thailand and the Zimbabwe Forestry Commission (Forest Research Centre). Other ACIAR projects have been developed subsequently in China, Indonesia, Malaysia, Pakistan, Fiji, Western Samoa and New Caledonia. The projects have been of a biotechnical nature and have not attempted to embrace wider socioeconomic considerations, which are usually very location-specific, or extension to the user which is seen more properly as the role of the extension service of the collaborating institution.

Since 1962 the Australian Tree Seed Centre of CSIRO's Division of Forestry and Forest Products has provided a valuable service exporting tree seed to many countries around the world. Some of this seed has not been used effectively due to lack of expertise in techniques of establishing valid tests with this unfamiliar material. The ACIAR input has enabled the development of a well-organised network of field trials for comparative assessment. It has utilised the combined skills of Australian scientists and their counterparts in other countries to test selected species in well-designed trials to give statistically valid results.

The aim of this chapter is briefly to trace the broad development of our strategies in testing lesser-known Australian species in both Australia and in each of our collaborating partner countries. Particular attention is given to the choice of collaborating countries, species, testing sites and designs. It is hoped that this experience will be of benefit to other groups embarking upon similar projects.

Selection of Countries

The selection of partner countries must be viewed in terms of Australia's foreign aid policy, and more specifically ACIAR's primary geographic focus in Southeast Asia and the Pacific Islands, and to a lesser extent in China, South Asia and East Africa. South American countries are excluded from Australian aid activities as a matter of policy.

In determining partner countries for ACIAR projects the following criteria apply:

(1) The research must be a high national priority;
(2) The collaborating institution must be of sufficient standard and have the capacity to provide an effective partnership; and
(3) The local environment(s) should be sufficiently representative of the region to enable considerable spillover of results to neighbouring countries.

It is clear that when these criteria are applied it is not necessarily the country with the greatest needs that becomes the partner, but rather the country with a strong commitment to the project and the financial and personnel resources to maximise the chances of success. The concept of spillover of results is particularly important where the research involves expensive field trials over a long period of time. The testing of many lesser-known, often totally unproven, species can only be justified in a small number of representative sites. The more promising species from the trials can then be recommended for testing more thoroughly in other countries with similar environmental conditions. Thailand, for example, was selected as a tropical country with acidic soils and a range of rainfall regions comparable to many parts of Indonesia, Philippines, Malaysia, Laos and Vietnam. In other words, the potential for spillover in the region was very high.

In pre-project activities, an Australian consultant, Professor L.D. Pryor, and others travelled widely, discussed the ACIAR objectives, sought agreements and secured approvals in principle to undertake the work. The CSIRO Division of Forestry and Forest Products and the Queensland Department of Forestry were contracted by ACIAR as

Commissioned Agents to undertake the program. ACIAR and DFFP/CSIRO staff negotiated agreements which led to Records of Understanding with collaborating institutions. Australian financial support was directed mainly towards employing Australian scientists to coordinate the project and conduct research in Australia, supplying technical equipment and seeds and funding reciprocal visits usually on a bilateral basis. The non-Australian partner organisation was largely responsible for funding personnel and establishing, maintaining and measuring the trials. This simplified approach minimised administrative problems, particularly in financial management.

The program relies heavily on the commitment of all parties and the scientific development of the project has always been one of having joint goals achieved through differing routes. One of the strengths of the project has been the scientific staff development and joint sharing of skills and experiences through an active reciprocal visits program involving study tours and hands-on training. Formal training leading to postgraduate degrees was provided for project staff from China, Thailand and Zimbabwe under the ACIAR Associated Fellowship program. Joint publication of results by the collaborating scientists where appropriate was encouraged.

It was recognised that it would be unrealistic to expect developing country scientists to test lesser-known Australian species when so little was known about them in Australia. Consequently a priority task was to record and publish all available information on selected species and to commence research in Australia on their nursery and silvicultural requirements. The former task was accomplished through the publication of an ACIAR monograph (Turnbull 1986). To address the latter task, the Queensland Department of Forestry was contracted by ACIAR to establish complementary species trials in selected sites in Queensland, to examine nursery and establishment requirements, and to conduct some small management trials (e.g. coppice, biomass, etc.). It was intended that the Australian trials should be the cornerstone of the overall ACIAR field-testing program and should serve as a field study laboratory and demonstration area for ACIAR project scientists visiting Australia, trainees of other foreign aid agencies, Australian Government staff at all levels and for university students and staff. It was also intended that the existence of these extensive trials would stimulate interest in the lesser-known Australian trees and shrubs and encourage further research on them. It was recognised that field trials are temporary, and that for lasting benefits to accrue an active program for the publication of results was an essential adjunct.

Selection of Sites Within Countries

Much has been written about general principles involved in the selection of sites for species trials (e.g. Burley and Wood 1976; Boland 1986). The aim of this section therefore is not to review past literature but to concentrate on those issues considered important in the context in which we developed the program.

The accessibility and the security of tenure of the trial sites were paramount considerations in the site selection within the chosen climatic zones. Trials were mainly on Government-controlled land and usually located on forest or agricultural research stations where trials could be protected and maintained, plants assessed, and where meteorological records had been or could be kept. The general location of each trial was the primary responsibility of the collaborating country scientist, but the actual site chosen was the joint decision of the Australian scientist and his/her counterpart following a field inspection.

Efforts were made to locate trials near areas where there was a perceived need for fuelwood or other tree products and benefits, but no detailed assessment of the representativeness of the soil type at the trial site for the region as a whole was made. Greater emphasis was given to the broader climatic conditions of the site. Locations of all experimental sites reported in this book are given in Fig. 1(a and b).

In general terms we sought to locate sites of uniform topography and establish trials in areas that were highly visible and could be visited easily by a wide range of interested people. In this sense the demonstration value and local spillover benefits of the trials was high. For example in Thailand a major site was developed at Ratchaburi near Bangkok which, because of its accessibility to international air travellers, has become the most visited trial site in the ACIAR network.

At most sites the chemical and physical properties of the soil were determined. This was the responsibility of the collaborating partner and the output reflected the in-country expertise and soil classification systems. No effort was made to obtain a standard set of analyses over all sites and countries. By contrast climatic parameters (at the macro level) have been standardised and are presented in Chapter 4. The impact of site microclimates is reported in the accompanying reports on field trials where appropriate.

The control of the in-country field experiments rested in the hands of individual scientists (e.g. David Gwaze — Zimbabwe, Paul Ryan — Australia, Khongsak Pinyopusarerk — Thailand, Sam Kaumi and later Patrick Milimo — Kenya). In China Bai Jiayu (tropical eucalypts, acacias and

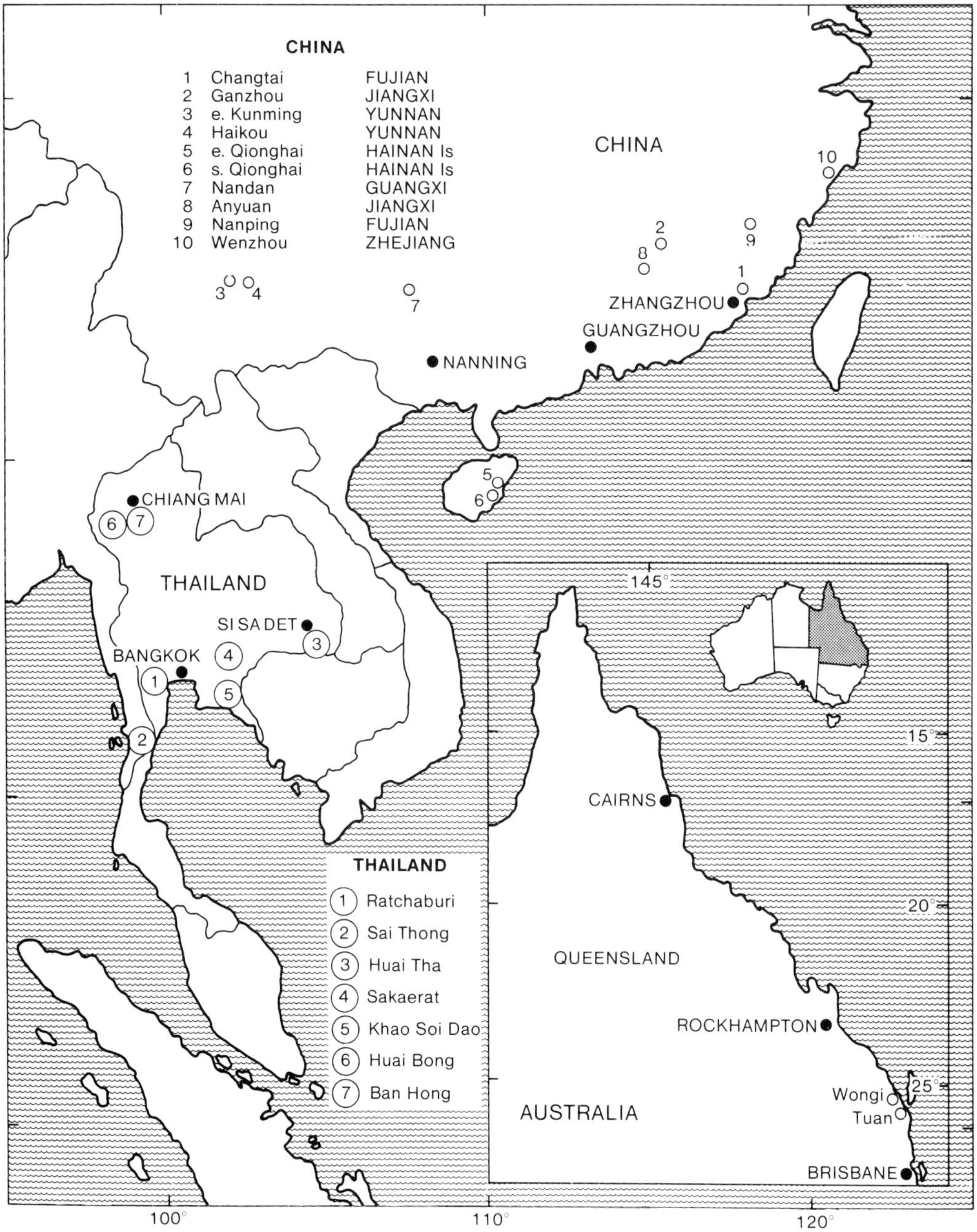

Fig. 1(a). ACIAR field trial sites in Southeast Asia, People's Republic of China and Queensland, Australia.

Fig. 1(b). ACIAR field trial sites in Zimbabwe and Kenya, Africa.

casuarinas) and Gao Chuanbi *(Acacia mearnsii* provenance trials) were the scientists in charge. Naturally, because of the large number of trials established in China, these leaders sought help from other local scientists. Subleaders included Wang Huoran (temperate and subtropical eucalypts and casuarinas), Yang Minquan (tropical acacias), Zhou Wenlong (tropical eucalypts and casuarinas — north Hainan), and Wu Kummin (tropical eucalypts and acacias — south Hainan). Each scientist efficiently coordinated in-country personnel and supervised assessment procedures, while a workshop at Gympie, Australia, in August 1986 served also to bring most of the ACIAR trial leaders together for discussions for the first time. The results were published as No. 16 in ACIAR's Proceedings Series.

Australia

The lesser-known species selected for trial grow naturally in a wide range of environments in northern Australia. These include the humid coastal lowlands *(Acacia mangium, A. oraria*, etc.) in northern Queensland, seasonally dry tropical woodlands in the country south of the Gulf of Carpentaria *(Parinari nonda, Terminalia* spp., *Grevillea* spp., *Melaleuca* spp., etc.) and semi-arid woodlands and shrublands of central Australia *(Acacia ammobia*, etc.). With such a range of material from diverse climatic regions a logical approach would have been to locate several appropriate trial sites to test selected suites of species in appropriate climatic regions in northern Australia. While such reasoning is sound, there are difficulties in the implementation, funding and administration of such an exercise, particularly in sparsely populated northern Australia. Consequently, a decision was made to forego growth data from a diversity of environments in Australia, in favour of concentrating the genetic material in a convenient location where it could be managed properly and assessed and be available for study. Two sites were chosen near Queensland's major forestry research station at Gympie which is readily accessible from Brisbane. This enabled an effective planting program where nursery, glasshouse, planting machinery and a whole range of technical and scientific backup could be brought together. In addition the location of the trials near a major technical training centre at Gympie and its close proximity to the Brisbane international airport ensured its relative accessibility as a field teaching and training laboratory for both local and international trainees.

The decision to consolidate near Gympie did, however, have some serious technical limitations. It meant, for instance, that most species being tested would be cultivated either outside their natural climatic range or on the southern fringes of their distribution. Exposure to light frosts caused damage to some species from frost-free localities on occasions. However, on the positive side, the sites provided an indication of those species with the capacity to tolerate wide environmental amplitudes.

Two major climatically similar (see Chapter 4) sites were chosen near Gympie (Tuan/Toolara State Forest and Wongi State Forest). Their soil types are dissimilar (e.g. Tuan/Toolara has deeper sandy loams while Wongi has a shallow profile and contains more clay — Ryan et al. 1987). Strategically, however, the duplication of the trials served also as insurance against disaster occurring to either.

Recognising the environmental limitations in the planting sites in South Queensland to provide suitable conditions for all the species, the Department of Forestry, and Division of Forestry and Forest Products CSIRO did attempt to establish small, arboretum-type plantings at several sites in northern Australia (Dalby, Atherton, Mareeba, Darwin, Cardwell and Broome). These trials met with some success (Applegate and Nicholson 1987) but were affected by administrative and logistic difficulties.

China

In the China experiments for project 8457: 'Introduction and Cultivation of Australian Broadleaved Trees in China', three broad climatic zones — tropical, subtropical and temperate — were selected. Within these zones planting sites were selected on tropical Hainan Island (Hainan Province), subtropical Zhangzhou (Fujian Province) and temperate Kunming (Yunnan Province) by the Tropical Forestry Research Institute, Chinese Academy of Forestry (CAF), in Guangzhou and the Research Institute of Forestry, Beijing.

The provenance trials of *A. mearnsii* (project 8458: 'Wattle Silviculture and Utilisation of Tannin Extracts') were managed by the CAF Subtropical Forest Research Institute in Fuyang. Priority was given to having secure land tenure and eight provenance trials were established in cooperation with forest farms, agricultural universities and research stations, and provincial and country forestry bureaus. Strategically, trials were located where black wattle is either grown currently or was anticipated to be established in commercial plantations in the future. The Chinese had a special interest in promoting black wattle cultivation in colder areas near the limits of its climatic tolerance.

The trials were managed usually by staff on forest farms in provincial and county forestry bureaus (since the Academy has no control of land) in collaboration with staff at the Tropical Forestry Research Institute in Guangzhou, the Subtropical

Forestry Research Institute, Fuyang, and the Forest Research Institute in Beijing.

Thailand

All planting sites in Thailand were under the direct control of the Royal Forest Department (RFD), and a regional forest research station under the control of the Department's Silvicultural Division was located quite close to each trial site. Trial sites were dispersed around Thailand from wet (e.g. Sai Thong) to seasonally very dry sites (Chiang Mai). In most instances there was a perceived need for fuelwood in the area (e.g. the Ratchaburi area had a need for small-diameter logs for fuel for pottery kilns), or the sites were chosen to extend the climatic range of the testing sites (e.g. Sakaerat). One site, Si Sa Ket, was located on the main agroforestry research station in Thailand. Climatic conditions at the trial sites are covered by Booth (Chapter 4).

Zimbabwe

The administration of all trials was controlled by the Forest Research Centre, Zimbabwe Forestry Commission. High priority was given to obtaining land controlled by forestry (e.g. Matopos) or agricultural researchers (e.g. Makoholi). High priority was also given to the assured availability of local labour, ready access to nursery facilities and security of the planting sites.

The trials were located at sites which covered a range of native vegetational associations reflecting soil types and moisture availability. One serious limitation was that no trials were located in the very dry western and southwestern parts of the country, mainly because of local security problems. Climatic conditions at the trial sites are covered by Booth (Chapter 4). The aim of the trials was to select species for use by villagers in highly populated and deforested communal lands.

Kenya

Three trial sites were selected and managed directly by staff of the Kenya Forestry Research Institute. They were located near the Institute's small regional stations at Turbo (near Eldoret) on good soils and high rainfall, at Loruk on the dry floor of the Rift Valley and near Gede on the humid coast of Kenya near Malindi. Each site had a history as a testing centre for exotic species. Climatic conditions at the sites are covered in Chapter 4. The lack of good nursery facilities proved a serious problem at Loruk.

Choice of Species

In April 1983 ACIAR sponsored a meeting in Canberra of Australian foresters familiar with forestry problems in other countries, and botanists conversant with the woody flora of Australia. These scientists nominated about 170 species of trees and shrubs with potential for planting in a range of environmental conditions for fuelwood or other community forestry uses. Emphasis was given to species with a tropical or subtropical distribution, especially those adapted to infertile soils. Only eucalypts that had been little-tested as exotics and could be considered as 'lesser-known' were considered. This meeting debated the merits of the nominated species and selected 108 species that deserved increased recognition and research. Most of the selections were suitable for fuelwood for individual family needs rather than for cultivation in larger plantations, and are little-known in traditional forestry. Some are short-lived, crooked, multistemmed shrubs rather than the more persistent tall straight forest trees, but nevertheless may meet the requirements for small-scale village use or soil conservation. In selecting the species the meeting aimed for:

(1) plants capable of providing products and services in addition to fuelwood;
(2) adaptable plants that are easily established and maintained; and
(3) plants capable of growing in extreme environments including arid and humid tropical zones, infertile soils, heavy clays, saline, highly alkaline or waterlogged sites or exposed coastal situations.

Other characteristics considered desirable were: an ability to fix atmospheric nitrogen, a capacity for rapid growth, an ability to coppice, and good burning properties.

The species used in the trials in Australia, Kenya, Thailand and Zimbabwe reflected strongly the recommendations of the Canberra meeting. Other species that were identified during later field reconnaissances as having potential value were included in the trials also (Chapter 3). Although a major seed collection effort was mounted some species were excluded from the trials due to lack of seed.

The nitrogen-fixing acacias, of which some 850 species are indigenous to Australia, have been underexploited, and acacias have formed a major part of most of the ACIAR trials. Although many Australian acacias are fast-growing and some, notably *A. auriculiformis, A. mangium, A. mearnsii* and *A. saligna* have been widely planted as exotics, little is known of their provenance variations. The ACIAR trials have included a range of provenances of the more promising species. Particular efforts have been made to investigate provenance variation in *A. auriculiformis* in cooperation with a USAID-sponsored forestry and fuelwood project (F/FRED) in Asia, and *A. mearnsii* provenances have been tested widely in China as part of ACIAR project 8458 ('Wattle Silviculture and Utilisation of Tannin Extracts').

In ACIAR Project 8320 ('Australian Hardwoods for Fuelwood and Agroforestry') consideration was given to maintaining the same species at all sites irrespective of the environmental conditions, but this was rejected due to the enormous variation amongst sites. A more flexible approach was adopted with the choice of species at each site dependent to a large extent on their perceived ecological requirements. In Kenya, for example, species suitable for the semi-arid, wet highlands and seasonally dry coastal conditions at the three planting sites were chosen. The trials still provide considerable potential for site/genotype interaction studies, and some of these have been conducted already in Thailand (Chapter 14).

In selecting provenances of tropical/subtropical species to plant outside Australia, we made use of the matching homoclime approach of Booth et al. (1987), in which geographic areas in Australia having an approximately similar climate to the planting site were determined. This approach was first developed and used for Ratchaburi, Thailand, and used extensively in later trials. Chapter 4 contains locations of appropriate climatic matches in Australia for each of the trial sites. Because many of the species used in the trials were lesser known, only a limited number of provenances of each was included. Additional provenances were used when genetic variation in the species was expected to be substantial. It was intended that detailed provenance studies would commence after particularly promising species were identified. Such studies (e.g. provenance trials of *Acacia auriculiformis* (with F/FRED), *A. crassicarpa* and *A. holosericea*) will constitute part of the program in 1989–91.

Choice of Design

Choice of design caused considerable debate in Australia. Eventually we decided to use a simple robust design comprising randomised complete blocks with square plots of 25 or 36 trees and with 3–4 replications. The aim was to restrict treatments to about 25 seedlots but more were used in some instances.

Choice of design must reflect the aims of the experiment. Our aim was to test a range of lesser-known Australian tree species over a range of locations for growth and survival over a 6–10-year period. Large plot sizes provide some protection against interplot competitive effects, and allow for the possiblity of thinning as trees mature (this was necessary after about 3 years for several fast-growing acacia species). Nevertheless there are statistically sound reasons for using incomplete block designs with 5–10 tree line plots if trials are only elimination trials of lesser-known species. This approach uses fewer resources in environmentally more difficult sites where chances of failure are high (arid sites with termites, etc.), and will be used during the next phase of the project. Greater numbers of seedlots can be tested and the trials repeated in successive years to cope with erratic climatic conditions (especially unreliable rainfall). Such trials have only a short life (2–3 years) but are very economical in terms of the reduced planting area and low maintenance requirements.

Because of the emphasis on assessing growth potential of these lesser-known species in the first phase of the program, every effort was made to prepare the trials to reduce extraneous environmental conditions that might have interfered with seedlots achieving their full potential. Consequently, most sites were ploughed, fertilised and kept weed-free until canopy closure. Termiticides and herbicides were used also where appropriate. With one notable exception, the trials in Zimbabwe were not protected from termite attack and this resulted in some change in the aims of these trials. The impact of this approach led to new trials evaluating the effect of termites on species survival as detailed in Chapter 23.

Choice of Assessment Procedures

Much thought and effort went into standardising assessment procedures for all the trials. The procedures developed by Paul Ryan at Gympie served as a basic model for the general attributes about which information was needed (e.g. survival, heights, diameters, crown densities, etc.). Particular difficulty was experienced in describing tree-form characteristics and this was not adequately resolved. Phenological assessments of flowering, fruiting, etc. were also devised. The overall aim was to develop procedures compatible with TREDAT data recording procedures.

The assessment procedures devised at Gympie were sometimes adopted in their entirety, but in most countries trial leaders selectively incorporated particular elements of them in their own assessments. The time involved in recording some attributes and in the assessment of characters by subjective scores caused some problems in applying the procedures.

Chapter 2

Acacia mearnsii: Its Past and Potential Use with Reference to the Development of Plantations in the People's Republic of China

W.E. Hillis

Abstract

A brief history is given of the use of *Acacia mearnsii*, to serve as a general background for the ACIAR forestry program development. The program involves a number of multidisciplinary studies to improve the yield and utilisation of the species in the People's Republic of China. Past research in the Republic of South Africa on *A. mearnsii* has already led to one of the most significant developments in contemporary forestry. Requirements for the selection of plantation species to provide the utilisation needs of different countries will increasingly involve versatile species such as *A. mearnsii*. The coordination of recent developments may again lead to other significant developments in forestry through the planting and use of this species. Brief details are provided of the ACIAR program on *A. mearnsii* in China.

Introduction

The rapid developments in the People's Republic of China (PRC) since 1976 have resulted in, among other things, increased demand for leather. About 40 tannin extract factories produce about 30 000 t/annum of mainly hydrolysable tannins, often from low-quality resources and in amounts inadequate to tan the large number of pigskins available. At the same time China has an urgent need for tree species capable of growing on poor quality soils so as to assist soil protection and improvement, and provide wood and other products as well as employment in rural districts. Species that could provide versatile condensed tannins in a relatively short time include *Eucalyptus astringens* (Brockway and Hillis 1955; this species requires climatic conditions not found in China), mangrove species (Hillis 1956; these species are limited to particular coastal regions) and rapidly growing *Acacia* species.

From the earliest times different *Acacia* species have satisfied various human needs. More recently the use of spindly stems of acacia regrowth for building huts in the early days of British settlement of Australia (about 1800) led to the adoption of the common name of 'wattle' (Sherry 1971). Different Australian species were planted in India in the mid 19th century to provide fuelwood. Apart from their use as decorative trees, wattles or mimosas are perhaps most widely known as a source of tanning agents. *Acacia nilotica* remnants have been found in a 5000-year-old tan-yard in Upper Egypt (White 1956). Fifteen years after settlement, William Goff established the first European-style tannery in Australia at No. 8 Pitts Row (Pitt Street) in the heart of Sydney, probably using wattle bark from the surrounding districts. Wattle bark was an export commodity before 1821 (Sherry 1971) and bark collectors preceded European settlers to various parts of the southeastern coast of Australia. The amount of bark exported from the colony of Victoria rose to 11 378 t in 1878 by which time indiscriminate stripping of immature trees (very largely *A. mearnsii*) was widespread and the quality of the bark supply deteriorated. A similar sequence of events occurred in New South Wales, which exported smaller quantities of bark, and in Tasmania, the largest exporter with an average of

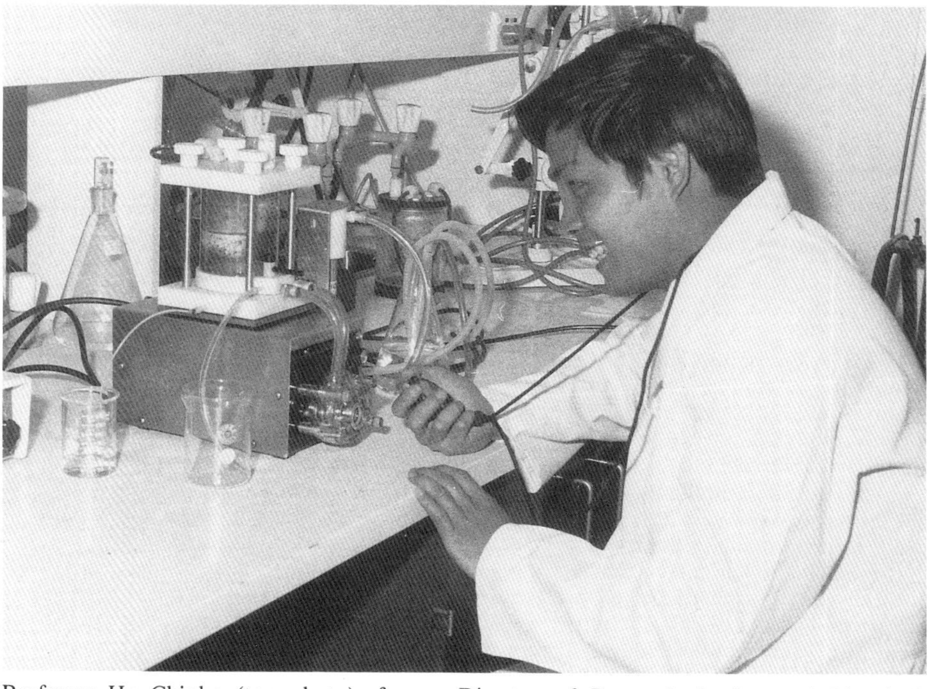

Professor Ho Chinko (top photo), former Director of Research Institute of Chemical Processing and Utilization of Forest Products, Chinese Academy of Forestry, Nanjing (left), and Dr W.E. Hillis, formerly Division of Wood Technology CSIRO, Melbourne, were mainly responsible for conceiving and developing ACIAR project "Wattle Silviculture and Utilization of Tannin Extracts", and in doing so renewed a research contact and friendship which commenced in 1947 in the Division of Forest Products CSIR, Melbourne.

Mr Zheng Guangcheng (bottom) from the Research Institute of Chemical Processing and Utilization of Forest Products, C.A.F., Nanjing, working on the ultrafiltration of *Acacia mearnsii* tannin extracts at the Division of Forestry and Forest Products CSIRO, Melbourne, Australia, in 1987 (photo: Y. Yazaki).

40 000 t of bark a year in that period. Today the natural occurrence of *A. mearnsii* in Australia has been greatly reduced, but its reputation as a tanning agent is well established.

Acacia mearnsii in Cultivation

The Vanderplank brothers were possibly the first to grow *A. mearnsii* in South Africa in 1865 to provide ornamental trees, shelter and fuel. The origin of this seed is thought to be Bicheno (Tasmania). A tanner who examined in 1884 the barks of *A. mearnsii* and *A. dealbata* for Sir George Sutton found the former species to be the most valuable. Following the submission of samples to a London exhibition in 1886, the first commercial bark was exported from South Africa in 1887, and then the first plantations anywhere specifically for the production of tanbark were established. A large industry was subsequently established with the plantation area reaching over 360 000 ha in 1960 (Sherry 1971). Considerable attention was given to raising the production of high-quality tanbark, and a yield of 53% tannin in moisture-free bark has been obtained, with a range of 44–48% not being unusual. Special attention was given to the production of high-quality extracts that would convert hides and skins into the light-coloured leathers required by European markets; to achieve this objective other *Acacia* spp. were excluded from plantation regions. *Acacia mearnsii* plantations and farmlots have also been established in other countries such as Zimbabwe, Kenya and notably Brazil. Sherry (1971) prepared a comprehensive account of *A. mearnsii* up to 1970, showing that it is the fastest biosynthesiser of tannin known.

The appointment of I.J. Craib in 1928 to study the stagnation of growth in black wattle plantations in South Africa became an event of great importance. He condemned the existing practice of intense mutual competition of trees in early life and proposed drastic thinnings in the first year of growth. The extent of the thinnings was determined by the length and density of the crown and its vigour. The continued success of this revolutionary approach in his subsequent work on pines established the foundations of forestry practices for fast-growing plantations (Craib 1933). His work has resulted in one of the most significant developments of contemporary forestry, with the increasing importance of industrial plantations of different species for the production of wood.

Acacia mearnsii can meet needs in addition to the tannin for which industrial plantations were originally established, and with the silvicultural foundations established by Craib attention can be given to these. A coordinated application of recent developments with the aid of modern computing and other techniques could lead to broad significant developments to extend the foundations of forestry. This will provide the particular needs, such as the more effective use of land, of different countries from a particular species.

Tannin Yield of Different Provenances

No comprehensive examination of the provenances throughout the range of *A. mearnsii* in Australia has been made, nor of the genetic variation within and between populations. The continuing decrease of the formerly extensive natural distribution of *A. mearnsii*, because of the clearing of land for agricultural and other purposes in Australia, means that seed must be collected from remaining provenances as soon as possible.

An early (1928) interest in New South Wales in the improvement of the quality of black wattle later led to the plan to establish a seed production stand of high tannin-producing provenances ('strains') collected in that State. At the completion of the last set of trials, Humphreys and Johnstone (1957) concluded that insufficient seed samples were taken to establish differences between the provenances. There were, however, highly significant differences between the mean tannin content at four different ages (from 28.0% at 2.75 years to 37.3% at 10.08 years) and a regression equation was derived. However, whereas in a South African study only an 8.3% increase was found with barks of 4 and 8 years of age (Sherry 1971), factors in addition to age may influence tannin content. The exact location of the sources of the 19 seedlots of *A. mearnsii* collected by S.P. Sherry in 1957 is unknown. When planted in South Africa and harvested after 8 or 10 years' growth, significant differences were found between the Australian seedlots for diameter at breast height and stem form. There were also differences in tannin content of bark samples, bark thickness and weight of bark per tree. In general, bark yields per hectare were lower from the Australian seedlots than from the progeny of selected South African parent trees (Anon. 1967, 1969). There appeared to be positive correlations between tannin content, bark thickness and tree diameter.

Tannin Analysis

The internationally accepted method of determining tannin is by means of its removal with an approved hide powder, previously prepared under standard procedures, from an aqueous extract obtained under controlled conditions from the bark. The method is highly empirical and relies strongly on close control of the conditions of analysis and the quality and physical form of the hide powder.

In addition a minimum of 30 g of bark is required for duplicate analyses, involving specialised extraction equipment over a period of 4–5 days. A faster method utilising smaller samples is needed to monitor biological practices aimed at obtaining maximum tannin yields. The rapid spectrophotometric methods developed by Roux (1951, 1957a, 1957b, 1957c) produce results showing a close relationship with those determined by the hide powder method. There are some disadvantages (Gordon-Gray 1957) with these methods and results may vary with changing composition of the raw material. A more direct basis for an analysis is the reaction of polyphenols in extracts with formaldehyde in the Stiasny reaction (Wissing 1955). The development of a method involving a satisfactory extraction procedure, the Stiasny reaction, readily available low cost equipment and a minimum of 3 g of bark for duplicate analyses now enables 10 samples to be analysed daily to provide closely reproducible results (Zheng and Yazaki 1988). A close linear relationship has been found between the Stiasny value and tannin content (by hide powder) of *Pinus radiata* bark (Bayfield et al. 1952).

Acacia mearnsii in China

Under conditions of financial restraints China has begun programs to employ its large and mainly rural population in the development of commodities in a situation of limited energy resources. In order to provide more tannin for leather manufacture, *Acacia mearnsii* has been grown since about 1950 in the Zhejiang, Fujian, Jiangxi, Guangdong, Guangxi, Yunnan, Sichuan and other provinces in China, with an estimated total area of 10 400 ha. The trees are mostly grown in small areas whereas efficient commercial operations require much larger plantation areas. Moreover the genetic history of seed for these plantations is uncertain and the quality of the trees is inferior. With the current plans to rapidly increase the plantation area of *A. mearnsii*, there is an opportunity to apply the most effective forestry and utilisation practices.

Acacia mearnsii can serve more purposes in China than the primary one of supplying tan-bark for which it has been planted. Requirements for the limited areas of better soils for food production for a growing population favour the introduction of undemanding tree species. (Between 1957 and 1980 one-third of the present agricultural area in China was lost to buildings — Richardson 1986.) In this regard it is of considerable importance that Australian acacias are pioneer species and can adapt to a variety of sites. They can symbiotically fix atmospheric nitrogen and thereby improve soil conditions, provide environmental protection and a component in agroforestry operations (Boland 1987). The selection of optimum provenances from the natural range of the species would maximise the possibilities of obtaining the most adaptable trees for the proposed sites for plantations, and provide, amongst other attributes, resistance to frost damage (Anon. 1963) and to the different causes of gummosis of bark.

In addition to converting hides and skins into leather, other uses for wattle bark extract have been extended or developed. These uses include the control of viscosity in clay-water mixes used in oil-well drilling or for ceramics manufacture, and anticorrosive compounds. Wood adhesives and bonding agents to improve the utilisation of wood are the most significant of these new uses.

Measures are being undertaken by China to reduce the severe shortage of wood, the consumption of which is less than one-tenth per capita than that in Australia. Furthermore, more pulpwood is required to help supply an expected 6% annual growth rate in paper consumption. Although *A. mearnsii* grows quickly its production of wood is not as rapid as that of a number of other *Acacia* species. Although a number of factors (genetic, spacing, soil characteristics, temperature, rainfall) can influence bark thickness, the ratio of wood to bark production increases with age (Sherry 1971). Accordingly, in addition to the selection of the most suitable seed source for the site, economic and social studies will be required to ascertain the optimum harvesting age to provide tan-bark of specified purity, as well as the fuelwood, pulpwood or building materials which may be needed for various regions.

The wood (air-dry density 650–750 kg/m^3) from the small-diameter trees from plantations has found many uses. It is very hard and tough and although the pale sapwood is susceptible to *Lyctus* attack it readily absorbs preservatives. The finely textured wood with a light brown heartwood is moderately easy to work, with moderate shrinkage, it polishes well and is very suitable for furniture when appropriate drying schedules are used to avoid checking (Bolza and Keating 1972). Plantation-grown *A. mearnsii* wood is being used commercially to produce different chemical pulps in good yields with good properties (Logan 1987; Hannah et al. 1977). Improved utilisation of all wood resources is achieved with increased production of panel and laminated products, which in turn increases the consumption of adhesives that are significant cost items of the processes. In a country with a rapidly expanding technological base, and increasing demands on the relatively small but enlarging supplies of chemicals and energy, there are advantages in supplying appropriate chemicals for adhesives from low-energy-demanding biosynthetic sources. Extracts of condensed tannins can provide the basis for adhesives and *A. mearnsii* has been used commercially for this purpose since 1959. As with other condensed tannins having a polyflavanoid structure, *A. mearnsii* tannin adhesives have the potential to form highly moisture-resistant and waterproof bonds comparable with those produced by phenol- or resorcinol-formaldehyde adhesives. Hydrolysable tannins, consisting of gallic acid and its derivatives esterified with glucose or other sugars, are unsuitable substrates for adhesives.

The development of high-quality adhesives from *A. mearnsii* requires extracts of uniform high quality, in which carbohydrates and other nonreactive components do not exceed a certain proportion. The initial production of extracts in China will be from plantations that differ in locality, age and degree of gummosis. It is necessary to have procedures capable of refining these extracts if required into sufficiently large quantities with the requisite and uniform quality. The continuing development of a range of membranes to increase commercial applicability of ultrafiltration in several industries could be extended to raise the quality of not only *A. mearnsii* but also other tannin extracts (such as from spruce bark) when necessary. Moreover, the distribution of molecular size in an extract, assessed by membrane filtration, provides (in addition to Stiasny value) data from which to predict gluing properties.

ACIAR Program in China

In 1985 ACIAR, through the Division of Forest Research CSIRO, and the Chinese Academy of Forestry, commenced a 3-year development program on *A. mearnsii* titled 'Wattle Silviculture and Utilisation of Tannin Extracts.' The program had two main components. The first involved the genetic improvement of plantations through the introduction and breeding of new seed sources that would result in higher yields of tannin extracts and wood. The second involved a program that would lead to the development of tannin-based wood adhesives. For the first component the Division of Forest Research CSIRO worked directly with the Subtropical Forest Research Institute in Fuyang and, for the second, the CSIRO Division of Chemical and Wood Technology, Melbourne, worked with the Research Institute of Chemical Processing and Utilisation of Forest Products, Nanjing. Events leading up to these collaborative arrangements are summarised in Action China.

To improve the genetic resources of *A. mearnsii*, provenance seed collections were made across the entire range of the species in Australia. This was the first seed collection program made, on a systematic basis, for the species. Subsequently, provenance trials were established in several centres across southern China to evaluate the growth performance of local Chinese seed sources against improved South African sources and new nonimproved Australian sources (see Chapter 9). Complementary studies were made in a glasshouse in Australia on geographic variation in seedling morphology (Bleakley and Matheson 1988), and a major study is in progress on determining levels of frost resistance in natural populations. Seedling seed orchards have also been established at two sites in China based on a breeding plan prepared by an Australian forest geneticist (Raymond 1987).

Bark samples were also collected from trees providing the provenance seed, and the highest content of extractives has been found in those barks from provenances in southern Victoria and Tasmania. With the assistance of a rapid method of analysis developed during the program, the purity or proportion of reactive components in some of those provenances was higher than elsewhere. This work was conducted both in Australia and China and involved several reciprocal scientific visits. It is realised that the results from uncultivated trees involve a confounding of genetic, age and environmental influences and that the work should be repeated in China on even-aged stands at near rotation age and growing in typical plantation environments.

Different extracts of *A. mearnsii* bark have been examined by an ultrafiltration technique. It was found that, if necessary, the extracts could be enriched by this technique although more work is needed for its application on a commercial scale. Moreover, the technique and the Stiasny analysis has been used as the first assessment of the suitability of extracts for adhesive preparation (Zheng and Yazaki 1988). Also, Chinese workers have successfully prepared wattle tannin-formaldehyde adhesives on a laboratory scale. Further work on other development and commercial applications will assist more extensive use of local forest resources.

Using a throwing rope to collect seed from *Albizia procera* north of Cairns, North Queensland (photo: S.D. Searle).

Chapter 3

Seed Collections of Lesser-Known Trees and Shrubs in Queensland, Australia

S.D. Searle

Abstract

A summary of seed collections funded by ACIAR and undertaken in Queensland (by CSIRO Division of Forestry and Forest Products) during a 30-month period (November 1983 to May 1986) is presented. The emphasis of these collections was on tropical and subtropical lesser-known Australian tree and shrub species with potential for fuelwood and agroforestry. About half of the 112 species and 194 provenances collected were acacias and melaleucas. This sampling enabled previously unavailable species from many genera to be tested in field trials. In addition to information on seed viabilities, field observations of flowering and seeding, coppicing and suckering, collection difficulties and seed cleaning techniques employed are summarised.

Introduction

In April 1983 ACIAR sponsored a meeting in Canberra between foresters, botanists and ecologists with experience in tropical and subtropical Australia. These scientists nominated Australian tree and shrub species with potential for planting in a range of environmentally difficult conditions for fuelwood and other community uses. On the basis of these selections, a book was written summarising knowledge of 100 species (Turnbull 1986), and seed collections were undertaken in Queensland.

The CSIRO Australian Tree Seed Centre was chosen to undertake this program to ensure accurate identification of these lesser-known species, high standards of seed collection and to facilitate follow-up activities.

A team was based at the Atherton CSIRO regional station in North Queensland from November 1983. In May 1985 they transferred to Samford CSIRO regional station in southeast Queensland to undertake collections in subtropical Australia. The collections were concentrated on species drawn from nominations made at the Canberra meeting, but the team had the flexibility to identify and collect other species worthy of inclusion in the program. The collection program was terminated in May 1986.

As well as gathering, processing and documenting the seed collections, the team also collated information on species to be described in Turnbull (1986); they photographed the selected species, recorded species and population characteristics, ecological and phenological details, commented on potential utilisation and sampled wood and foliage.

This chapter presents summaries of seed viabilities, observations of flowering, seeding, coppicing and suckering, difficulties encountered collecting the species and the seed cleaning methods employed. These observations and practices have been included as a guide to those making seed collections from these species in the future.

Methodology

Sampling Strategy

These collections were intended for species screening trials. Given the limited time available and the absence of performance information, it was

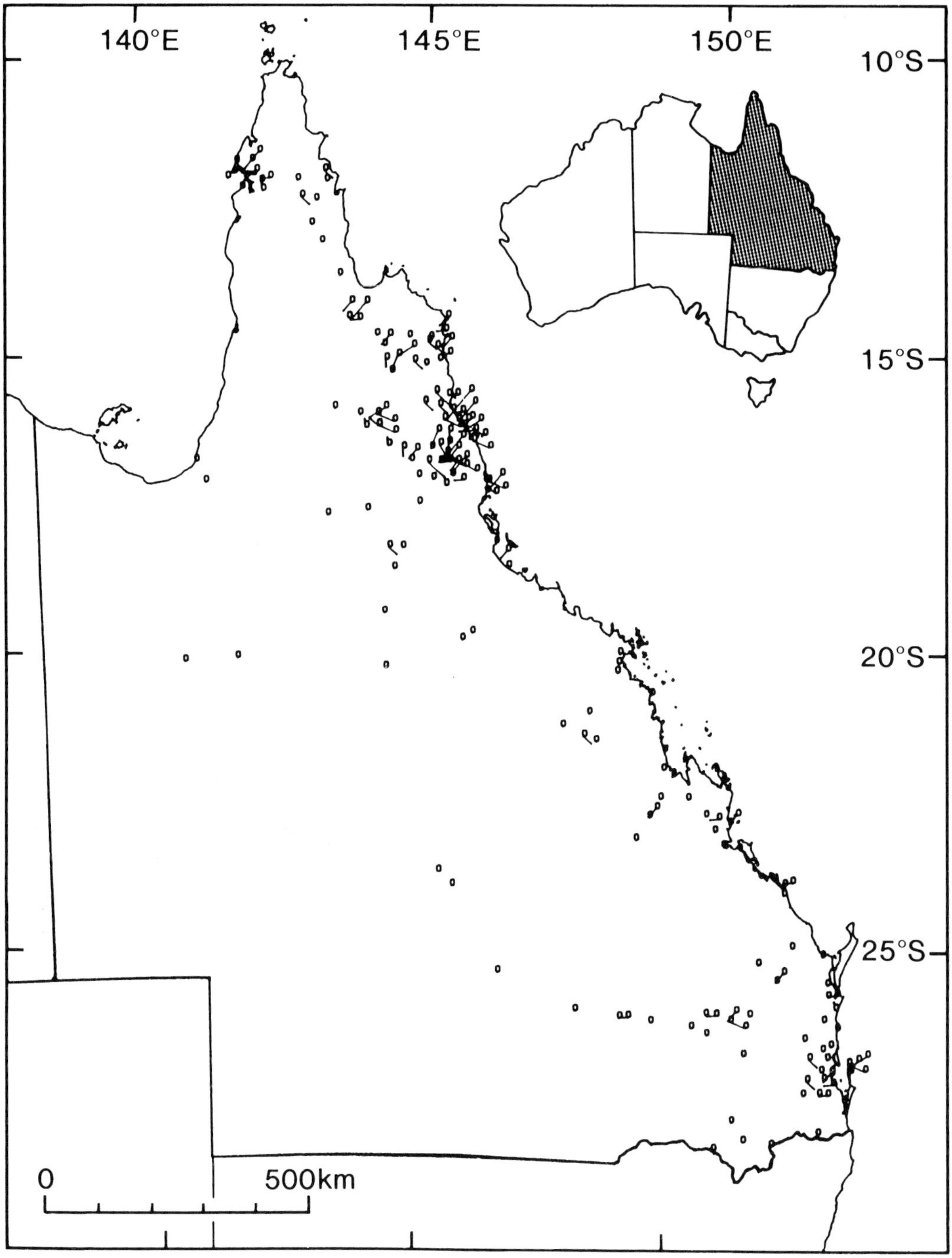

Fig. 1. Map of Queensland indicating seed collection sites from Nov. 1983 to May 1986.

considered more important to concentrate on seed collections from as many species as possible. Therefore seed from two provenances, from differing environments, for each species was considered adequate. It was envisaged provenance collections would be initiated following demand engendered by the performance of these species in field trials.

The main aim, therefore, was to sample genetic variation within populations and, to a lesser extent, between populations of the target species. Bulk or individual collections were made according to the number and density of individuals, the area they covered and the size of the seed crops present. For example, where the population extended over some distance and it was possible to sample individuals more than about 100 m apart, individual tree collections were made. Where a population was confined to a small area, or seed crops were small and individual collections would have resulted in very small amounts of seed, bulk collections were made from as many trees as possible.

Seed Processing

All seed collected was cleaned in the field or at the regional station where the team was based. A portable electric (2 hp) seed thresher and a cement mixer were used to thresh and scarify fruits when required prior to sieving. Seed was then sent to the Australian Tree Seed Centre in Canberra for germination testing, storage and despatch to trial sites.

Results

Seed from 32 genera, 112 species and 194 provenances was collected during a 30-month period. A list of these species is presented in Table 1 together with average seed viabilities for each species, observations of their flowering and seeding, and vegetative reproduction capabilities. Locations of the collections are presented in Fig. 1.

The field observations were limited by the relatively short period of time the team could spend on each species. Timing of flowering and seeding of many tropical species can also vary considerably from year to year, and Table 1 should therefore be considered a guide only. Further details of many of these species can be found in Turnbull (1986). A summary of collection difficulties encountered for those species which proved particularly elusive is given in Table 2, and these should be noted for future provenance collections.

Many of the species sampled were tested for seed viability and stored by the Australian Tree Seed Centre for the first time. With few or no guidelines to follow, the fleshy-fruited species from the genera *Planchonella* and *Persoonia* proved difficult to clean and, together with *Terminalia*, *Melia*, *Petalostigma* and *Alphitonia*, difficult to germinate. The Centre conducted long-term glasshouse trials (6 months) to determine germination requirements for these genera. Viabilities for species in these trials are included in Table 1. Seed cleaning methods employed are also summarised in Table 1.

Table 1. Seed viabilities, field observations and seed cleaning methods for species collected in Queensland.

		CSIRO Tree Seed Centre results average seed viability/10 g^{++}	Field observations														
			Flower and seeding months												Vegetative reproduction		
Species	Provenances	Individual and bulked tree collections	J	F	M	A	M	J	J	A	S	O	N	D	Ability to coppice	Ability to sucker	Seed cleaning method
Mimosaceae																	
Acacia aulacocarpa	1	520		O	O	O		O	O		X	X	X	X	C	-	T*
" *bancroftii*	1	185										X			-	-	T
" *bidwillii*	3	39	X		X	X					O	O	O	O	-	-	T
" *blakei*	1	838											X		-	-	T
" *brassii*	1	1117							X	X	X		X		-	-	T
" *burrowii*	1	1125									O		X		-	-	T
" *cambagei*	1	-					O						X		C	S	Sieve
" *concurrens*	1	1097											X		-	-	T
" *deanei* ssp *deanei*	1	439									O		X	X	-	-	T
" *falcata*$^{+}$	2	584						O	O			X			-	-	T
" *falciformis*	2	264		OX								X		OX		-	T
" *farnesiana*$^{+}$	2	75					O				OX	X			-	-	T
" *fasciculifera*	1	136												X	C	S	T
" *fimbriata*	2	991										X			-	-	T
" *flavescens*	2	228 ± 65		O	O	O	O	O		X	X	X	X		C	-	T
" *glaucocarpa*	1	446											X		C	-	T
" *hammondii*$^{+}$	2	1035					O				X	X	X	X	-	-	T

(*Continued*)

Table 1. (*Continued*)

Species	Provenances	CSIRO Tree Seed Centre results average seed viability/10 g[++]: Individual and bulked tree collections	Field observations: Flower and seeding months: J	F	M	A	M	J	J	A	S	O	N	D	Vegetative reproduction: Ability to coppice	Ability to sucker	Seed cleaning method
" *harpophylla*	1	134							O	O	O	OX	X		C	S	Sieve
" *hylonoma*	1	427											O	OX	C	S	T
" *julifera* ssp *gilbertensis*	1	214									X				C	-	T
" *julifera* ssp *julifera*	5	547 ± 330					O					X		X	C	S	T
" *juncifolia*[+]	1	833											X		-	-	T
" *leiocalyx*	3	660											X		-	-	T
" *leiocalyx* vel aff.[+]	1	894										X			C	S	T
" *leptocarpa*	2	826 ± 167					O	O	O	O	X	X	X		-	-	T
" *leptoloba*	1	224		X	X						O				-	-	T
" *leucoclada*	1	789			O								X	X	-	-	T
" *maidenii*	2	633										X	X		C	-	T
" *melanoxylon*	1	1204 ± 199	O								X		OX	OX	-	-	T
" *oraria*	2	315	O		O			O			X	X			C	-	T
" *oswaldii*[+]	1	61												X	-	-	T
" *penninervis* var *longiracemosa*[+]	1	295											X		-	-	T
" *penninervis* var *penninervis*[+]	1	204									X	X	X		-	-	T
" *platycarpa*	2	60	O		O	O				X	X	X		O	C	-	T
" *pustula*[+]	1	449											X		C	-	T
" *rothii*	2	26									X	X	X	X	C	-	T
" *salicina*	1	244							O		O	O		X	C	-	T
" *simsii*	3	814	O		OX	X	X	X	X		X	X			-	-	T
" *spectabilis*	1	260											X		-	-	T
" *stenophylla*	1	77									X				C	-	T
" *tephrina*	1	241											X		C	-	T
" *torulosa*	3	367 ± 193									X	X	X	X	C	-	T
" *victoriae*	2	258										O		X	C	S	T
Adenanthera abrosperma	2	23	O							X	O	O	OX	OX	C	-	T
Albizia procera	2	151								X	X	X			-	-	T
Rhamnaceae																	
Alphitonia excelsa	2	32 ± 23	X		O	O					X			O	C	-	T
" *petrei*	2	524		X	X						O				-	-	T
Atalaya hemiglauca	2	107										O	X	X	C	-	See Footnote[1]
Proteaceae																	
Banksia integrifolia var. *compar*	1	706	X												-	-	S**
" *serrata*	1	80			OX										-	-	S**
Sterculiaceae																	
Brachychiton populneus ssp *populneus*	1	-			X										-	-	T
Proteaceae																	
Buckinghamia celsissima	1	NIL		O									X		-	-	Sieve
Caesalpiniaceae																	
Cassia brewsteri	2	60	X	X											C	-	T
Casuarinaceae																	
Casuarina cunninghamiana[+]	2	7131			X	X									-	-	Sieve
" *cristata* ssp *cristata*[+]	2	1516 ± 621						X							C	-	Sieve
" *equisetifolia*[+]	11	2222 ± 1408	X		X	X	X								C	S	Sieve
" *glauca*	1	1250				X									C	S	Sieve
Mimosaceae																	
Cathormion umbellatum	1	-													C	-	T
Fabaceae																	
Dendrolobium umbellatum	1	16	O	O	O					X					C	-	T***

(*Continued*)

Table 1. (*Continued*)

		CSIRO Tree Seed Centre results average seed viability/10 g⁺⁺	Field observations														
			Flower and seeding months												Vegetative reproduction		
Species	Provenances	Individual and bulked tree collections	J	F	M	A	M	J	J	A	S	O	N	D	Ability to coppice	Ability to sucker	Seed cleaning method
Myrtaceae																	
Eucalyptus argophloia	1	13733 ± 8343	X												-	-	Sieve
〃 *raveretiana*⁺	1	24667 ± 11501				X									-	-	Sieve
Rutaceae																	
Flindersia maculosa⁺	1	568												X	-	-	Sieve
Geijera parviflora	1	-												X	-	-	T
Verbenaceae																	
Gmelina dalrympleana	1	9			X								O	O	C	-	MS
Proteaceae																	
Grevillea glauca⁺	2	184	O		OX	X	OX	O						OX	C	-	Sieve
〃 *parallela*	1	293			O		O	O	O	O	O	O	X	X	-	-	Sieve
〃 *pinnatifida*	1	480	X										O	O	-	-	Sieve
〃 *pteridifolia*	2	200	O		O	O	O	O		OX	OX	X	O	O	C	-	Sieve
Myrtaceae																	
Leptospermum flavescens	3	9951				X	X		X		OX	X			C	-	Sieve
〃 *flavescens* vel. aff.	1	8150							X						C	-	Sieve
〃 *flavescens x petersonii*⁺	1	3560						X							C	-	Sieve
〃 *liversidgei*	1	2700			X										-	-	Sieve
〃 *longifolium*	2	12925 ± 7958								O	O		X	X	C	-	Sieve
〃 *petersonii*⁺	1	24000							X	X	X				C	-	Sieve
Lophostemon confertus	2	3062			X												Sieve
〃 *suaveolens*	2	3000	X								O		O	X	C	S	Sieve
Melaleuca acacioides ssp. *acacioides*	1	25998 ± 10555								X	X		X		C	S	Sieve
〃 *angustifolia*	1	1000										X			C	-	Sieve
〃 *arcana*	2	76818 ± 31279								X	X				C	-	Sieve
〃 *argentea*	2	6919 ± 4165	X							O	O	O		X	C	-	Sieve
〃 *bracteata*	3	56200	X	X	X					O	OX	O	O		C	S	Sieve
〃 *brassii*	1	3000											X	X	-	-	Sieve
〃 *cajuputi*	1	52091 ± 13838							O			X	X		C	S	Sieve
〃 *citrolens*	1	20000									X	X			-	-	Sieve
〃 *decora*	1	46000	X									X			-	-	Sieve
〃 *'fluviatilis'* MS	1	6225									X		X	X	C	-	Sieve
〃 *lanceolata*	1	12400		X	X								X		C	-	Sieve
〃 *leucadendra*	3	21654 ± 13809			X	X	O	O	O	O			OX		-	-	Sieve
〃 *linariifolia*	3	21478 ± 10782	X			X				O	O	OX			C	-	Sieve
〃 *nervosa*	1	56100 ± 19105					O			O	O		X	O	C	-	Sieve
〃 *quinquenervia*	4	27762		X	X			X					X	X	C	-	Sieve
〃 *saligna*	2	49482 ± 16973		O						X	X	O			C	-	Sieve
〃 *stenostachya*	1	26000											X		-	-	Sieve
〃 *symphyocarpa*	2	4200 ± 1079				X							X		C	-	Sieve
〃 *viridiflora*	4	21196 ± 10790	O	O	OX	OX	O	O		OX			X	O	C	S	Sieve
Meliaceae																	
Melia azedarach var. *australasica*	2	3					X	X			O	O			C	S	MS
Myrtaceae																	
Metrosideros tetrapetala	1	-								X	X				C	S	Sieve
Neofabricia myrtifolia	2	2345											X	X	C	S	Sieve
Neofabricia sp. aff. 〃 *myrtifolia*	1	1850										X			-	S	Sieve
Chrysobalanaceae																	
Parinari nonda	1	0.2								O	OX	O	OX	O	-	-	MS
Proteaceae																	
Persoonia falcata	3	NIL	O		X	X	X			O	O		X	OX	-	-	MS

(*Continued*)

Table 1. (*Concluded*)

Species	Provenances	CSIRO Tree Seed Centre results average seed viability/10 g[++]: Individual and bulked tree collections	Field observations: Flower and seeding months: J	F	M	A	M	J	J	A	S	O	N	D	Vegetative reproduction: Ability to coppice	Ability to sucker	Seed cleaning method
Euphorbiaceae																	
Petalostigma pubescens	3	5	X		X				X	X				X	C	-	Sieve
Sapotaceae																	
Planchonella pohlmaniana var. *vestita*	5	0.5								X	X			X	-	-	MS
Anarcardiaceae																	
Rhodosphaera rhodanthema	3	12			X				X	X					C	-	T
Myrtaceae																	
Syncarpia hillii	1	671 ± 482			X										-	-	Sieve
Syzygium suborbiculare	1	-									O		X		C	-	See Footnote [2]
Combretaceae																	
Terminalia arenicola	3	1.2						X	X						C	-	MS
" *muelleri*	4	0.7						X	X						C	-	MS
" *platyphylla*	1	1.0	O			O								X	C	-	MS
" *platyptera*	1	NIL								X				X	-	-	MS
Rhamnaceae																	
Ventilago viminalis	1	88												X	C	S	See Footnote [1]
		- = No test; NIL = No germination	J	F	M	A	M	J	J	A	S	O	N	D			

Family nomenclature follows the Australian standard (Cronquist 1981).
[+] Additional species to original list included as a result of opportunistic collections.
[++] Standard deviations for seed viabilities were determined for species with 5 or more individual tree or bulk tree values

O = Flowering
X = Seeding

C = Coppice observed
S = Suckering observed
- = C and/or S not observed at the sites visited

Seed Cleaning Methods

Sieve — Fruits were dried in the sun and sieved from the seed.
T — Fruits were threshed using a portable electric (2hp) seed thresher and then sieved from the seed.
MS — Moist scarification was achieved efficiently with the use of a concrete mixer and varying combinations of sand, rocks and water.
* — With some species threshing is not necessary if the pods have matured and opened sufficiently.
** — This species requires high temperatures to open follicles and release seed.
*** — Pods break into sections with the seed remaining inside and are stored and germinated in this state.

[1] *Atalaya hemiglauca* and *Ventilago viminalis* cannot be threshed to remove the wings without damaging the seed. Therefore they are handled in an entire state.
[2] *Syzygium suborbiculare* has a very limited storage life as it usually germinates at or shortly after fruit maturity. Cleaned seed can be stored for short periods in peat moss or vermiculite with fungicide in plastic bags.

Table 2. Summary of species which presented particular seed collection difficulties.

	Level of difficulty	
	Some difficulties	Difficult
Dry Fruit		
Capsular and other dry fruit		
Alphitonia excelsa	L	
Lophostemon suaveolens	F,W	
Melaleuca angustifolia	T	
" *argentea*	T,W	
" *brassii*		R
" *dealbata*	T,W	
" *saligna*	R	
Metrosideros tetrapetala		L,R,S,T
Petalostigma pubescens	E,F,L	
Leguminous fruit		
Acacia brassii	R,T	
" *hylonoma*		C,CD,T
" *julifera* ssp. *gilbertensis*	S,T	
" *shirleyi*		E,R,T
Adenanthera abrosperma	L	
Albizia toona		CD,E,S
Cassia queenslandica		C,E,S
Cathormion umbellatum		L,R
Follicular fruit		
Banksia integrifolia var. *aquilonia*		S
Buckinghamia celsissima		CD,F,L,T
Grevillea parallela		F,L,S,T
" *pinnatifida*	F,L,T	
Fleshy Fruit		
Gmelina dalrympleana		L,R,S,T,W
Parinari nonda		L,R
Persoonia falcata		L,S
Pouteria sericea		E,L,S
Syzygium suborbiculare		F,L,R,S
Terminalia platyphylla	E,L	
" *sericocarpa*	E,L	

KEY

C Immature crops were prone to cockatoo attack.
CD Collection difficulties were experienced as a result of the height of trees and associated dense canopy in rainforest.
E Erratic seeding from year to year was observed for this species.
F This species flowers and fruits over a relatively long period and therefore at any one time it was difficult to collect reasonable quantities of mature seed.
L Fruits of this species are relatively large and it is therefore difficult and time consuming to collect large quantities of seed to meet trial requirements.
R Remoteness of species populations made it difficult to regularly monitor crop maturity.
S Only small amounts of seed were available in any one year.
T Timing of collection is critical.
W Stands can be inundated by water when they are seeding.

Conclusion

The ACIAR seed collection program has made available for field trials previously uncultivated species which are adapted to a wide range of difficult sites in tropical and subtropical Australia. Early trial results are demonstrating the potential of many of these for fuelwood and agroforestry purposes. As a result, provenance collections of the most promising species are being initiated. In North Queensland, these are being undertaken by a CSIRO Australian Tree Seed Centre team recently based again at the regional station in Atherton.

This program was also the first concentrated thrust by the Australian Tree Seed Centre into the collection of Australian genera other than *Eucalyptus*, *Casuarina* and *Acacia*, and the experience gained has led to the problems of cleaning, storing and germinating recalcitrant species becoming a research priority for the Centre.

The knowledge gained of these lesser-known species will assist future collections and their promotion for a range of uses and environments. There are, of course, many other Australian species to be sampled and assessed for their adaptability, growth and utilisation. For this to be accomplished there must be a continuing commitment to similar scientifically based seed collection programs.

Acknowledgments

This seed collection program would not have been as successful without the dedication, hard work and taxonomic skills of Jim Moriarty who worked with the program for its full duration. My thanks also for his contributions to Table 1. Our sincere thanks to Vince and Billie Moriarty (Australian Tropical Plant Supplies, Dimbulah, Qld) and John Clarkson (Queensland Department of Primary Industries, Mareeba, Qld) for their enthusiastic support during the collections in North Queensland. Their accurate species identifications, population locations and information concerning the timing of flowering and seeding of a number of species were invaluable.

I would also like to thank Tim Vercoe who headed the collection team for the final 4 months of the program in southeast Queensland. At all times the team was supported by CSIRO Australian Tree Seed Centre staff and in particular by Jerry Cole and John Doran.

Our colleagues from the CSIRO Atherton and Samford regional stations also provided support and information which contributed to the success of the program as did botanists working with the Queensland Department of Primary Industries, ecologists working with mining companies and the Queensland National Parks and Wildlife Service, and foresters with the Queensland Department of Forestry.

Field Trials

Chapter 4

Climatic Conditions at Trial Sites in China, Kenya, Thailand and Zimbabwe Compared to Similar Regions in Australia

T.H. Booth

Abstract

This paper summarises climatic conditions at 19 multipurpose tree trial sites in Australia, China, Kenya, Thailand and Zimbabwe. The best available information was obtained for monthly mean values of maximum temperature, minimum temperature and precipitation at each site. From these 36 values 18 indices were calculated which represent annual and seasonal conditions. The 18 indices for each site were compared with conditions in a regular grid across Australia. This comparison indicated the parts of Australia that experience similar climates to the trial sites. The 18 indices for all trial sites were also analysed together to show how similar each site was to the others.

Introduction

For many years climate has been an important consideration when selecting tree species for trials. All the major sources of information on species' requirements consider climate in some detail (e.g. Streets 1962; U.S. National Research Council 1980; FAO 1981; Webb et al. 1984). Significant new advances in climatic analysis have been made as part of the work reported in this monograph (see also Booth 1988). These new methods enable useful information to be obtained from climatic analyses of a species' natural distribution (Booth 1985), its performance in trials (Booth et al. 1988) or the conditions at a trial site (Booth et al. 1987).

Trials are expensive, so it is important to make the best possible use of results. Climatic analysis, along with evaluation of other important factors, such as soil attributes, can help generalise the results of trials. In this way, recommendations of suitable species can be made for other areas. Improving these techniques will be a major objective for continuing analyses as part of the ACIAR forestry program. In the meantime, some basic information on climatic conditions at the trial sites is presented here. The purpose of this description is to indicate the range of climatic conditions at the trial sites, how they relate to conditions in Australia and how similar they are to each other.

Climatic Conditions at Trial Sites

Monthly mean values of average daily maximum temperature, average daily minimum temperature and total precipitation were obtained for 19 trial sites. Conditions at the Australian sites were estimated using interpolation surfaces, which allow mean climatic conditions to be assessed for any location in Australia (see Hutchinson 1988 in press). Record lengths for sites in other countries were often short, as trial sites were usually some distance from major meteorological stations. The names and locations of the sites, along with the number of years of meteorological records, are shown in Table 1. Figure 1a–s shows the maximum temperature, minimum temperature and precipitation data for these sites.

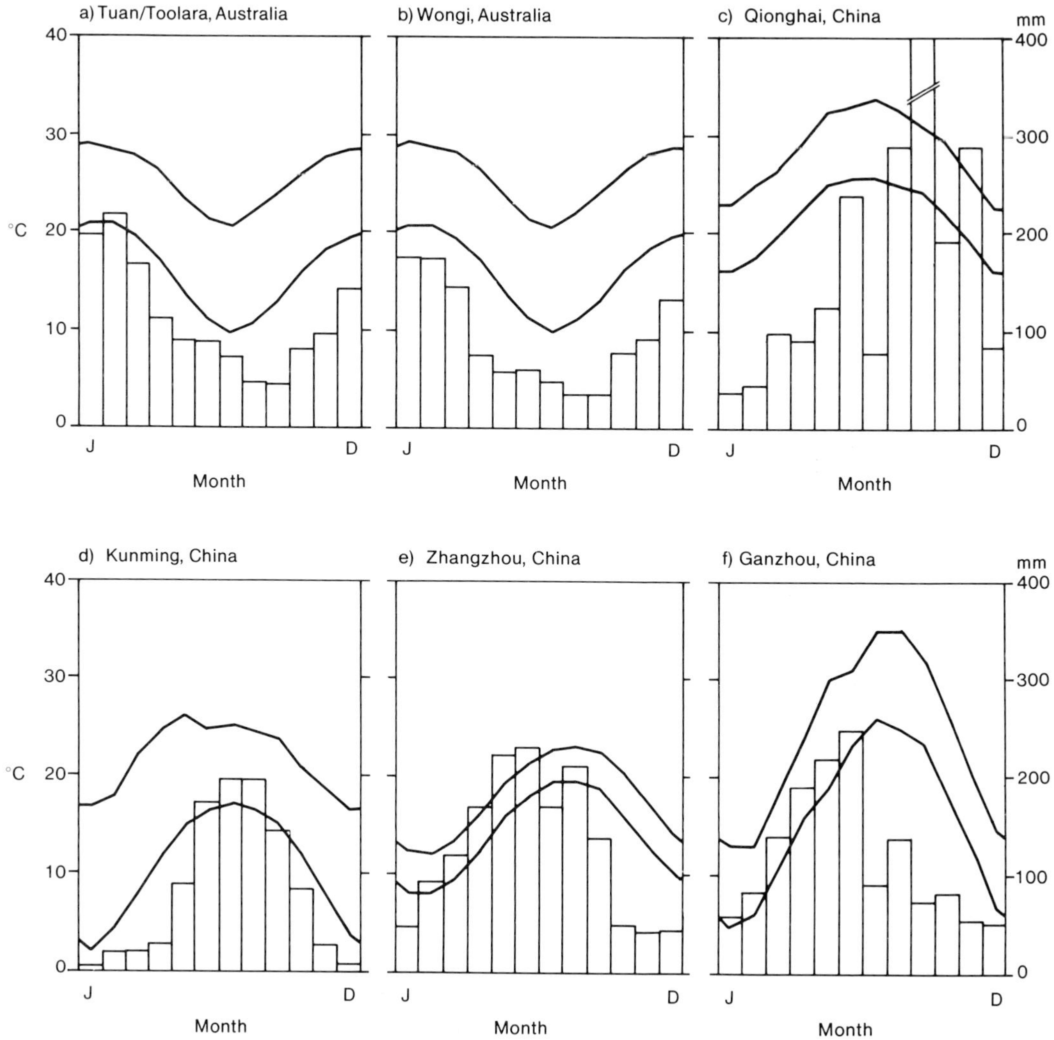

Fig. 1. Climate diagrams for trial sites showing monthly mean values of average daily maximum temperature, average daily minimum temperature and total precipitation.

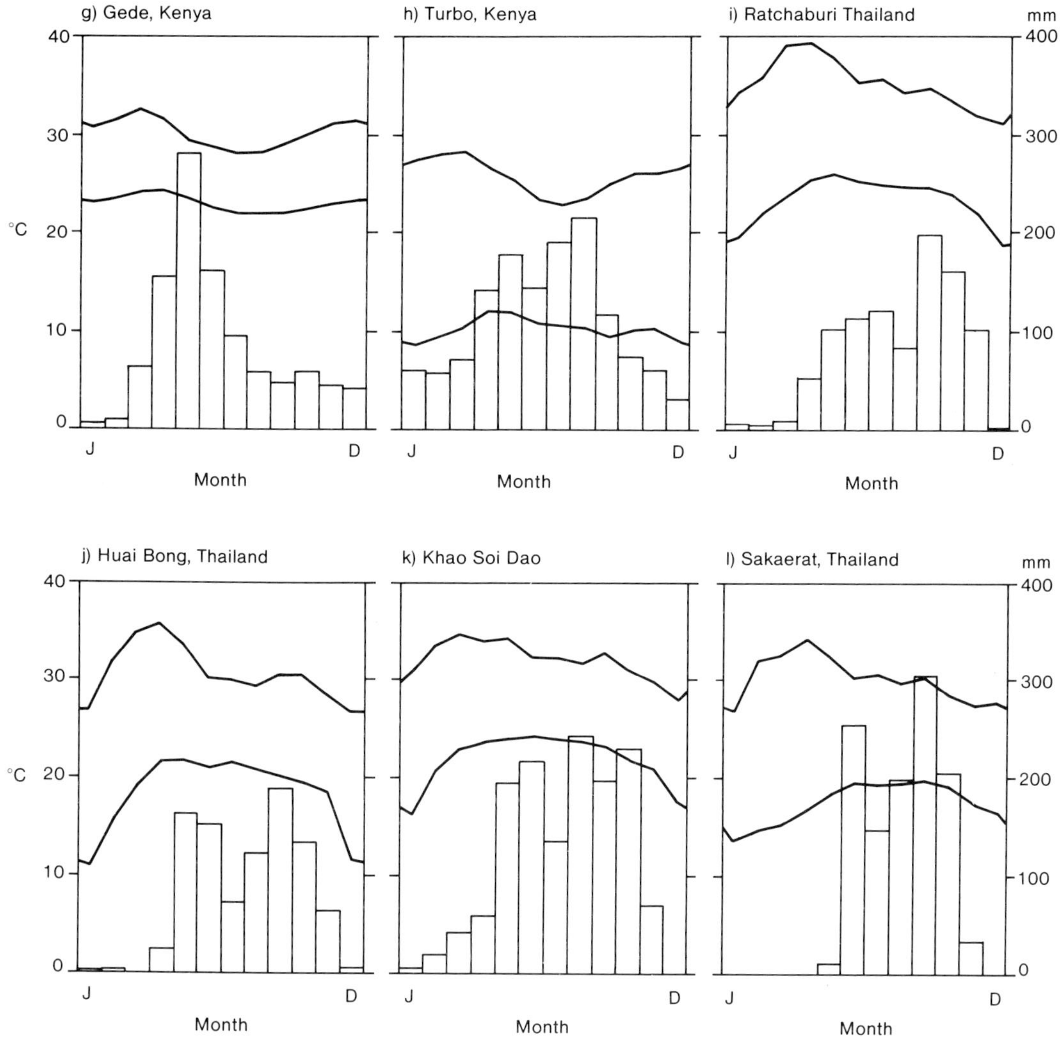

Fig. 1 *(continued).* Climate diagrams for trial sites showing monthly mean values of average daily maximum temperature, average daily minimum temperature and total precipitation.

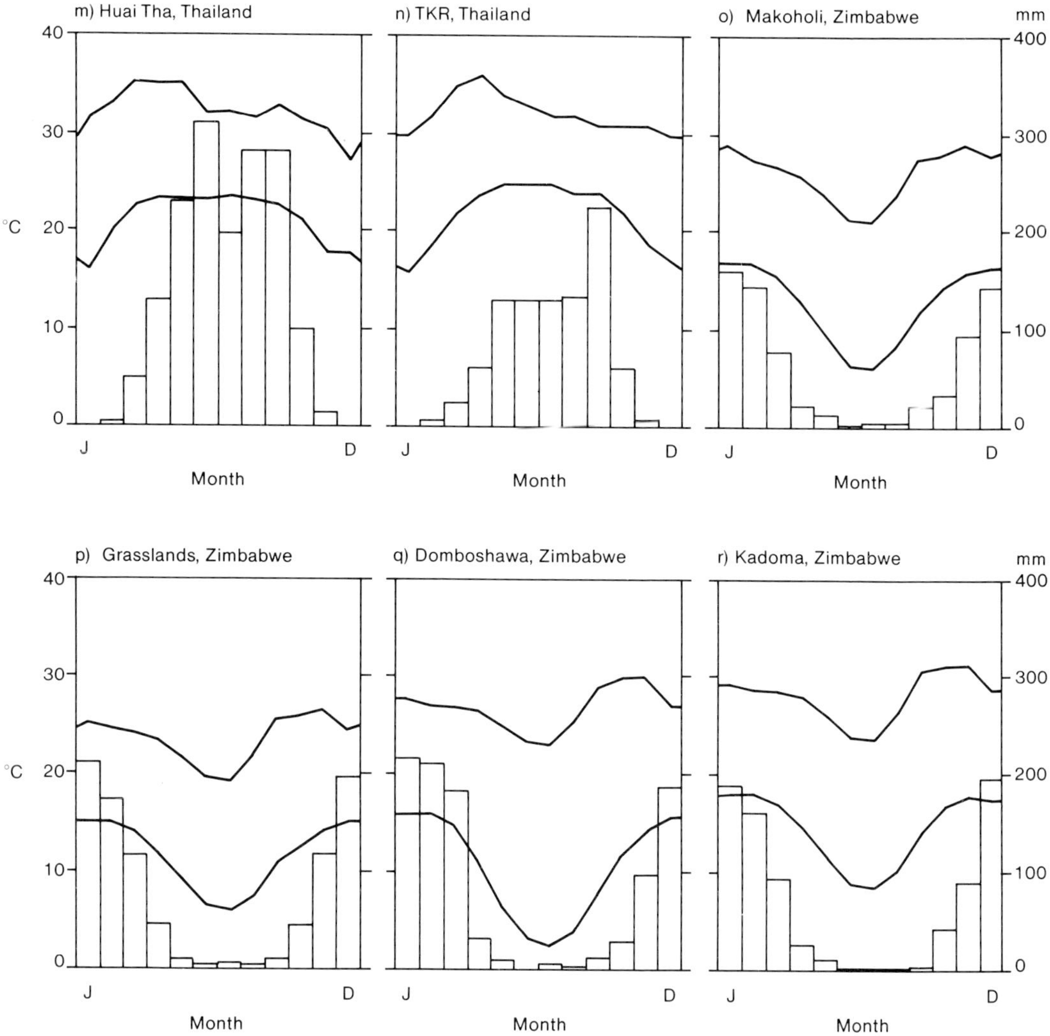

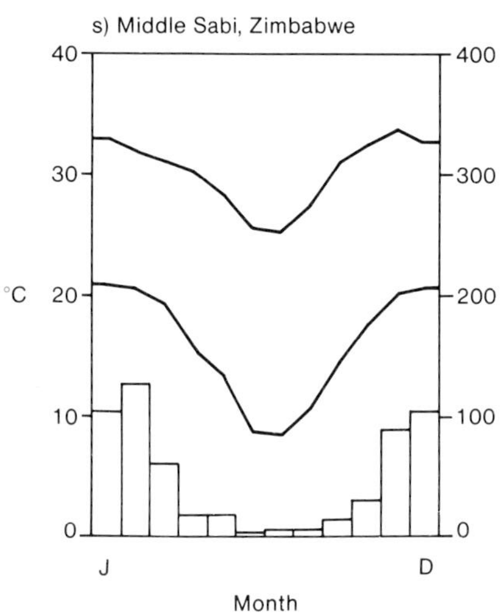

Fig. 1 *(concluded).* Climate diagrams for trial sites showing monthly mean values of average daily maximum temperature, average daily minimum temperature and total precipitation.

Fig. 2 Areas within Australia most climatically similar to conditions at trial sites outside Australia. Most similar areas are shown in dark shading. A note at the top right of each map indicates whether the best match was very good, good, fair or poor.

i) Ratchaburi Good

j) Huai Bong Good

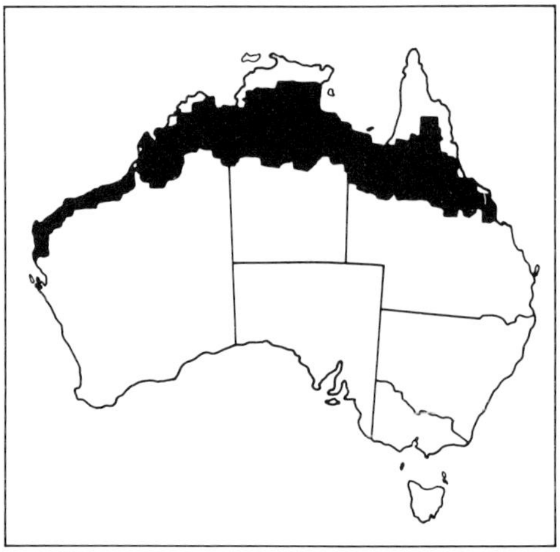

k) Khao Soidao Good

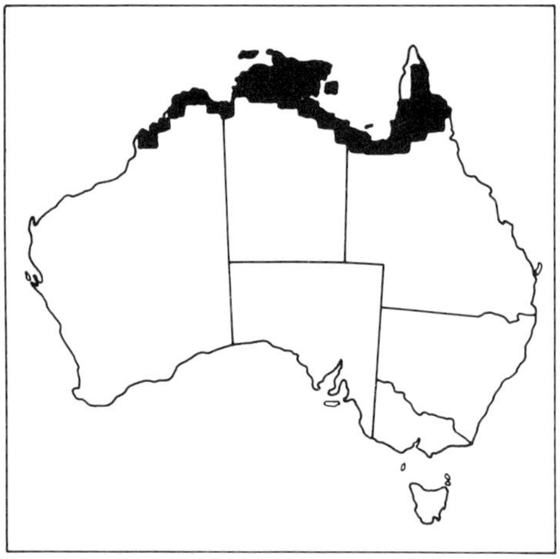

l) Sakaerat Good

m) Huai Tha Very good

n) T.K.R. (Roi Et) Very Good

o) Makoholi — Very good

p) Grasslands — Good

q) Domboshawa — Good

r) Kadoma — Very good

s) Middle Sabi — Very good

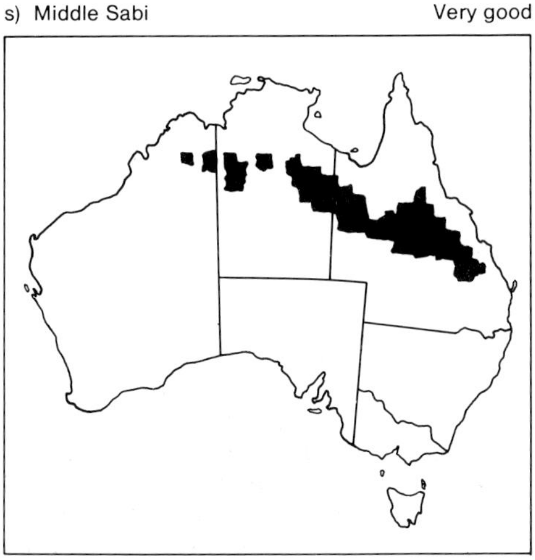

Fig. 2 *(concluded).* Areas within Australia most climatically similar to conditions at trial sites outside Australia. Most similar areas are shown in dark shading. A note at the top right of each map indicates whether the best match was very good, good, fair or poor.

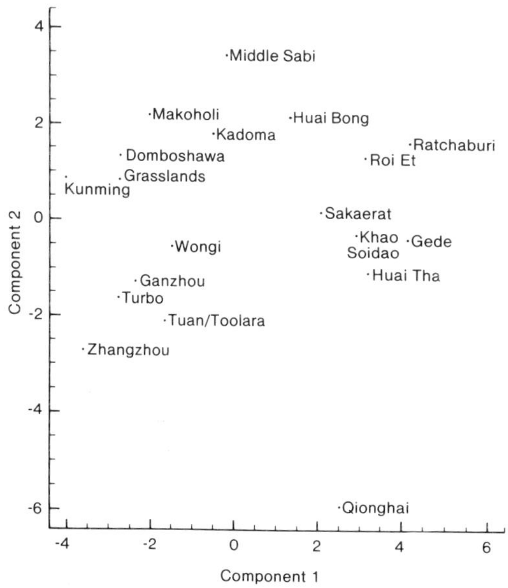

Fig. 3 Principal component analysis of 19 trial sites based on 18 climatic factors.

Comparison of Climates

From the 36 values shown for each site in Fig. 1, 18 indices were calculated which summarise the variation in mean climatic conditions. The indices, which are measured in degrees C, millimetres or dimensionless units, are as follows:

1. Annual mean temperature
2. Coldest month minimum temperature
3. Hottest month maximum temperature
4. Annual temperature range (i.e. index 3 - index 2)
5. Wettest quarter mean temperature
6. Driest quarter mean temperature
7. Annual mean precipitation
8. Wettest month mean precipitation
9. Driest month mean precipitation
10. Annual precipitation range (i.e. index 8 - index 9)
11. Wettest quarter mean precipitation
12. Driest quarter mean precipitation
13. Warmest quarter mean precipitation
14. Coldest quarter mean precipitation
15. Warmest quarter mean temperature
16. Coldest quarter mean temperature
17. Annual precipitation range/(annual mean precipitation/12)
18. Annual temperature range/annual mean temperature

These values for each site are tabulated in Table 2. It should be noted that these indices can be calculated from data for northern or southern hemisphere sites without having to define summer or winter seasons.

Using these values each site was compared with climatic conditions at 2795 locations in a regular grid across Australia using the method described by Booth et al. (1987). However, to show a large number of comparisons in a small space, the maps have been simplified here to show only the most climatically similar areas (Fig. 2).

It is not always possible to find a good match between the locations across Australia and the sites outside Australia. The similarity measure for the most similar Australian site is printed out by the program to indicate how good a match has been found. A note at the top right of each map in Fig. 2 indicates whether the best match was very good, good, fair or poor.

Climatic Similarity

It is useful to compare conditions at all the sites to appreciate the range of conditions where provenances are being tested. The results of a principal component analysis of the data shown in Table 2 are shown in Fig. 3 (see Jeffers 1978 for simple explanation of PCA). The first two components shown accounted for 70% of the variance in the data.

Figure 3 shows how similar one site is to another. Locations that are close together on the graph experience similar climatic conditions and those that are far apart experience markedly different conditions. The first component was largely associated with temperature. Cool sites, such as Kunming and Zhangzhou, are at the left of the graph. Warm sites, such as Ratchaburi and Gede, are at the right of the graph. The second component was mainly associated with rainfall-related factors. Dry sites, such as Middle Sabi, are at the top of the graph. Wet sites are towards the bottom of the graph. Qionghai is shown to be considerably different from the other sites, because it is not only the wettest site in terms of annual mean precipitation, but also is markedly different in terms of wettest month and wettest quarter precipitation (see Table 2).

Discussion

The climatic factors described here provide an introduction to the range of conditions experienced at the trial sites. The short record lengths available for most of the sites mean that the data shown here should be treated with some caution. Nevertheless,

Table 1. Location of trial sites included in climatic analysis.

Site*		Latitude	Longitude	Elev (m)	Record length (years)
Australia					
a)	Tuan/Toolara	25°47′S	152°50′E	45	n/a
b)	Wongi	25°26′S	152°32′E	70	n/a
China					
c)	Qionghai (Hainan Island)	19° 0′N	110°30′E	45	8
d)	Kunming (Yunnan)	25° 2′N	102°43′E	1893	10
e)	Zhangzhou	24°30′N	117°39′E	30	10
f)	Ganzhou	25°51′N	114°50′E	124	15
Kenya					
g)	Gede	3°19′S	40° 3′E	40	6
h)	Turbo	0°37′N	35° 5′E	1800	16
Thailand					
i)	Ratchaburi	13°25′N	99°50′E	30	7
j)	Huai Bong, nr Chiang Mai	18°12′N	98°25′E	790	4
k)	Khao Soidao, Chanthaburi	13°00′N	102°15′E	200	5
l)	Sakaerat Thai/Japan Proj.	14°13′N	101°55′E	550	1
m)	Huai Tha, Si Sa Ket	14°53′N	104°27′E	130	3
n)	TKR, Roi Et	15°50′N	103°20′E	110	2
Zimbabwe					
o)	Makoholi	19°50′S	30°47′E	1210	14
p)	Grasslands	18°10′S	31°30′E	1646	14
q)	Domboshawa	17°36′S	31° 8′E	1552	14
r)	Kadoma	18°19′S	29°54′E	1157	14
s)	Middle Sabi	20°21′S	32°20′E	448	14

*For location of trial sites see Fig. 1 in Chapter 1 of this Monograph.

they provide a first impression of the range of climatic conditions covered by the trials. Climatic data are being collected at or near the sites during the course of the trials and these data will be used later in more detailed analyses.

The climate diagrams in Fig. 1 differ from the well-known diagrams of Walter (1970) as they show both maximum and minimum temperatures. Figures 1d and 1e for Kunming and Zhangzhou show how different these may be, even though average monthly temperatures remain similar.

The comparisons of climatic conditions at trial sites with conditions in Australia suggest general regions from which successful species and provenances might come. For example, *Eucalyptus camaldulensis* is widely successful in Thailand (Thailand Royal Forest Department 1987). The comparisons in Fig. 2 show how climatically similar these sites are to much of the distribution of northern provenances of *E. camaldulensis*. However, the comparison is only a quick guide and there are exceptions. For example, sites which show closest similarity with sites in southwestern Australia are often found to be more suited to southeast *Eucalyptus* species. This is probably because relatively few Western Australian *Eucalyptus* species have proved successful in plantations (FAO 1981).

Figure 3 gives an impression of the range of conditions covered by the trials. Examination of Fig. 3 and Table 2 suggests that the coverage of climatic conditions might be improved by addition of more low rainfall sites, as well as sites around 24°C annual mean temperature and 1700 mm annual mean precipitation.

The data gathered in the field trials described in this book will help us understand how lesser-known Australian tree species react to climate and other

Table 2. Bioclimatic indices for 19 ACIAR trial sites.

	*Index no.																	
	1 ANN TEMP	2 MIN MIN	3 MAX MAX	4 TEMP RANG	5 TEMP WET Q	6 TEMP DRY Q	7 ANN PREC	8 WET MTH PREC	9 DRY MTH PREC	10 PREC RANG	11 WET Q PREC	12 DRY Q PREC	13 WARM Q PREC	14 COLD Q PREC	15 WARM Q TEMP	16 COLD Q TEMP	17 PREC SEAS	18 TEMP SEAS
a) Tuan/Toolara	20.7	9.8	29.0	19.2	24.5	16.8	1362	219	47	172	583	168	559	209	24.6	16.0	1.5	0.9
b) Wongi	20.9	9.9	29.4	19.5	24.7	16.9	1113	176	36	140	496	121	484	144	24.8	16.2	1.5	0.9
c) Qionghai	24.9	15.7	33.6	17.9	25.1	19.8	2070	526	34	492	1006	158	603	158	29.1	19.8	2.9	0.7
d) Kunming	16.3	2.0	26.1	24.1	20.8	10.3	992	197	3	194	567	31	567	31	20.8	10.3	2.3	1.5
e) Zhangzhou	16.0	8.0	23.0	15.0	19.5	12.5	1533	230	41	189	622	131	522	257	21.0	10.6	1.5	0.9
f) Ghanzhou	20.1	5.0	35.0	30.0	23.8	12.0	1431	249	51	198	658	162	304	190	29.3	9.8	1.7	1.5
g) Gede	26.4	21.7	32.3	10.6	26.5	27.0	988	279	2	277	591	47	220	191	27.7	25.0	3.4	0.4
h) Turbo	17.9	8.4	28.2	19.8	17.0	18.0	1315	214	28	186	546	140	264	517	18.9	17.0	1.7	1.1
i) Ratchaburi	29.4	19.0	39.5	20.5	28.6	26.9	964	198	3	195	465	14	164	114	31.9	26.4	2.4	0.7
j) Huai Bong	24.6	11.2	35.9	24.7	25.1	23.3	934	189	0	189	446	3	192	70	27.8	20.6	2.4	1.0
k) Khao Soidao	27.0	16.2	34.6	18.4	27.4	24.5	1421	241	1	240	670	28	298	78	28.9	23.9	2.0	0.7
l) Sakaerat	23.8	13.5	34.1	20.6	24.4	22.4	1146	305	0	305	706	0	263	32	25.2	21.5	3.2	0.9
m) Huai Tha	26.6	15.7	35.2	19.5	27.5	24.0	1586	309	0	309	787	3	403	13	29.0	23.2	2.3	0.7
n) TKR, Roi Et	27.0	16.0	36.0	20.0	28.0	23.8	949	228	0	228	526	7	324	7	29.5	23.8	2.9	0.7
o) Makoholi	19.2	6.1	29.0	22.9	22.3	14.5	719	159	2	157	446	10	398	10	22.5	14.5	2.6	1.2
p) Grasslands	17.4	5.9	26.5	20.6	19.6	13.4	925	208	2	206	576	10	521	10	19.9	13.4	2.7	1.2
q) Domboshawa	18.4	2.5	29.8	27.3	21.4	13.5	973	214	0	214	609	6	496	6	21.7	13.5	2.6	1.5
r) Kadoma	21.2	8.5	31.3	22.8	23.3	17.0	812	198	1	197	544	4	330	4	24.0	17.0	2.9	1.1
s) Middle Sabi	22.9	8.2	33.5	25.3	26.4	17.5	555	125	1	124	326	10	287	10	26.6	17.5	2.7	1.1

*See text for list of climatic indices.

environmental factors, such as soil conditions. As we begin to understand these relationships we can replace simple comparisons, such as those shown in Fig. 2, with detailed descriptions of environmental requirements for particular species and provenances.

Aerial photograph of the 1985–86 ACIAR species trials at Tuan/Toolara State Forest near Gympie, Qld, Australia, managed by the Queensland Department of Forestry. Photograph indicates the block plantings of species showing variable survival. The whole trial is surrounded by routine plantings of *Pinus caribaea*. Photograph courtesy Queensland Department of Forestry.

Chapter 5

Growth, Coppicing and Flowering of Australian Tree Species in Trials in Southeast Queensland, Australia

P.A. Ryan and R.E. Bell

Abstract

Early data on growth, coppicing ability and flowering are presented for 148 Australian species derived from a range of environments and established in trials in southeast Queensland from 1984 to 1986. Some information is provided also on potentially destructive sources of damage, especially insects. Generally, the most successful species were those originating from wetter and warmer areas while those from cool, dry environments failed. Nevertheless, many species derived from dryland areas have performed well in cultivation under the moist subtropical conditions of the region. A number of species previously unknown in cultivation have shown very fast growth rates comparable with species used currently in commercial plantings. There has been substantial variation in performance between provenances for some species highlighting the importance of assessing provenance as well as species performance. Within-provenance variation has been substantial also in some species. In such cases a tree improvement program may be warranted in the longer term to realise fully their potential utility.

Coppicing ability has been consistent across many of the genera but varied widely within the acacias, ranging from complete failure to abundant regeneration through root suckering following cutting. The capacity to spread by regeneration by root suckering in some species, or by prolific seeding in others, indicates high potential for weediness. Introduction of such species into foreign environments needs to be treated with considerable caution.

Introduction

Field trials of 177 Australian species, many described in Turnbull (1986), comprising a total of 306 seedlots were established between 1984 and 1987. The rationale for the trials owes much to Boland and Turnbull (1981) who identified the potential role for Australian species in assisting to alleviate fuelwood shortages in developing countries, discussed the ecological, botanical and management criteria for species selection and pointed out potential constraints to use. The aim of these trials is to:

(1) Gather information on growth rates and general performance of these taxa in cultivation;

(2) Assemble qualitative information on biological attributes of the species; and

(3) Provide a resource for studies into the utilisation of the species.

A basic premise underlying the conduct of the trials was that constraints to productivity should be minimised to enable valid evaluation of potential performance. Thus the management of the trials aimed at providing reasonably high levels of inputs (cultivation, weed control, fertilising, insect control) where practicable. The need for various degrees of cultural inputs can be determined once base performance data from the initial screening process have been obtained and the more promising species identified.

This paper outlines the techniques used in the

trials and provides preliminary summary information for those species (148) and seedlots (276) that were tested in the first three sets of trials (1984-86).

Methods

Location

The field trials were located on sites in the Tuan/Toolara (25°47′S, 152°50′E, 45 m ASL) and Wongi (25°26′S, 152°32′E, 70 m ASL) State Forests near Gympie in southeast Queensland (see Fig. 1 in Chapter 1 of this Monograph). This area has a subtropical climate of warm wet summers and cool mainly dry winters. Climate for the two regions is comparable, with the sites at Wongi State Forest receiving slightly lower rainfall (see Chapter 4 for details).

Soils are of low fertility, loamy sands in the upper horizons increasing in texture to sandy clay loam with depth. The soils at Tuan/Toolara are deeper with generally better drainage than those at Wongi, though seasonal saturation to the soil surface may occur at all sites.

The Tuan/Toolara sites originally carried tall, open forest while the vegetation at the Wongi sites was woodland (Specht et al. 1974). Vegetation on all sites was dominated by an overstorey of *Eucalyptus* species.

Site Preparation and Establishment

Standing vegetation was cleared by crawler tractors, heaped into windrows and burnt. Sites were ploughed to about 30 cm depth and reploughed prior to planting for the first 2 years. Subsequently, the second ploughing was replaced by the construction of small mounds to overcome drainage problems resulting from depressions created by ploughing. Plots established for the growth trials avoided ash heaps left from burning.

Planting stock was raised at the Toolara nursery in small (50–70 ml) tubes or net pots while refill stock was raised in larger (200 ml) tubes (see Ryan et al. 1987 for details). Refilling was carried out where necessary within 2 months of planting. Planting in all cases was by the use of metal bars designed to punch holes in the ground of identical dimensions to the seedling root ball.

Design and Treatments

Over half the 276 entries in the first three sets of trials were acacias, the other major genera being the eucalypts and melaleucas (Table 1). Though most of the material was derived from subtropical and tropical areas, the selections covered a diverse range of climatic origins within Australasia (Table 2 — details of individual seedlots are covered in Appendices 1–4). The large numbers of seedlots under test necessitated establishment of trials over several years and on different sites at each location. To enable an assessment of the effects of uncontrolled variation introduced as a consequence, seedlots of a number of species were repeated in each year's plantings to act as standards in the comparisons of the performance of all species. In addition, several well-documented species (*Eucalyptus camaldulensis, E. cloeziana* and *E. grandis*) were included in the 1986 planting.

The field trials for each of the first three planting years consist of two replicate growth plots and a single replicate coppice plot at each location for each seedlot. The growth study plots consist of 36 trees planted at 3 m × 2 m, the middle 16 trees being measured. The coppice study plots are 20 tree line plots with 1.5 m spacing between trees and 3 m spacing between plots. Cutting treatments were applied to the 1984 coppice plots at 3 years of age and to 1985 and 1986 plots when 2 years old. Within each plot, trees were cut at 0.1, 0.5 or 1.0 m above ground with generally 6–7 trees per cutting height. All branches were removed from 0.1 and 0.5 m stumps, but at least one viable branch retained on each 1.0 m stump where possible. Where there were 10 or fewer surviving trees, all were cut at 0.5 m.

Management

Trials were fenced to exclude cattle where necessary while net fencing was erected around some sites to exclude small herbivores following considerable browsing damage to the first year's planting.

Weed control aimed to maintain a 1 m radius around each tree free of competition for about the first 18 months. In the first year this was achieved by chipping with hoes, but subsequently a guarded application of glyphosate was the major method used with some hand weeding around the base of the plant.

Nitrogen and phosphorus fertilisers were applied in three split dressings at increasing rates over the first 2 years to supply totals of 150 kg/ha P (as triple superphosphate, 19.2% P) and 235 kg/ha N (as ammonium sulfate, 20.5% N and ammonium nitrate, 34% N). Potassium (50 kg/ha as potassium chloride, 50% K and potassium sulfate, 29.7% K) as well as copper, zinc and boron (1 kg/ha of each element) was applied in 1987 to all trials after limited foliar analysis revealed low levels of foliar potassium and marginal levels of some of the trace elements. A total of 140 kg/ha S and 125 kg/ha Ca have been added as incidental elements in the fertilisers.

Table 1. Genera, families and numbers of entries of each planted in ACIAR species trials 1984–86.

Genus Code	Genus	Family	Numbers of entries
Aca	*Acacia*	Leguminosae (Mimosoideae)	152
Adn	*Adenanthera*	Leguminosae (Mimosoideae)	2
Alb	*Albizia*	Leguminosae (Mimosoideae)	2
Alo	*Allocasuarina* (syn *Casuarina*)	Casuarinaceae	9
Alp	*Alphitonia*	Rhamnaceae	2
Ang	*Angophora*	Myrtaceae	2
Ata	*Atalaya*	Sapindaceae	2
Ban	*Banksia*	Proteaceae	1
Cal	*Callitris*	Cupressaceae (Gymnospermae)	3
Cas	*Casuarina*	Casuarinaceae	10
Csa	*Cassia*	Leguminosae (Caesalpinioideae)	1
Des	*Dendrolobium* (syn *Desmodium*)	Leguminosae (Papilionoideae)	1
Dod	*Dodonea*	Sapindaceae	3
Euc	*Eucalyptus*	Myrtaceae	29
Gre	*Grevillea*	Proteaceae	7
Lep	*Leptospermum*	Myrtaceae	4
Lop	*Lophostemon* (syn *Tristania*)	Myrtaceae	3
Mel	*Melaleuca*	Myrtaceae	28
Mla	*Melia*	Meliaceae	2
Nau	*Nauclea*	Sterculiaceae	1
Neo	*Neofabricia* (syn *Leptospermum)*	Myrtaceae	2
Par	*Parinari*	Rosaceae	1
Pet	*Petalostigma*	Euphorbiaceae	2
Pla	*Planchonella*	Sapotaceae	1
Syz	*Syzygium* (syn *Eugenia)*	Myrtaceae	2
Ter	*Terminalia*	Combretaceae	3
Ven	*Ventilago*	Rhamnaceae	1

Insect control was carried out initally on one replication per site by applying acephate (Orthene) on a regular basis after planting. Spraying was carried out in the second replication only when potentially high levels of damage were threatening. There was little indication that the intensive regime of insect control resulted in overall growth improvement, though a few species suffered significant damage. Consequently, the spraying regime was relaxed and insect control has been carried out when needed and only when trees were small enough to spray in safety. No disease control measures have been applied.

A small but significant number of trees in the 1986 planting at Tuan/Toolara were damaged by a native rat (*Rattus tunneyi* ssp *culmorum*) which has caused significant damage also in plantations of *Araucaria cunninghamii* (hoop pine) in southeast Queensland (Kehl 1980). Baits of 1080 on sweet potato coated with linseed oil were laid twice, the first time at 2 kg/ha and the second at 6 kg/ha. Successful control was not achieved though populations were halved.

Assessments and Analysis

Annual measures of height and diameter at ground level were the major growth parameters recorded, while crown width measures and an assessment of health were carried out concurrently. General characteristics of individual species are recorded annually to provide information on foliage density, presence and abundance of thorns and spines, effects on understorey growth and the occurrence of natural regeneration. In addition, general observations of the phenological patterns of individual species and of damage due to insects, disease, wind, frost or animals are recorded during monthly inspections.

Coppice plots were assessed at the time of cutting when height and diameter at ground level were measured and stump health, number of branches, foliage density and health were assessed subjectively. Monthly assessments were recorded of the type of coppice, abundance of coppice shoots, vigour and health, while cause and severity of any

Table 2. Numbers of entries in trials planted in 1984-86 by climate of origin (temperature, rainfall and rainfall distribution).*

	Mean annual rainfall (mm)	Mean annual temperature (°C)					
		<18	18-20	21-22	23-25	≥26	Total
Summer rainfall	<300		1	20	9	3	**33**
	301-500			6	18	10	**34**
	501-700		11	7	4	8	**30**
	701-900	2	9	10	3	7	**31**
	901-1200		3	9	13	7	**32**
	1201-1500	1	9	3	9	7	**29**
	>1500		4	1	16	27	**48**
Winter rainfall	<300	1	2	4	3		**10**
	301-500	4	4				**8**
	501-700	3	3				**6**
	701-900	2	1				**3**
	901-1200	5					**5**
	1201-1500	1					**1**
	>1500	2					**2**
All rainfall	<300	1	3	24	12	3	**43**
	301-500	4	4	6	18	10	**42**
	501-700	3	14	7	4	8	**36**
	701-900	4	10	10	3	7	**34**
	901-1200	5	3	9	13	7	**37**
	1201-1500	2	9	3	9	7	**30**
	>1500	2	4	1	16	27	**50**

*Climatic data provided by T.H. Booth, CSIRO Division of Forestry and Forest Products.

Table 3. Some better performers in ACIAR plantings in 1984-86 by origin mean annual rainfall.

>1100mm	*Aca aulacocarpa*	*Aca leptocarpa*	*Cas cunninghamiana*
	Aca auriculiformis	*Aca mangium*	*Euc cloeziana*
	Aca brassii	*Aca mearnsii*	*Euc grandis*
	Aca cincinnata	*Aca melanoxylon*	*Lep flavescens*
	Aca crassicarpa	*Aca platycarpa*	*Lop suaveolens*
	Aca elata	*Aca podalyriifolia*	*Mel cajuputi*
	Aca falciformis	*Aca rothii*	*Mel leucadendra*
	Aca flavescens	*Aca torulosa*	*Mel quinquenervia* (vel aff)
	Aca holosericea	*Aca trachyphloia*	*Mel saligna*
	Aca hylonoma	*Alo littoralis*	*Mel viridiflora*
	Aca julifera ssp. *gilbertensis*	*Ang costata*	*Mla azedarach* v. *australasica*
700-900 mm	*Aca deanei*	*Aca neriifolia*	*Aca simsii*
	Aca falcata	*Aca parramattensis*	*Aca storeyi*
	Aca fimbriata	*Aca penninervis*	*Aca torulosa*
	Aca glaucocarpa	*Aca plectocarpa*	*Cas cunninghamiana*
	Aca leptoloba	*Aca saligna*	*Euc camaldulensis*
500-700 mm	*Aca aneura*	*Aca decurrens*	*Aca podalyriifolia*
	Aca blakei	*Aca difficilis*	*Aca shirleyi*
	Aca concurrens	*Aca julifera* ssp. *julifera*	*Gre robusta*
	Aca crassa ssp. *crasa*	*Aca plectocarpa*	
<500mm	*Aca ammobia*	*Aca torulosa*	*Euc melanophloia*
	Aca tumida		

damage to stumps or shoots were noted also. The diameter and length of the largest coppice shoot were measured at the final assessment 10 months after cutting.

Statistical analysis to date has been confined to determination of means and estimates of variance for each seedlot by plot and site for the growth study trial only. More detailed analysis of particular subsets of the data may be undertaken on the completion of the first phase of each trial at age 4.5 years.

Results

Growth

Six seedlots only were not outplanted due to total failure in the nursery. Outplantings have been classified as failures where survival is negligible or, progressively from about 18 months, where overall survival, health and vigour are poor. The application of the latter category is conservative to allow the maximum amount of information to be collected for each seedlot. Of the outplanted seedlots, 65 had been classified as failures by September 1987 (Appendix 1).

There is a discernible pattern of failure rate in relation to the climate of seedlot origin. Virtually all seedlots originating from winter rainfall (generally cooler) areas receiving less than 500 mm rain annually have failed. In summer rainfall areas of less than 500 mm, failure rate appears to decrease as mean annual temperature (MAT) increases. Thus, where MAT is less than 23°C, most seedlots have failed but the proportion of failures decreases as MAT increases above 23°C. The failure level for seedlots derived from areas receiving in excess of 500 mm/year is low.

Species derived from dry through to wet zones have all been among the best performers (Table 3) though the majority are from those areas receiving in excess of 1100 mm of rain annually. Some lesser-known species have been outstanding, including *A. cincinnata, A. crassicarpa, A. deanii, A. flavescens, A. plectocarpa* and especially *A. neriifolia* and have been comparable with some of the better known commercial species such as *A. mearnsii, A. melanoxylon, E. camaldulensis* and *E. grandis*.

Nevertheless, within these groupings, not all species could be classified as highly successful. For example, *A. auriculiformis* and *A. aulacocarpa* have sustained continual leaf pathogen infestation from about age 2 years; *A. elata* is highly variable as are *A. ammobia* and *A. aneura; Angophora costata* and *Melia azedarach* suffer frequent and extensive defoliation by insects; *C. cunninghamiana* (fertilised but not inoculated with *Frankia*) developed severe nitrogen deficiency symptoms after cessation of fertiliser applications (see Chapter 22). On the other hand, some species (e.g. *A. simsii* and *Leptospermum flavescens*) are inherently small but have performed well.

In general, performances vary between provenances within species with few exceptions (*A. cincinnata, A. plectocarpa, E. melanophloia, M. cajuputi*), though species failures have been across provenances (e.g. *A. pendula, A. pruinocarpa, Allocasuarina decaisneana* and *E. gamophylla*). In general also, relative provenance performance has been consistent across sites though there are some exceptions (e.g. *A. rothii* and *A. torulosa*).

Provenance variation in some species appears to be related to the level of similarity between provenance climate and site climate (e.g. for *A. melanoxylon* and *M. viridiflora*). In these cases variation in performance may be related more to climatic requirements of the provenances than to inherent differences in vigour. In other instances, this appears not to be the case. In particular, a number of Papua New Guinea provenances have performed better than their North Queensland counterparts (e.g. *A. auriculiformis, A. crassicarpa, A. leptocarpa* and *A.mangium*). There are also a number of instances where provenances are geographically close but differ markedly in performance (e.g. *A. oraria* and *A. platycarpa*). The greatest level of variation between provenances has been for *A. holosericea, A. neriifolia, A. mangium, A. melanoxylon, A. oraria* and *A. platycarpa*.

Overall results for the Wongi sites have been better than for the Tuan/Toolara sites for the 1984 and 1986 plantings, but inferior for the 1985 planting. The last is probably attributable to very wet post-planting conditions in 1985 and the poorer drainage of the Wongi site. However, there has been no consistent pattern in performance differences between the Wongi and Tuan/Toolara sites — some taxa have been considerably better on the first, others considerably worse.

Similarly, there is no consistent trend in the year-to-year performance of the standards, though it appears that generally the performance of stock planted in 1986 is better than that planted in 1984, which in turn is better than that planted in 1985 (Fig. 1). This trend may, in part, be due to improvements in stock quality and silviculture with increasing experience in the management of this type of material. The effects of the very wet post-planting conditions in 1985 may be a factor also. The lack of any consistent trends in performance between sites and planting years highlights the usefulness of the methods of Booth et al. (1987, 1988).

The pattern of development of trees within the trials can be illustrated by the pattern of height growth of a few selected examples from the 1984 planting (Fig. 2). All species were slow to develop

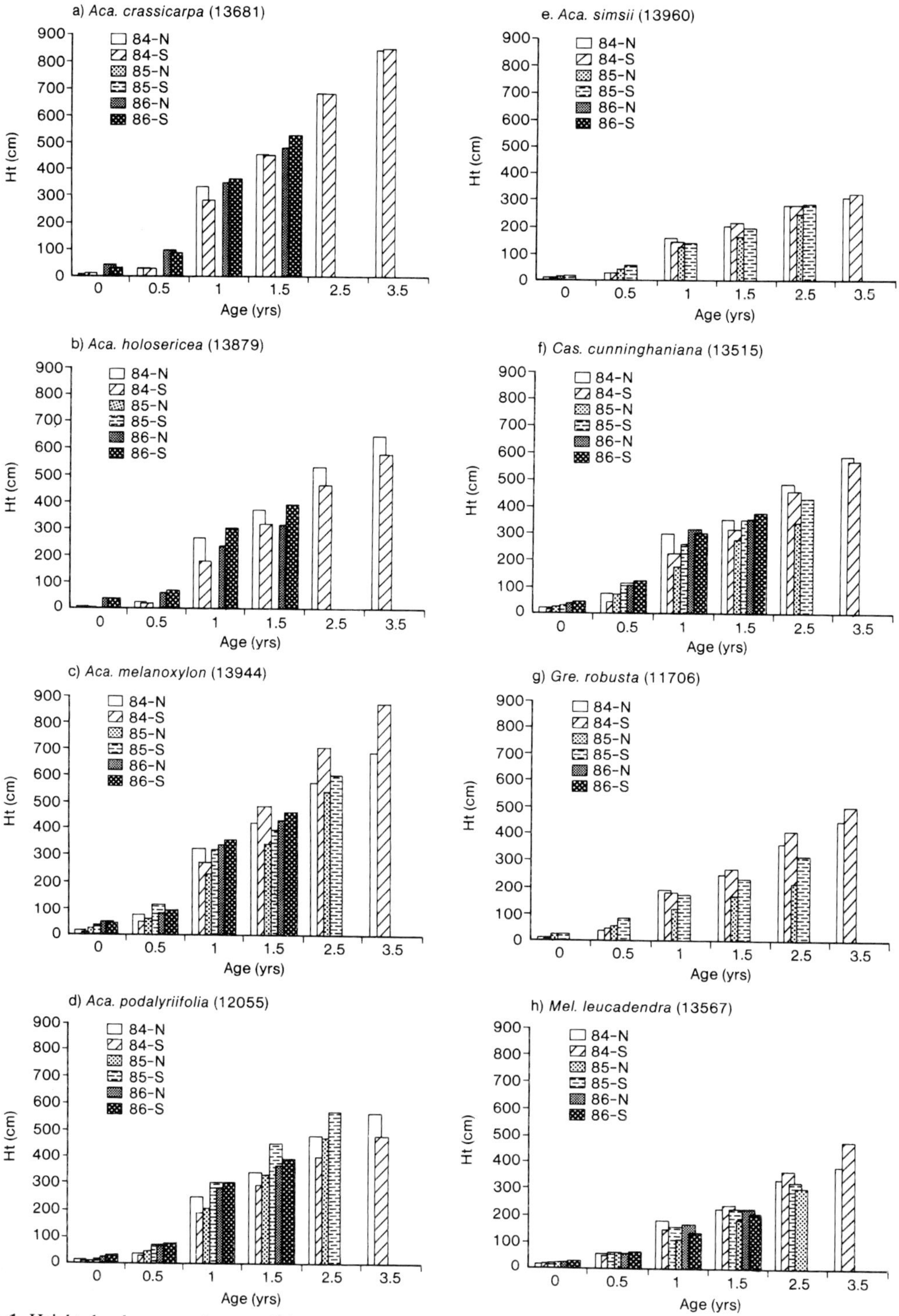

Fig. 1. Height development of some of the standards by site and year of planting.
(*N = Wongi State Forest trial; S = Tuan/Toolara State Forest).

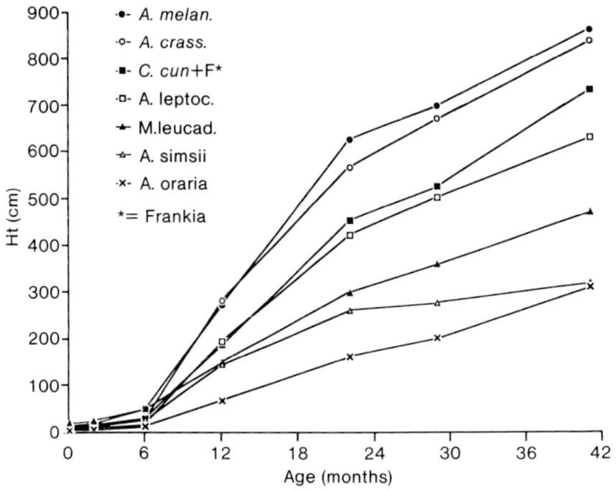

Fig. 2. Pattern of height development of some selected species planted in the 1984 trials at Tuan/Toolara.

for the first 6 months after planting. Most then entered into their most rapid growth phase over the next 18 months before settling into a steady phase of sustained growth. However, some (e.g. *A. simsii*) have matured early and have grown little after age 2 years. Others (e.g. *A. oraria*) were slow starters but have maintained consistent growth rates, and their rate of growth may be increasing steadily as they age.

Coppicing

Albizia procera, Alphitonia excelsa, Petalostigma pubescens, Syzygium suborbiculare and *Terminalia platyphylla* coppiced well with little difference in response to treatment height. Species in the genera *Angophora, Banksia, Eucalyptus, Grevillea* and *Lophostemon* coppiced very well with little stump mortality. Although there was some variation between treatments, there was no distinct trend to indicate the best cutting height for production of the most vigorous coppice. *Melaleuca* and *Leptospermum* species responded well to all treatments. *Casuarina cunninghamiana* coppiced over all treatments but generally had the best coppice at 1 m.

There was a great deal of variation between *Acacia* species in response to cutting:

(a) Some species coppiced extemely well (e.g. *A. saligna, A. rothii*) at all cutting heights with little stump mortality;

(b) Others (e.g. *A. melanoxylon*) coppiced over all treatments but shoots were most vigorous from the 1 m treatment;

(c) Some species (*A. mearnsii, A. monticola, A. tumida, A. simsii, A. podalyriifolia* and *A. plectocarpa*) coppiced when cut at 1 m with little or no response from the lower cut heights and produced shoots mainly from the branches providing there was at least one healthy live branch retained. Those with little or no foliage left after cutting eventually died;

(d) Some species which had little or no foliage in the lower part of the crown at the time of cutting (e.g. *A. holosericea, A. mangium, A. cowleana* and *A. tumida*) responded poorly to all treatments;

(e) A number of species (*A. aulacocarpa, A. auriculiformis, A. cincinnata, A. crassicarpa, A. polystachya,* and *A. oraria*) showed a high degree of variability in responses within treatments, with a high percentage of stump mortality but with some stumps coppicing vigorously. Some stumps began to sprout but the shoots died (probably killed by frost) and did not reshoot.

Differences in sprouting ability between provenances of some species were noted. For example, coppice production from Papua New Guinea (PNG) provenances of *A. aulacocarpa, A. auriculiformis* and *A. crassicarpa* was poor and was generally inferior to that from Queensland provenances. However, there was little difference between these provenances of *A. leptocarpa*.

Variation in the coppicing performance of some of the standards planted in 1984 (coppiced at age 3 years) and 1985, 1986 (coppiced at age 2 years) suggest that coppicing ability of at least some species may decline markedly with increasing age. Younger material tended to produce a greater number of coppice shoots of greater vigour with lower stump mortality (e.g. *M. leucadendra*).

There were some differences in coppicing success between sites. Variations in stump mortality, number of coppice shoots produced and shoot vigour were noted particularly for *A. hylonoma, A. flavescens, A. leptocarpa, A. rothii* and *Grevillea robusta*.

Root suckering following cutting was recorded for several species, most notably *A. melanoxylon* (Queensland provenances) and *A. storeyi*, while seedling regeneration was noted also for several species (e.g. *A. simsii* and *A. podalyriifolia*).

Flowering

The levels and particularly the seasonal patterns of flowering and fruiting are tentative only at this stage, and should be treated with caution. Our data appear to indicate that for some species, flowering patterns and development into mature fruit may vary from year to year depending on weather conditions. Data from observations over a number of years after first flowering are required before more precise patterns can be determined for each species.

Damage

The brown hare (*Lepus capensis*) caused considerable damage to the newly established 1984 plantings at Tuan/Toolara by nipping off seedling stems. *Casuarina cunninghamiana* in particular suffered high levels of nipping.

A native rat caused extensive damage to established plantings of some species in the 1985 and particularly the 1986 plantings at Tuan/Toolara. Feeding burrows and runways were widespread, but particularly noticeable in plots of *A. julifera* ssp. *julifera, A. torulosa* and *A. difficilis*, and the roots of the first two were eaten extensively. Burrowing and root cutting resulted in affected plots being very susceptible to windthrow. The rats were feeding also on acacia seed, the two most obvious being *A. julifera* and *A. penninervis*.

Wind damage has been noted in several species in addition to those predisposed to windthrow through the activity of rats. In general this has been minor with the exception of the PNG provenances of *A. crassicarpa* and *A. simsii*. In the former case, damage results primarily from the high foliage biomass, large size and weak junction points where stems bifurcate and where major branches join the stem. In the case of the *A. simsii*, stems were weakened by wood moths.

Although a wide range of insects has been collected and identified from the trials (Appendix 5), only a few have been potentially serious. These included Scarabaeidae (especially on *Angophora costata*), Chrysomelidae, Limacodidae (especially on eucalypts) and Cossidae (on acacias).

Detailed information on pathogenic fungi is not yet available. Pathogenic fungi isolated and identified to date include:

(a) Powdery mildew (*Oidium* sp.) — severe on Western Australia and Northern Territory provenances of *A. holosericea*, and to a lesser extent on *A. aulacocarpa, A. auriculiformis, A. mangium* and *A. polystachya*;
(b) Leaf spots (*Glomerella cingulata*) — *A. simsii* (relatively minor); common on acacias in the nursery;
(c) Rust (*Uromycladium robinsonii*) — *A. melanoxylon* (minor);
(d) Stem canker (*Cytospora* sp.) — some melaleucas and some casuarinas (?); and
(e) A root rot — *Casuarina obesa* (severe).

Discussion

Consistent patterns in performance for different species, or even different provenances of the same species from similar climatic zones, are generally difficult to detect and highlight the need for such techniques as climatic analysis (Booth et al. 1987, 1988) in determining the suitability of species for introduction into particular climatic regions.

Though the majority of successful species generally are those derived from wetter regions, there are a number of species from very dry areas that have survived, grown well and that are in good health under the moist and humid environment of these sites. Good performances by these taxa may indicate that they possess some degree of environmental adaptability, a very useful trait for broadscale species introductions. There is also an indication that species from warmer areas may adapt more readily to relatively cooler areas than do species from cooler areas to relatively warmer areas. Again, more detailed analysis is required to determine whether this is the case.

The high level of variation in the performance of some species (e.g. *A. aneura*) suggests that, while these species may not be suitable in the short term for broadscale use, they may have a role in the longer term following a program of tree improvement. This is particularly so where species have potentially high utilisation value. In contrast, there is probably little to be gained from a tree improvement program on species with low variation in performance (e.g. *A. simsii*), unless these species are shown to have high levels of variation in other useful attributes.

Size and rate of development may be important criteria for determining the potential usefulness of species, but other factors (e.g. adaptability, range and usefulness of products, ease of establishment and management) also need to be considered. Thus some small species with fast, early growth and rapid maturity (e.g. *A. simsii*) may be useful in particular situations (e.g. around garden plots as a source of mulch, as part of a mixed planting with larger but slower-growing species or in rehabilitation of degraded areas). Similarly, slow-growing species (e.g. *A. aneura*) may be very useful if they have a high utility value, can tolerate environmentally difficult situations or have other desirable attributes.

Biological traits and form of management will influence the selection of species also. Some species have shown definite potential to become weeds due to their capacity for abundant regeneration either as seedlings (e.g. *A. simsii*) or root suckers (e.g. *A. melanoxylon* and *A. storeyi*). These characteristics may be advantageous in some situations and undesirable in others. In some cases potential weediness may be relatively easy to control, e.g. in the case of *A. simsii*, either by lopping before seed maturity or by cutting regeneration. However, both *A. melanoxylon* and *A. storeyi* coppice well when cut and would be potentially difficult to control unless utilisation pressure is high.

Coppicing is a useful management tool enabling regular harvesting of material without having to replant. The ability to coppice is determined genetically and the coppicing study has provided some preliminary indicative information on coppicing ability of most of the species and provenances included in the trials. However, the success of coppicing may be affected by a variety of factors:

(a) The ability of stumps to produce sprouts can decline with increasing age due to a reduction in the number of dormant buds (Busgen and Munch 1929; Kramer and Koslowski 1979; Evans 1982) and increasing stump diameter (Evans 1982; Sharma 1985);
(b) Season of felling, since heavy frosts or inadequate soil moisture may reduce or delay shoot production (Clarke 1975; FAO 1979; Evans 1982);
(c) Poor felling techniques may reduce the rate of callus development increasing the risk of infection by wood-rotting fungi (Schonau 1975; Evans 1982);
(d) Failure to clear stumps of slash and branch material (Schonau 1975).

While some of these factors (particularly weather conditions) may have affected the coppicing response by some species, it is unlikely that those that failed to reshoot have the ability to coppice strongly. Such species are not appropriate where the use of coppicing as a management tool is desirable. However, they may be amenable to lopping or pollarding provided at least one healthy, well-foliaged branch is retained.

Acknowledgments

We wish to thank the Queensland Department of Forestry and many of its staff for their substantial contributions to the study. In particular Mark Podberscek, Cliff Raddatz and Dave Taylor for management of the trials, Ross Wylie, Murdock de Barr, Bruce Brown, John Tierney, John Kehl, Chris Corben and Greg Cooper from the Biology Section of Queensland Department of Forestry, Mark Nester and Tony Burridge for data handling and analysis, Clair Ross for typing and Gary Bacon the project leader. Seed and valuable information were provided by the CSIRO Australian Tree Seed Centre; Trevor Booth from the CSIRO Division of Forestry and Forest Products provided climatic information and many others from that Division have provided valuable inputs. Our thanks also to ACIAR for financial support.

(*Appendix Tables follow*)

Appendix 1. ACIAR trials planted 1984–86 — failures and seedlots details.

Plnt yr	Seedlot	Species	Origin			Climate		Failure level[a]		
			Lat ° ′	Long ° ′	Alt (m)	MAT (°C)	MAR (mm)	Gmn	Nrs	Fld
84	13834	*Aca ancistrocarpa*	20 12	137 30	300	26	310	F	S	X
84	13493	*Aca bidwillii*	23 17	144 21	250	23	432	F	F	X
84	13485	*Aca cambagei*	24 01	143 48	210	23	391	S	X	X
84	13487	*Aca cambagei*	25 22	141 43	110	23	220	S	S	X
84	13767	*Aca coriacea*	22 18	130 52	600	23	320	F	F	X
84	13768	*Aca coriacea*	20 15	130 02	380	25	379	P	P	X
84	13775	*Aca cowleana*	19 58	129 42	450	25	398	F	F	X
84	13629	*Aca dealbata*	37 48	146 07	900	10	1686	F	S	X
84	13739	*Aca ligulata*	23 15	132 52	620	22	244	F	S	X
84	13740	*Aca ligulata*	22 18	130 52	600	23	320	F	X	
84	13621	*Aca mangium*	3 04	128 12	150	na	na	F	F	X
84	13807	*Aca mearnsii*	34 00	150 00	500	14	783	F	S	X
84	13773	*Aca monticola*	22 21	131 18	700	22	317	F	F	X
84	13781	*Aca murrayana*	25 13	130 53	580	21	263	F	X	
84	13782	*Aca murrayana*	25 12	130 48	590	21	265	F	X	
84	13482	*Aca pendula*	25 51	146 36	380	21	507	F	X	
84	13962	*Aca pendula*	31 40	148 18	200	18	481	F	P	X
84	7859	*Aca pruinocarpa*	26 37	120 15	520	22	209	S	X	
84	7947	*Aca pruinocarpa*	26 00	118 00	490	23	194	S	X	
84	13488	*Aca stenophylla*	25 06	142 50	120	23	283	F	F	X
84	13961	*Aca stenophylla*	30 56	147 52	200	19	432	F	F	X
84	13271	*Aca victoriae*	25 51	146 35	310	21	491	S	P	X
84	13494	*Aca victoriae*	21 32	139 15	240	25	324	S	P	X
84	13164	*Alo campestris* ssp *campestris*	32 03	117 23	290	17	387	F	X	
84	13225	*Alo campestris* ssp *eriochlamys*	29 56	121 07	420	20	227	F	S	X
84	13226	*Alo campestris* ssp *grossa*	32 00	121 40	280	17	275	F	X	
84	13201	*Alo decaisneana*	25 18	131 42	440	21	236	F	F	X
84	13204	*Alo decaisneana*	23 45	132 41	580	22	241	F	P	X
84	13171	*Alo huegeliana*	32 05	118 50	380	17	320	F	S	X
84	13172	*Alo huegeliana*	32 28	118 53	360	17	327	F	X	
84	9496	*Cal endlicheri*	31 00	148 00	290	18	455	P	X	
84	8052	*Cal macleayana*	29 00	153 00	375	17	1461	S	X	
84	13753	*Dod augustissima*	23 52	132 33	650	21	257	F	S	X
84	13754	*Dod augustissima*	25 20	131 47	500	21	244	F	S	X
84	12336	*Euc annulata*	33 38	119 51	300	17	442	F	P	X
84	11468	*Euc brevifolia*	15 42	130 07	20	28	777	F	F	X
84	7034	*Euc gamophylla*	22 00	118 00	300	26	329	S	X	
84	10499	*Euc gamophylla*	22 50	133 25	670	22	263	F	X	
84	13158	*Euc gamophylla*	22 57	118 38	500	25	278	S	X	
84	12839	*Euc gongylocarpa*	28 32	122 15	460	21	211	F	X	
84	10700	*Euc normantonensis*	20 20	138 50	300	25	391	F	F	X
84	8583	*Euc ochrophloia*	27 39	143 49	140	22	268	S	P	X
84	12507	*Euc ochrophloia*	26 08	145 40	302	21	444	F	F	X
84	7228	*Euc oxymitra*	24 39	132 18	540	21	241	P	S	X
84	12262	*Euc sheathiana*	31 17	119 52	340	18	306	F	S	X
84	12776	*Euc socialis*	31 32	143 34	20	20	222	F	S	X
84	13792	*Euc socialis*	23 34	132 31	600	22	259	S	S	X
84	13759	*Euc trivalvis*	23 34	132 31	600	22	259	F	S	X
84	13760	*Euc trivalvis*	25 12	130 48	580	21	264	F	X	
84	13751	*Mel lasiandra*	22 18	130 52	600	23	320	F	F	X
84	13440	*Mel nervosa*	21 33	145 50	300	23	554	F	F	X
84	11354	*Nau orientalis*	4 42	151 47	20	na	na	S	P	X

Plnt yr	Seedlot	Species	Origin			Climate		Failure level[a]		
			Lat ° ′	Long ° ′	Alt (m)	MAT (°C)	MAR (mm)	Gmn	Nrs	Fld
85	14204	*Aca bidwillii*	17 34	145 13	780	20	851	P	S	S
85	14039	*Aca calcicola*	25 13	130 20	530	22	256	F	S	X
85	13485	*Aca cambagei*	24 01	143 48	210	23	391	P	S	X
85	13487	*Aca cambagei*	25 22	141 43	110	23	220	P	P	X
85	13779	*Aca latzii*	24 34	132 47	450	22	212	F	X	X
85	13782	*Aca murrayana*	25 12	130 48	590	21	265	F	S	S
85	13482	*Aca pendula*	25 51	146 36	380	21	507	P	F	X
85	7859	*Aca pruinocarpa*	26 37	120 15	520	22	209	P	P	X
85	7947	*Aca pruinocarpa*	26 00	118 00	490	23	194	P	P	X
85	12541	*Euc gamophylla*	22 57	118 38	500	25	278	P	P	X
85	14044	*Euc gamophylla*	25 05	130 03	610	21	269	F	S	X
85	11731	*Euc ochrophloia*	26 53	144 20	180	22	319	P	P	X
85	14153	*Par nonda*	12 33	141 52	10	26	1663	P	F	X
85	14179	*Pla pohlmanniana* v *vestita*	18 42	146 17	na	24	2040	X		
85	14506	*Ter ferdinandiana*	12 27	130 50	30	28	1565	S	F	X
85	14182	*Ter platyptera*	16 40	143 59	202	23	1460	X		
86	14596	*Aca farnesiana*	23 33	145 18	265	23	466	X	X	
86	14100	*Cas obesa*	26 34	120 03	550	22	212	F	P	X
86	14501	*Mla azedarach* v *australasica*	18 05	144 52	780	21	825	X		

[a]Abbreviations: Gmn, germination; Nrs, post-germination to planting out; Fld, post-planting; F, few; P, partial; S, substantial; X, total.

Appendix 2. ACIAR trials planted April 1984 — seedlot details and results (to 41 months).[a]

		Origin			Climate		Growth Performance: Wongi					Growth Performance: Tuan/Toolara					Coppicing performance: Cut ht. (m)				Age first buds	Flowering & fruiting: Level		Flowering & fruiting: Season	
Seedlot	Species	Lat ° ′	Long ° ′	Alt (m)	MAT (°C)	MAR (mm)	Ht. (cm)	CV (%)	DGL (cm)	CV (%)	Srv. (%)	Ht. (cm)	CV (%)	DGL (cm)	CV (%)	Srv. (%)	0.1	0.5	1.0	Notes	(mo)	Flw	Mat seed	Flowers	Fruit
13794	*Aca ammobia*	25 20	131 12	580	21	261	427	17	6.4	41	38	436	25	6.2	36	34					15	3	2	Mr-Oc	Jn-De
13480	*Aca aneura*	27 53	148 43	210	20	503	340	35	7.8	44	34	229	43	4.8	57	41		2		s					
13481	*Aca aneura*	26 25	146 17	300	21	454	249	60	4.6	60	38	235	55	4.9	58	38		1		s					
13719	*Aca aneura*	22 12	130 55	600	23	321	177	17	5.0	23	91	170	42	4.1	45	69	2	2	2	s	31	1		De-Mr	
13687	*Aca aulacocarpa*	8 41	141 29	35	26	1804	499	13	12.9	24	97	549	15	12.4	25	88	1	1	1	L					
13865	*Aca aulacocarpa*	17 09	145 37	720	21	1612	502	27	15.1	32	100	547	12	13.5	33	100	1	1	1	L					
13866	*Aca aulacocarpa*	16 40	145 18	400	23	1317	434	25	11.3	28	94	422	18	9.0	37	91	2	2	2	f,V	33	1	1	Ja-My	Ap-No
13686	*Aca auriculiformis*	8 41	141 29	50	26	1804	523	19	11.7	25	88	559	33	10.8	41	75	1	1	1	f,V,L					
13854	*Aca auriculiformis*	12 20	133 04	35	28	1355	416	39	10.6	40	88	511	26	11.6	37	84	1	1	2	V					
13861	*Aca auriculiformis*	15 50	144 55	500	23	1624	562	25	11.8	34	88	452	26	9.2	27	84	1	1	2	V					
11690	*Aca baileyana*	32 18	148 35	260	17	537	437	35	9.8	42	88	309	33	7.4	38	25					15				
13864	*Aca cincinnata*	16 57	145 38	440	23	1629	740	6	18.2	19	100	667	19	13.3	31	84	1	2	2	V					
13878	*Aca cincinnata*	16 35	145 25	410	23	1639	828	8	15.5	19	100	731	16	13.0	25	94	1	2	2		26	1	1	Jn-Jl	Au-No
13774	*Aca cowleana*	22 18	130 52	600	23	320	399	19	6.1	29	84	328	50	4.2	53	41	1	1	1	F,L	15	2	2	De-Au	My-De
13681	*Aca crassicarpa*	8 40	141 45	30	26	1848	830	16	16.7	14	100	840	18	16.3	21	88	1	1	2	V,F					
13863	*Aca crassicarpa*	16 57	145 38	440	23	1629	622	10	15.6	22	91	699	10	14.4	19	100	2	2	3	V	27	1	1	My-Jl	Jn-No
9972	*Aca elata*	33 04	151 35	120	17	1162	743	8	20.6	15	44	552	32	11.9	46	41		2		V					
13958	*Aca flavescens*	17 19	146 00	0	24	3941	622	20	14.2	24	81	692	17	14.3	28	97	3	3	3		26	1		Ap-Au	
13853	*Aca holosericea*	12 50	132 50	50	28	1386	461	15	10.1	29	88	425	17	7.8	37	88	0	1	1	B,L	15	2	2	Ap-Au	My-Oc
13879	*Aca holosericea*	16 46	145 15	380	23	1090	640	12	9.8	23	94	575	22	9.9	37	88	0	1	1	B,L	15	2	2	Mr-Au	My-No
13652	*Aca leptocarpa*	12 45	143 15	60	26	1912	581	26	11.9	29	97	600	23	11.6	30	91	3	3	3	r,f	16	1	1	Ma-Au	Jn-No
13691	*Aca leptocarpa*	8 52	143 03	30	26	2017	646	13	12.7	25	91	630	20	11.8	25	81	3	3	3	f	15	1	1	Ma-Au	Jn-No
13460	*Aca mangium*	8 50	143 08	10	26	2090	498	38	11.7	51	81	632	28	13.7	40	81	1	1	1	F,L					
13846	*Aca mangium*	16 31	145 24	60	25	1977	602	27	13.7	31	53	422	49	9.8	54	56	0	0	1	F,L					
13630	*Aca melanoxylon*	38 25	146 30	550	11	1277	447	25	11.5	36	84	423	29	11.4	42	63	2	2	2						
13944	*Aca melanoxylon*	26 36	153 02	100	20	1886	681	11	14.1	21	97	864	11	16.3	26	100	2	2	3	R	15	2	2	No-Au	Ma-No
13654	*Aca oraria*	14 16	144 26	180	26	1272	131	52	6.0	56	100	156	57	4.9	56	94	2	2	2	V	24	1	1	Ap-Jn	Jn-De
13867	*Aca oraria*	15 48	144 56	150	25	1390	280	22	9.4	29	88	313	29	8.6	38	94	2	2	2	f,V	24	2	2	Ap-Jn	Jn-De
9094	*Aca plectocarpa*	16 20	126 50	410	26	760	571	12	11.3	22	100	551	12	11.8	28	47	0	1	1	B,L	15	2	2	Ja-Se	Fe-De
12055	*Aca podalyriifolia*	24 50	152 40	100	21	1136	559	18	8.6	29	88	476	21	8.0	51	88	0	0	2	B	<15	3	3	Jn-Oc	Jl-No
13500	*Aca polystacha*	13 42	143 18	360	25	1183	336	16	7.9	24	97	353	23	6.9	35	84	1	1	2	V,F	35	1		Ap-Au	
13871	*Aca polystacha*	16 58	145 37	480	22	1576	273	31	9.3	43	91	255	43	7.3	38	94	1	1	2	V,F					
13599	*Aca retinodes*	37 18	142 46	854	10	1051	413	31	7.4	30	19										<15	2	1	My-De	Jn-Mr
13501	*Aca salicina*	26 34	149 08	310	21	585	226	33	4.6	49	88	192	56	3.4	79	91	2	2	2	V	22	1	1	Fe-Au	Ap-De
13651	*Aca saligna*	31 45	115 48	20	18	812	568	13	13.8	29	100	663	13	13.9	25	91	3	3	3		15	3	3	Au-Oc	Se-De
13690	*Aca simsii*	8 42	141 32	30	26	1817	318	21	9.1	26	38	456	22	9.4	30	56					<15	3	3	Fe-Au	Ap-No
13960	*Aca simsii*	17 19	145 13	700	21	768	305	11	7.8	19	94	321	10	7.3	18	97	0	1	2	B	<15	3	3	No-Jn	De-Oc
13843	*Aca torulosa*	17 33	133 32	210	26	489	390	15	8.3	37	100						1	1	1	V,L					
13332	*Ang costata*	30 03	153 04	250	18	1372	718	18	14.0	15	97	789	19	14.2	19	97	3	3	3		31	1		No-De	
9711	*Cal intratropica*	11 32	132 56	30	27	1330	248	18	8.3	22	81														
13134	*Cas cunninghamiana*	26 20	152 41	200	20	1263	543	20	11.6	23	100	606	20	11.6	25	94	2	2	3		26	1	1	Jn-Au	Au-Jn

Seedlot	Species	Origin			Climate		Growth Performance										Coppicing performance				Age first buds (mo)	Flowering & fruiting			
							Wongi					Tuan/Toolara					Cut ht. (m)					Level		Season	
		Lat ° ′	Long ° ′	Alt (m)	MAT (°C)	MAR (mm)	Ht. (cm)	CV (%)	DGL (cm)	CV (%)	Srv. (%)	Ht. (cm)	CV (%)	DGL (cm)	CV (%)	Srv. (%)	0.1	0.5	1.0	Notes		Flw	Mat seed	Flowers	Fruit
13511	*Cas cunninghamiana*	23 49	150 18	120	22	759	553	13	12.3	20	97	609	22	13.6	22	88	2	2	3		27	1	1	Jn–Au	Au–Jn
13515	*Cas cunninghamiana*	17 04	145 28	400	23	1087	577	18	11.7	21	97	559	22	11.6	25	100	2	2	3		27	1	1	Jn–Au	Au–Jn
13713	*Euc argophloia*	26 20	150 40	300	19	684	342	41	7.6	38	72	306	52	6.8	43	59	3	3	3						
13158	*Euc melanophloia*	26 25	146 13	330	21	459	376	31	8.6	31	97	419	28	7.8	25	97	3	3	3		20				
13588	*Euc melanophloia*	22 46	147 31	300	22	596	370	32	8.7	36	91	344	43	8.1	51	94	3	3	3		20	1		No–De	
10945	*Euc normantonensis*	16 27	144 47	460	23	1281	278	42	6.9	48	88	323	22	7.6	35	50	3	3	3		22	2	1	Fe–De	Ap–No
11706	*Gre robusta*	31 42	148 42	280	18	532	440	21	11.1	27	91	492	34	11.9	23	91	2	2	3						
13955	*Lep flavescens*	35 02	150 36	40	17	1165	246	15	12.2	24	91	268	23	11.1	34	84	2	2	2		17	3	3	Oc–De	No–Jn
13529	*Lop grandiflorus*	13 46	143 08	170	26	1144	314	23	8.9	28	97						3	3	3		31	1	1	De–Fe	Ja–Jn
11935	*Mel dealbata*	12 35	131 18	20	28	1492	233	27	9.0	29	100	297	24	10.7	25	94	2	2	2		42	1	1	Oc–No	No–Fe
13752	*Mel lasiandra*	20 15	130 02	380	25	379	136	36	3.8	31	34	118	37	3.6	36	81					<15	3	1	Fe–Au	Ap–Oc
13532	*Mel leucadendra*	12 42	143 20	40	26	1909	336	23	14.9	16	100	407	23	15.0	21	97	2	2	2	V	28	2	2	Jn–Ja	Oc–Jn
13567	*Mel leucadendra*	17 00	145 30	500	22	1145	377	21	15.0	21	88	472	18	15.0	22	97	2	2	2	V	27	2	2	Mr–Au	No–Jn
7717	*Mel stypheloides*	32 42	151 47	70	na	na	346	21	9.6	24	97	411	12	10.2	13	94	2	2	2		43	1	1	No–De	De–Jn
13530	*Mel viridiflora*	12 42	143 20	60	26	1906	180	51	4.2	49	56	113	85	3.6	132	63									
14157	*Syz suborbiculare*	12 39	141 50	2	27	1675	266	22	8.3	34	63	128	57	3.5	53	88	3	3	3						

[a]Abbreviations for Appendices 2–4:

Growth Performance
- **Ht** Mean Height
- **DGL** Mean tree diameter at ground level [$D=\sqrt{\Sigma (d*d)}$ for multiple stems]
- **CV** Coefficient of variation
- **Srv** Survival

Flowering & Fruiting Codes
- **1** light
- **2** moderate
- **3** abundant

Coppicing Codes
- **0** none
- **1** poor
- **2** fair
- **3** good
- **B** reshooting mainly from branches
- **f** shoots damaged by frost
- **F** shoots killed by frost — no reshooting
- **L** Little or no foliage remaining on 1m tmt after cutting
- **r** root suckering
- **R** abundant root suckering
- **s** slow to coppice
- **S** seedling regeneration
- **SS** abundant seedling regeneration
- **V** variable responses within treatments.

Appendix 3. ACIAR trials planted February1985 — seedlot details and results (to 31 months).

Seedlot	Species	Origin			Climate		Growth Performance										Coppicing performance				Age first buds	Flowering & fruiting			
							Wongi					Tuan/Toolara					Cut ht. (m)					Level		Season	
		Lat ° ′	Long ° ′	Alt (m)	MAT (°C)	MAR (mm)	Ht. (cm)	CV (%)	DGL (cm)	CV (%)	Srv. (%)	Ht. (cm)	CV (%)	DGL (cm)	CV (%)	Srv. (%)	0.1	0.5	1.0	Notes	(mo)	Flw	Mat seed	Flowers	Fruit
14093	*Aca ancistrocarpa*	24 26	125 06	140	24	221	118	28	2.8	40	13	166	12	4.8	16	22		0			10	3	2	De–Jl	Ja–Oc
14187	*Aca bidwillii*	23 50	149 57	119	22	710				Fail		55	52	1.4	56	69									
14488	*Aca farnesiana*	22 15	131 47	600	23	279				Fail		77	41	1.2	46	78		2							
14175	*Aca flavescens*	16 40	145 18	167	23	1317	424	32	9.6	33	72	592	10	12.0	21	97	3	3	3		25	2	1	Fe–Jl	Ap–Oc
14197	*Aca hylonoma*	17 01	145 18	167	25	2387	337	21	7.8	30	97	381	12	7.8	24	100	3	3	3	r					
13740	*Aca ligulata*	22 18	130 52	600	23	320				Fail		140	26	2.6	33	50									
12986	*Aca melanoxylon*	41 00	145 00	200	11	1501	296	37	7.0	40	100	415	21	9.0	26	100									
13944	*Aca melanoxylon*	26 36	153 02	100	20	1886	561	23	12.3	33	97	617	10	12.0	14	100	2	2	3	R	21	2	2	De–Au	Ja–Oc
14176	*Aca melanoxylon*	17 17	145 26	1022	19	1428	465	24	10.9	32	100	515	18	10.9	25	100	2	2	3	R	33	1			
14008	*Aca monticola*	18 50	121 40	25	27	447	218	23	5.0	33	72	236	19	5.2	23	63	0	0	1	B,L	9	3	3	De–Au	Ja–Oc
13781	*Aca murrayana*	25 13	130 53	580	21	263	123	26	2.4	23	19	152	57	3.0	53	38		0			17	3	1	Oc–No	No
14003	*Aca plectocarpa*	15 45	128 40	50	28	748	495	15	10.7	35	78	508	11	10.0	24	100	0	1	3		23	2	2	Fe–Ap	Ap–No
14004	*Aca plectocarpa*	15 50	128 40	50	28	729	486	9	10.7	25	88	506	9	9.7	23	100	0	1	3		23	3	3	Fe–Ap	Ap–No
12055	*Aca podalyriifolia*	24 50	152 40	100	21	1136	494	22	7.5	37	91	597	8	9.9	20	97	0	1	2	B,S	9	3	3	Jl–Oc	Au–No
14140	*Aca rothii*	14 17	143 26	210	26	1209	345	21	7.3	28	94	308	25	6.8	29	84	3	3	3						
14160	*Aca rothii*	12 32	141 51	10	26	1662	204	36	4.8	28	44	354	18	7.0	21	94	3	3	3						
13960	*Aca simsii*	17 19	145 13	700	21	768	254	12	6.8	26	97	297	5	6.5	21	100	0	1	2	SS	10	3	3	No–Ju	De–Oc
14183	*Aca torulosa*	16 41	144 02	275	25	945	422	37	7.5	47	81	530	14	8.6	13	100	1	1	2		23	2	2	Ap–Jn	My–No
11505	*Aca tumida*	20 08	119 23	110	27	299	232	27	6.2	36	31	328	23	7.9	39	75	0	0	1	BL	8	3	2	De–Au	Mr–No
11514	*Aca tumida*	21 41	117 45	480	25	357	345	16	7.6	37	88	361	18	8.0	20	84	0	0	1	B,L	8	3	3	De–Au	Ap–No
14489	*Aca victoriae*	22 08	133 02	552	23	277	76	32	1.4	27	44	85	51	2.3	58	53									
14180	*Adn abrosperma*	16 30	143 21	108	26	998				Fail		69	85	1.2	83	44									
14213	*Alb procera*	16 34	145 30	30	25	1864	90	63	2.7	66	69	93	39	2.7	43	97	3	3	3						
14186	*Alp excelsa*	23 07	150 20	20	22	845	275	26	5.8	44	94	319	13	7.5	24	97	3	3	3		11	2	2	De–Jn	Mr–Oc

Seedlot	Species	Origin			Climate		Growth Performance										Coppicing performance				Age first buds (mo)	Flowering & fruiting			
							Wongi					Tuan/Toolara					Cut ht. (m)					Level		Season	
		Lat ° ′	Long ° ′	Alt (m)	MAT (°C)	MAR (mm)	Ht. (cm)	CV (%)	DGL (cm)	CV (%)	Srv. (%)	Ht. (cm)	CV (%)	DGL (cm)	CV (%)	Srv. (%)	0.1	0.5	1.0	Notes		Flw	Mat seed	Flowers	Fruit
14190	*Alp excelsa*	23 11	149 17	212	22	709	305	25	6.8	45	94			Not	planted		3	3	3		23	2	2	Fe–Ap	Mr–Jl
13332	*Ang costata*	30 03	153 04	250	18	1372	463	32	11.4	30	97	436	30	9.9	24	97	3	3	3						
14181	*Ata hemiglauca*	16 36	143 48	211	26	933				Fail		72	47	0.8	28	88									
14486	*Ata hemiglauca*	22 14	134 34	491	23	269				Fail		56	57	0.7	45	88									
14191	*Ban integrifolia* v *comp.*	23 13	150 48	5	22	1202	256	28	6.9	28	94	245	31	7.7	30	97	3	3	3						
13515	*Cas cunninghamiana*	17 04	145 28	400	23	1087	359	32	7.2	32	100	447	12	9.7	13	100	2	2	3		29	1	1	Jn–Au	No–Ap
14188	*Csa brewsteri*	23 35	149 03	195	22	637				Fail		82	56	1.7	48	84		2							
13713	*Euc argophloia*	26 20	150 40	300	19	684	213	54	5.3	46	72	346	32	7.6	30	78	3	3	3						
10945	*Euc normantonensis*	16 27	144 47	460	23	1281	181	43	4.6	52	81	234	25	5.7	36	94	3	3	3		22	1	1	My–Au	Jn–Ap
14164	*Gre glauca*	12 43	142 06	18	26	1588	139	45	4.9	42	56	120	93	4.3	63	72	3	3	3		24	1	1	Mr–Au	My–Jn
14177	*Gre glauca*	17 40	145 07	765	21	753	148	68	4.2	65	72	139	88	4.9	61	50	3	3	3						
11706	*Gre robusta*	31 42	148 42	280	18	532	225	35	6.9	42	97	328	31	9.7	26	100	2	2	3						
14144	*Lep longifolium*	12 40	142 06	10	26	1590	152	36	5.3	43	97	172	23	6.3	31	100	2	2	2		14	2	2	Jn–Ja	Jl–Jn
14185	*Lop suaveolens*	17 35	145 27	935	20	1197	408	20	10.1	18	100	414	15	10.2	23	100	3	3	3	r	7	2	2	Ap–De	All Year
14146	*Mel acaciodes*	12 43	142 05	2	26	1599	86	26	2.8	39	97	86	41	2.5	40	81		3							
14485	*Mel bracteata*	23 36	133 52	840	20	271	99	38	3.1	42	94	103	25			94		2			20	1	1	Oc–Ja	No–My
13567	*Mel leucadendra*	17 00	145 30	500	22	1145	338	21	13.5	21	100	317	21	11.7	22	100	2	2	2		28	2	2	Jl–Se	Au–Jn
14147	*Mel leucadendra*	12 31	141 48	10	26	1665	305	33	11.3	30	100	315	20	10.2	20	97	2	2	2		27	1	1	Jl–Se	Au–Jn
14149	*Mel saligna*	12 44	142 06	10	26	1590	272	16	7.9	23	97	274	17	7.5	29	97	2	2	2		26	1	1	Ap–Jn	Jl–Mr
14150	*Mel symphyocarpa*	12 31	141 48	10	26	1665	220	31	5.0	37	97	212	27	4.9	34	97	2	2	2		23	1	1	Fe–Jl	All Year
14170	*Mel symphyocarpa*	12 40	141 53	10	26	1690	207	32	5.1	31	84	189	29	4.7	37	97	2	2	2		23	1	1	Fe–Jl	All Year
14155	*Pet pubescens*	17 38	145 20	650	21	1138	133	36	3.2	49	100	212	31	4.0	38	97	3	3	2						
14189	*Pet pubescens*	23 11	149 17	192	22	702	145	49	3.2	60	97	199	31	4.7	36	100	3	3	2		19	1	1	Se–Oc	No–Jn
14178	*Ter platyphylla*	16 31	145 06	360	26	907				Fail		123	35	4.6	23	97									

Appendix 4. ACIAR trials planted March 1986—seedlot details and results (to 18 months).

Seedlot	Species	Origin			Climate		Growth Performance										Coppicing performance				Age first buds	Flowering & fruiting			
							Wongi					Tuan/Toolara					Cut ht. (m)					Level		Season	
		Lat ° ′	Long ° ′	Alt (m)	MAT (°C)	MAR (mm)	Ht. (cm)	CV (%)	DGL (cm)	CV (%)	Srv. (%)	Ht. (cm)	CV (%)	DGL (cm)	CV (%)	Srv. (%)	0.1	0.5	1.0	Notes	(mo)	Flw	Mat seed	Flowers	Fruit
14652	*Aca adsurgens*	19 14	127 46	340	26	348						203	15	3.9	35	19					11	2	2	Fe–Au	Ap–De
14668	*Aca ampliceps*	18 26	127 51	400	26	441	134	25	4.1	39	97	213	28	5.2	32	94	2	2	2	r	11	2	2	Mr–Jl	Jn–Se
14631	*Aca ampliceps*	17 26	130 56	230	27	493	196	24	5.3	28	94	250	15	6.2	19	97	2	2	2	r	11	2	2	Mr–Au	My–Se
14599	*Aca bidwillii*	20 44	140 16	340	25	428	35	36	0.9	45	69	44	42	0.9	46	53									
14754	*Aca blakei*	26 31	150 03	360	19	644	407	29	6.3	37	41	400	27	6.3	32	75					15	2	2	Jn–Au	Oc–No
14965	*Aca brassii*	13 44	143 07	165	26	1149	404	19	7.0	24	91	461	14	8.0	20	94	0	0	1	B,L	15	2	2	Ap–Jn	Jl–No
14732	*Aca concurrens*	27 39	151 29	100	20	588	502	19	8.3	28	97	547	12	5.4	19	100	1	1	1	r,L	17	1		Au	
14621	*Aca cowleana*	16 15	133 22	200	27	621	132	26	2.5	39	78	223	20	4.2	32	84	0	0	1	L	10	3	2	Ja–Au	My–Se
14745	*Aca crassa* ssp. *crasa*	27 51	151 02	380	18	643	444	34	6.9	40	88	513	13	7.6	14	100	1	1	1	B,L	16	1		Au–No	
13681	*Aca crassicarpa*	8 40	141 45	30	26	1848	485	9	9.2	20	100	530	6	8.4	14	100	0	1	3	V					
13682	*Aca crassicarpa*	8 50	143 10	20	26	2107	501	12	9.1	18	97	509	11	8.5	12	100	1	1	3	V					
14739	*Aca deanei*	26 55	151 49	550	17	772	572	27	9.3	28	97	613	12	9.0	18	97	0	1	2	B					
14726	*Aca decurrens*	34 53	149 17	685	13	683	322	27	6.2	25	97	430	23	7.5	27	100	1	1	1	L	25	1		Ap	
14619	*Aca difficilis*	17 24	133 30	250	26	505	391	15	7.6	20	94	317	22	6.7	24	100	1	1	2		9	2	2	Ja–Au	Jl–Fe
14623	*Aca difficilis*	16 21	133 22	235	27	618	410	11	8.1	18	100	399	12	8.4	19	100	1	1	2		9	2	2	Ja–Jl	Jn–Oc
14738	*Aca falcata*	26 46	151 54	500	18	822	277	21	5.2	21	97	305	24	4.6	23	97	0	1	2	B	8	2	3	Ap–Au	Au–Ja
14970	*Aca falcata*	17 36	145 28	890	20	1169	189	19	4.0	27	53	244	21	4.1	26	91	1	1	2		8	2	2	Jl–Au	Au–No
14981	*Aca falciformis*	17 31	145 26	1050	19	1721	332	24	6.4	30	66	432	12	6.9	18	100	1	1	2	r	7	2	2	All Year	Jn–Ja
14736	*Aca fimbriata*	26 46	151 50	500	18	804	355	22	6.7	30	97	324	15	5.6	19	100	1	1	2	B	12	1		Au	
14590	*Aca flavescens*	23 06	150 45	6	22	1327	399	20	8.6	24	75	416	14	8.0	19	97	2	2	2		24	1		Mr	
14763	*Aca glaucocarpa*	23 51	149 05	840	18	827	327	46	5.9	38	81	548	17	9.0	22	100	3	3	3		23	1		Mr–Jl	
14891	*Aca hammondii*	17 43	141 03	20	27	876	124	40	2.6	48	81	187	18	3.9	29	91		0			13	2	1	My–Mr	Jn–No
14657	*Aca hemignosta*	17 30	127 56	395	26	581	71	48	1.5	57	88	55	39	0.9	41	53									
13879	*Aca holosericea*	16 46	145 15	380	23	1090	314	23	5.5	25	100	393	11	6.1	17	97	1	1	2	B	13	2	2	My–Jl	Jl–No
14660	*Aca holosericea*	17 04	128 12	400	26	650	180	47	3.0	57	78	74	23	2.2	27	72					13	2	2	Jn–Au	Jl–No
14977	*Aca hylonoma*	17 01	145 50	110	25	2387	212	29	4.6	39	59	327	9	6.4	11	97	2	2	2						
14885	*Aca julifera* ssp. *gilbertensis*	16 47	144 08	280	25	922	307	33	6.5	38	100	354	21	6.3	25	97	3	3	3	r	13	2	2	My–Jl	Jn–No
14890	*Aca julifera* ssp. *julifera*	19 54	144 16	930	20	685	389	20	7.6	31	88	405	15	7.2	26	94	2	1	2	r	13	2	2	My–Jl	Jl–No
14974	*Aca julifera* ssp. *julifera*	20 13	145 53	330	24	663	371	22	6.6	22	88	376	23	5.8	27	94	2	2	2	r	14	2	2	My–Au	Jl–No
14758	*Aca juncifolia*	25 12	149 59	360	20	721	233	21	4.3	24	84	265	27	5.0	14	97	1	1	2	B	14	3	3	Jl–Au	Au–De
14139	*Aca leptocarpa*	16 40	145 18	400	23	1317	339	23	6.8	25	100	389	21	6.6	23	97	2	2	2	r	14	1	1	Jn–Jl	Au–No
14577	*Aca leptoloba*	17 23	145 14	780	20	804	325	13	6.7	17	88	350	18	6.2	15	100	2	2	2	r	10	2	2	No–Jn	De–Au
14676	*Aca maconochienana*	20 17	127 19	260	26	269	106	45	2.7	58	69	93	29	1.5	41	97					17	1		Ap/Au	
14398	*Aca mearnsii*	36 20	150 13	40	16	946	404	33	7.5	30	81	611	16	9.0	23	100	0	0	2	B					
14766	*Aca melanoxylon*	27 22	152 47	300	18	1236	436	28	8.9	32	94	465	14	8.3	18	97	2	2	2	r					
14735	*Aca neriifolia*	27 24	152 00	500	17	819	661	18	10.0	26	97	663	18	9.6	23	97	1	1	2	B,V	16				
14759	*Aca neriifolia*	23 51	149 04	860	18	839	307	26	5.7	28	94	420	14	6.4	21	97	1	1	2	B,V	13	2	1	Jn–Au	Se–No
14961	*Aca oraria*	16 41	145 35	5	25	1825	178	25	6.1	28	94	171	20	4.6	24	100	1	2	2	r					
14672	*Aca pachycarpa*	19 33	127 41	300	26	313	45	35	0.7	25	34	51	42	1.0	36	69									
14629	*Aca pallidifolia*	16 41	131 46	200	27	584	30	0	0.4	35	6	30		0.7		6									
14767	*Aca parramattensis*	34 42	150 02	550	13	704	339	31	6.4	40	88	336	35	5.9	27	97									

Seedlot	Species	Origin			Climate		Growth Performance										Coppicing performance				Age first buds (mo)	Flowering & fruiting			
							Wongi					Tuan/Toolara					Cut ht. (m)					Level		Season	
		Lat ° ′	Long ° ′	Alt (m)	MAT (°C)	MAR (mm)	Ht. (cm)	CV (%)	DGL (cm)	CV (%)	Srv. (%)	Ht. (cm)	CV (%)	DGL (cm)	CV (%)	Srv. (%)	0.1	0.5	1.0	Notes		Flw	Mat seed	Flowers	Fruit
14757	*Aca penninervis*	25 12	149 59	360	2[illegible]	721	418	21	6.5	32	84	487	10	6.8	14	97	1	1	2	B	8	3	1	All Year	All Year
14542	*Aca platycarpa*	14 35	132 30	190	2[illegible]	884	81	90	1.6	85	28	69	65	1.4	65	63		2		s					
14960	*Aca platycarpa*	16 47	144 08	280	25	922	243	45	4.8	53	69	358	13	5.9	22	97	2	2	2		12	2	1	Ap–Jn	Jl–Ja
14695	*Aca plectocarpa*	16 18	128 15	120	23	656	318	19	6.8	23	100	376	14	6.1	18	100	0	1	3		12	2	2	Mr–Jl	My–De
12055	*Aca podalyriifolia*	24 50	152 40	100	21	1136	367	28	7.3	44	56	396	29	5.8	36	97					9	2	2	Jl–Au	Se–De
14733	*Aca podalyriifolia*	27 38	151 29	100	20	589	341	23	7.4	39	69	443	12	6.9	15	100					9	2	3	Jl–Au	Se–De
14747	*Aca polybotrya*	27 57	151 08	420	18	647	215	36	5.6	39	47	176	19	4.5	17	44					14	1	1	Au	No–Ja
7915	*Aca pruinocarpa*	26 00	118 00	490	23	194	48	31	1.4	30	16					0									
14592	*Aca salicina*	24 36	149 54	105	21	652	184	34	3.8	28	97	233	20	4.9	22	88	2	2	3	r	25	1		Ap–	
14622	*Aca shirleyi*	16 19	133 23	225	27	620	227	37	4.9	45	41	181	60	3.0	63	72					25	1		Ap–	
14753	*Aca shirleyi*	26 40	150 15	360	[illegible]9	644	251	26	5.0	34	91	281	19	5.3	26	88	0	0	2	B	14	1		My–Jn	
14773	*Aca silvestris*	36 49	149 00	800	[illegible]1	566	235	45	6.4	66	6					0									
14670	*Aca stenophylla*	18 41	128 21	340	26	374	110	39	2.3	50	91	118	19	2.3	34	100									
14612	*Aca stipuligera*	18 38	133 57	305	26	406						168	10	3.3	12	50					11	3	1	Mr–No	My–Ja
14760	*Aca storeyi*	23 52	149 01	860	18	836	461	21	8.6	36	84	490	12	7.3	14	100	3	3	3	R	15	1	1	All Year	All Year
14975	*Aca tephrina*	20 50	144 18	333	24	457	47	52	1.2	48	84	47	37	1.1	39	94									
14141	*Aca torulosa*	12 39	141 49	2	26	1690	378	4	11.5	7	100	363	15	6.0	21	97	2	2	2						
14888	*Aca torulosa*	15 27	144 13	110	26	956	378	19	7.5	24	97	416	19	6.3	20	100	2	1	2	r					
14229	*Aca trachyphloia*	35 36	149 55	710	12	910	443	47	8.4	37	88	457	23	7.9	28	88	1	1	2	B	25	1			
14675	*Aca tumida*	20 10	127 34	260	26	266	173	35	4.6	34	81	205	11	4.5	25	75	1	1	2	B	2	2	3	De/Jn	Jn–No
14557	*Adn abrosperma*	16 59	144 18	220	25	868	74	52	1.1	40	22	54	41	0.7	32	25									
14959	*Alb procera*	16 50	145 41	10	25	1958	98	151	1.9	61	50	63	38	1.7	45	94									
13880	*Alo luehmannii*	16 49	145 23	380	23	1055	116	29	2.7	44	84	120	24	2.1	27	81									
13133	*Alo littoralis*	25 57	152 56	50	21	1411	293	19	8.9	23	97	254	13	8.0	11	100	2	2	3	s	12	1	1	Ap–Au	De–Jn
14843	*Cas cristata*	31 43	148 40	290	18	529	219	17	4.3	21	97	178	29	3.3	37	97	2	2	2						
13515	*Cas cunninghamiana*	17 04	145 28	400	23	1087	358	20	7.7	21	100	380	17	8.2	21	100	2	2	3		16	1	1	Jl–Au	No–Ap
14560	*Des umbellatum*	17 18	144 35	500	23	773	66	66	2.2	64	88	48	55	1.8	57	91		2			22	2	2	Ja–Ap	Fe–Jn
13755	*Dod viscosa* ssp. *spatulata*	23 45	133 28	720	21	241	175	31	3.1	25	66	164	18	2.7	22	100	2	2	2		11	2	1	Fe–Jn	Ap–No
13713	*Euc argophloia*	26 20	150 40	300	19	684	256	36	5.5	40	81	153	53	3.4	52	88	3	3	3						
14338	*Euc camaldulensis*	17 17	145 03	500	23	703	500	33	9.1	28	100	488	17	8.3	19	100	3	3	3						
14425	*Euc cloeziana*	26 18	152 48	100	20	1387	481	36	8.9	25	94	455	13	8.1	10	100	3	3	3						
13886	*Euc grandis*	30 18	153 03	60	19	1792	598	20	9.7	18	97	508	19	8.4	19	94	3	3	3						
13906	*Euc grandis*	26 18	152 47	60	20	1331	742	18	10.2	19	97	548	23	8.1	19	100	3	3	3						
13936	*Euc jensenii*	16 46	125 52	400	26	826	116	66	2.9	57	94	117	49	2.9	49	97	3	3	3		23	1	1	Fe–	Mr–
14143	*Gre parallela*	12 33	141 52	10	26	1663	63	62	1.9	65	63	69	54	2.1	71	78									
14980	*Gre pinnatifida*	16 34	145 22	415	23	1606	102	44	2.3	43	66	164	25	2.5	26	100	3	3	3						
14905	*Gre pteridifolia*	15 17	144 59	280	25	1687	174	30	5.4	18	25	253	26	6.5	22	100	3	3	3		17				
14900	*Lep longifolium*	15 26	144 11	90	26	944	127	22	5.6	23	91	112	23	3.8	24	100	3	3	3		10				
14555	*Lep petersonii*	17 21	145 24	935	20	1231	121	28	5.0	29	84	123	23	4.0	26	94	3	3	3		22				
14185	*Lop suaveolens*	17 35	145 27	935	20	1197	329	12	8.5	16	47	274	16	5.8	26	100	3	3	3		10	2	2	Oc–Ma	All Year
14866	*Mel arcana*	12 43	143 12	100	26	1840	194	19	5.2	21	94	203	19	5.3	20	100	1	1	2	V	23	1		Mr–My	
14876	*Mel arcana*	15 12	145 09	40	26	1713	200	17	6.7	23	97	196	15	5.2	18	91	1	1	2	V	24	2		Mr–My	

(Continued)

Appendix 4. *(Concluded).*

Seedlot	Species	Origin: Lat ° ′	Origin: Long ° ′	Origin: Alt (m)	Climate: MAT (°C)	Climate: MAR (mm)	Growth Performance – Wongi: Ht. (cm)	Wongi: CV (%)	Wongi: DGL (cm)	Wongi: CV (%)	Wongi: Srv. (%)	Tuan/Toolara: Ht. (cm)	Tuan/Toolara: CV (%)	Tuan/Toolara: DGL (cm)	Tuan/Toolara: CV (%)	Tuan/Toolara: Srv. (%)	Coppicing performance – Cut ht. (m): 0.1	0.5	1.0	Notes	Age first buds (mo)	Flowering & fruiting – Level: Flw	Level: Mat seed	Season: Flowers	Season: Fruit
14903	*Mel bracteata*	15 50	144 54	180	25	1378	97	30	4.8	27	94	109	19	3.7	28	100	2	2	2		7	2	3	Jn–Fe	Se–Ap
14550	*Mel cajuputi*	16 16	145 22	5	25	3067	204	22	7.7	21	94	218	13	7.0	16	97	2	2	2		14	2	2	Ap–Au	My–Se
14878	*Mel cajuputi*	16 16	145 23	12	25	3120	225	26	6.3	24	97	224	27	6.3	26	100	2	2	2		13	2	2	Mr–Au	Fe–Jn
13567	*Mel leucadendra*	17 00	145 30	500	22	1145	228	24	7.5	30	94	205	26	5.9	31	100	2	2	2						
14979	*Mel linariifolia*	18 56	144 30	550	22	659	105	26	4.6	26	97	109	19	3.6	23	100	2	2	2		14	2	2	My–No	My–Mr
13440	*Mel nervosa*	21 33	145 50	300	23	554	71	61	2.0	71	78	66	50	1.8	61	88		2							
14879	*Mel nervosa*	20 20	145 42	320	24	645	134	25	4.2	30	100	82	50	2.0	55	94									
14902	*Mel quinquenervia* (vel aff)	16 38	145 23	375	23	1473	228	16	7.1	18	84	231	15	6.0	19	100	2	2	2		17	1		Ap–Au	
14495	*Mel symphyocarpa*	13 45	130 42	8	27	1285	87	64	2.1	93	91	86	49	1.7	58	97	3	3	3		13	1	1	My–Jl	Jn–
14558	*Mel viridiflora*	16 36	144 07	265	25	938	195	26	5.9	33	91	180	30	4.1	26	100	2	2	2		13	1	1	Mr–Jn	My–Se
14589	*Mel viridiflora*	22 52	150 17	30	22	1140	227	33	6.8	25	100	223	23	5.6	26	100	2	2	2		13	1	1	Ap–Jn	Jn–No
14500	*Mla azedarach* v *australasica*	17 17	145 27	752	21	1171	244	29	5.4	40	100	286	33	5.2	36	100	3	3	3						
14889	*Neo myrtifolia* (sp ass n)	15 49	144 16	360	24	1141	50	28	0.8	39	16	34	49	0.5	68	56									
14896	*Neo myrtifolia* (sp ass n)	12 38	143 25	10	26	1845	135	21	3.6	23	91	111	22	3.0	30	84	2	2	2		13	2	2	Jn–Jl	Au–Fe
15194	*Syz suborbiculare*	10 45	142 36	5	na	na	106	36	3.1	34	100	92	29	2.8	30	97	3	3	3						

Appendix 5. Insect pests recorded in ACIAR trials in southeast Queensland.

Order	Family	Species	Host	Comment
Coleoptera	Cerambycidae	*Ancita marginicollis*	*Aca. crassicarpa* *A. aulacocarpa* *A. holosericea* *A. ampliceps*	Stem-borer. The adult beetle ringbarks branches.
		Ancita sp.	*A. cowleana*	Stem-borer. The adult ringbarks branches and twigs.
		Chlorophorus curtisi	*Lop. suaveolens*	Stem-borer. Minor.
	Chrysomelidae	*Cryptocephalus iridipennis*	*L. suaveolens*	Defoliator. Causes occasional severe damage to young eucalypts.
		Dicranosterna picea	*Aca. holosericea*	Defoliator. Generally minor.
		Monolepta australis	*A. trachyphloia*	Defoliator. Swarming leaf beetle; wide range of tree and plant species. Important.
		M. germari	*A. trachyphloia*	Defoliator. Usually minor.
	Curculionidae	*Myllocerus* sp.	*A. leptoloba* *A. holosericea* *A. deanii*	Defoliator. Generally Minor.
	Lagriidae	*Lagria grandis*	*Euc. grandis*	Defoliator. Minor.
	Scarabaeidae	*Anoplognathus boisduvali*	*E. camaldulensis* *E. grandis*	Defoliator. Important pest of eucalypts in Qld.
		Automolius (?) *vulgaris*	*E. normantonensis*	Defoliator. Swarming leaf beetle; causes occasional severe damage to eucalypts and other tree species.
		Liparetrus discipennis	*E. jensenii* *Lop. suaveolens* *Euc. cloeziana*	Defoliator. Swarming leaf beetle causes occasional severe damage to eucalypts.
		Liparetrus sp.	*E. jensenii*	Defoliator. (Members of this genus are important pests of eucalypts and other tree species).
		Repsimus aeneus	*E. grandis*	Defoliator. Important pest of eucalypts in Qld.
Hemiptera	Cicadellidae	*Tartessus* sp.	*Lop. suaveolens*	Sap-sucker. Minor.
	Coreidae	*Mictis profana*	*Aca. ampliceps*	Sap-sucker. Causes dieback of young shoots. Important.
	Eriococcidae	*Eriococcus coriaceus*	*A. torulosa*	Sap-sucker. Occasional severe pest of eucalypts and other tree species in Qld. Common scale on eucalypts.
	Margarodidae	*Monophlebulus* sp.	*Mel. viridiflora*	Sap-sucker. Range of plant species Usually minor.
	Membracidae	*Sextius* sp.	*Aca. platycarpa* *A. holosericea* *A. ampliceps*	Sap-sucker. Minor.
	Pentatomidae	*Cuspicona simplex*	*Euc. tereticornis*	Sap-sucker. Minor.
		Poecilometis gravis	*E. grandis* *Aca. mearnsii*	Sap-sucker. Minor.
	Psyllidae	Unidentified sp.	*A. holosericea*	Sap-sucker.
	Scutelleridae	*Coleotichus* sp.	*A. difficilis*	Sap-sucker. Minor.

Order	Family	Species	Host	Comment
Isoptera	Rhinotermitidae	*Coptotermes acinaciformis*	*Aca. mearnsii*	Root-feeder and stemborer. One of dominant subterranean termites in Qld. Attacks young and (more commonly) old living trees.
Lepidoptera	Cosmopterigidae	Unidentified sp.	*A. falcata*	Skeletoniser. Probably minor.
	Cossidae	*Xyleutes* (?) *liturata*	*A. decurrens* *A. glaucocarpa*	Stem-borer. A principal wood moth species in Qld. Larval tunnelling weakens stem and branches of young trees and may result in breakage.
		Xyleutes sp.	*A. mearnsii*	Stem-borer. Common in wattles and eucalypts in Qld.
	Limacodidae	*Doratifera casta*	*Euc. grandis*	Defoliator. Important pest of eucalypts in Qld.
	Lycaenidae	*Jalmenus daemeii*	*Aca. platycarpa*	Defoliator. Minor.
		Theclinesthes miskini	*Acacia* sp.	Defoliator. Minor.
	Notodontidae	*Teara contraria*	*Acacia* sp. *Aca. crassicarpa*	Defoliator. Common defoliator acacias in Qld. Important.
	Oecophoridae	*Zonopetala* sp.	*A. aulacocarpa*	Defoliator. May cause occasional severe damage on young trees.
	Pyralidae	*Epipaschia* sp.	*Lop. suaveolens*	Leaf-tier. Occasionally severe on young trees.
	Tortricidae	*Bathrotoma quiteaha*	*Mel. quinquenervia*	Leaf-tier. Important pest of young trees.
	Xyloryctidae	Unidentified sp.	*Aca. platycarpa*	Stem-borer. Ring-barks branches and occasionally the stems of small trees.
Orthoptera	Acrididae	*Adreppus* sp.	*Euc. cloeziana*	Polyphagous. Minor.
		Coryphistes sp.	*Aca. holosericea*	Polyphagous feeder on plant tissue. Minor.
		Valanga irregularis	*Euc. grandis*	Chews shoots and foliage of a wide range of tree species. Plague locust.
	Gryllidae	(?)*Hemiphonus* sp.	*Aca. ampliceps*	Leaf and stem chewer. Generally minor except in nursery.
	Tettigoniidae	Unidentified sp.	*Aca. difficilis* *Des. umbellatum*	Polyphagous insects attacking growing tips and leaves. Minor.

From: F.R. Wylie and M. de Baar (1988). A checklist of insects collected in hardwood plantations of the ACIAR project in Queensland. Unpublished report, Queensland Department of Forestry.

Chapter 6

Temperate Eucalypt Trials in Southwest People's Republic of China

Wang Huoran, Yan Hong and Zhang Rongqui

Abstract

Early results of two 18-month species/provenance trials of temperate Australian eucalypt species near Kunming, southwest China, are given. The results indicate better growth rate of *Eucalyptus globulus* ssp. *globulus* and ssp. *bicostata* over other blue gums. Promising new introductions include *E. nitens, E. viminalis, E. camphora* and *E. smithii*, whereas *E. badjensis* and *E. scoparia* deserve further study. Further evaluation is required before definite recommendations can be made.

Introduction

A collaborative ACIAR-supported research project (Introduction and Cultivation Experiments for Australian Broadleaved Tree Species) was established between the Chinese Academy of Forestry (CAF) and the CSIRO Division of Forest Research in October 1985. A component of this project involved species/provenance trials of potentially promising temperate-zone eucalypts which occur naturally in southeastern Australia. These trials were planted in October 1986 near Kunming, Yunnan Province, southwest China.

Eucalyptus globulus (blue gum) was first introduced into Yunnan Province about 100 years ago. The original source of the seed is unknown. It is now estimated that about 500 million trees of *E. globulus* (including a smaller number of *E. globulus* subsp. *maidenii*) have been planted in Yunnan Province. Large-scale planting programs apparently commenced in the 1960s because the species grows quickly and its wood can be used for construction purposes and pulp and paper. The leaves are widely used for the extraction of cineole. Although blue gums grow well in this part of China, it is believed that introduction of other seed sources and of seed of other species could improve overall productivity.

The objective of these trials is to evaluate species and provenances of blue gum plus other *Eucalyptus* species suitable for wood production in southwestern China. The wood is required for paper pulp, artificial board and for general construction purposes by rural communities. The aim of the present project was to select better provenances of both species and to test a wider range of lesser-known temperate eucalypts.

This chapter reports some preliminary results of the trials at 18 months after planting.

Materials and Methods

Trial Location

Two trials were planted near Kunming. One is a species/provenance trial of *Eucalyptus globulus* including three subspecies, *globulus, bicostata* and *maidenii*, and two closely related species, viz. *E. nitens* and *E. viminalis*, located at Jindian Experimental Forestry Farm, 8 km from Kunming. The second trial is a species elimination trial consisting of 30 eucalypt species and is located at Haikou State Forestry Farm, 50 km west of Kunming (see Fig. 1 in Chapter 1 for location of trial sites).

Fishing boats (top) moored on the lake near Kunming, Yunnan Province, People's Republic of China, with tall *Eucalyptus globulus* trees in the background. *Eucalyptus globulus* timber is commonly used in the construction of fishing boats for use on the lake.

Wang Huoran and Bai Jiayu (bottom), Trial Leaders of one of the ACIAR-supported projects, in the CAF/ACIAR species trial at Haikou forest farm, near Kunming.

Kunming, the capital city of Yunnan Province, lies in the central part of the province (25°01′N, 102°41′E)at an altitude of 1890 m. Climate in the Kunming area is considered subtropical, receiving 990 mm of precipitation per annum, with a very long dry season from December to April. Temperature records over 30 years indicate an annual mean temperature of 14.8°C with an absolute minimum temperature of -5.4°C and an absolute maximum temperature of 31.9°C. Booth (see Chapter 4) had difficulty finding a climatic match between the climate in Kunming with that of similar areas in Australia. In his analyses the match was with upland areas in northern New South Wales and Queensland. The Kunming area can experience environmental extremes which can be damaging to plant growth, with periodic cold blasts of wind from the north in winter and long dry periods in the summer. These extremes, together with the past success with *E. globulus* and *E. maidenii*, suggest species from southern Australia (especially species belonging to section *Maidenaria*) should be tried more extensively in Yunnan Province.

Site Preparation and Planting

Prior to clearing for the trials, both sites were occupied by failed plantations of *Pinus yunnanensis*. This species is important locally for timber but had been very badly attacked by tip-borer insects. After clearing, both sites were fully cultivated and holes 60 × 60 × 60 cm were dug. At both sites 100 g NPK compound fertiliser was put in each hole prior to planting and the same amount was added in the following year before the wet season. Garbage from Kunming was also used at Haikou site as a source of organic fertiliser. Both trials were established during the wet season in late July 1986. Seedlings were about 15 cm in height when out-planted, except for those of *E. nitens* which were 6–11 cm.

Experimental Design and Layout

Details of the seed sources of the species and provenances of *Eucalyptus* tested at Jindian and Haikou are given in Tables 1 and 2 respectively.

At Jindian, 49 provenances (from five species) were planted in a 7 × 7 balanced incomplete block design with eight replicates. Each plot consisted of nine trees arranged in a single line. Spacing was 2 × 2 m. At Haikou, 32 species/provenances were planted in completely randomised blocks with nine replications and single line plots each with nine trees. Spacing was 2 × 2 m. Two local seedlots of *E. globulus* ssp. *globulus* and *E. globulus* ssp. *maidenii* were used in both trials as controls.

Protection

Chemical insecticides were used shortly after planting because young plants were attacked by termites and yellow scarab beetle. *Eucalyptus viminalis* seedlings were more seriously damaged than those of other species.

A disease, possibly a virus, occurred on juvenile foliage of *E. globulus* ssp. *bicostata* and *E. nitens.* It did not seriously affect growth of the trees and no control measures were used.

Measurements and Data Analyses

After planting, measurements were taken at six-month intervals of tree height, diameter at ground level or at breast height, number of surviving trees per plot and crown diameter. Details of stem form and phenological characteristics were recorded.

Analyses of variance were performed using the GENSTAT statistical analysis program for data collected at 18 months on height, diameter at breast height, number of surviving trees per plot and crown diameter.

Results

Species/Provenance Trial at Jindian

Species Differences

Analysis of variance showed that there were significant differences between species in diameter at breast height and survival, but not for height and crown diameter (Table 3). Also of note was the significant difference in replicate effects for all attributes except height.

Eucalyptus globulus ssp. *globulus* and ssp. *bicostata* were the fastest growing (mean values) and they were closely followed by *E. viminalis. Eucalyptus nitens* was performing well given that its seedlings were much smaller at planting time (Table 4).

The two local seedlots, *E. globulus* ssp. *globulus* and ssp. *maidenii*, differed markedly in their growth. The former was among the tallest in height growth while the latter was the shortest in the whole trial (Table 4).

Provenance Differences

There were highly significant differences between provenances in all growth parameters measured (Table 3).

Eucalyptus globulus has shown great variation between provenances. For ssp. *globulus* the top six provenances in height growth were from Geeveston, Taranna, Leprena, Channel, Bruny Island and Rheban. All these provenances are located in the southeastern region of Tasmania (Table 4).

There was great variation in height growth and leaf morphology among geographic populations of

Table 1. Details of the seed sources of species and provenances of *Eucalyptus* tested at Jindian.

Species	CSIRO seedlot no.	Origin	Lat. (S)	Lon. (E)	Alt. (m)
E. globulus ssp. *globulus*	–	Bruny Is. TAS			180
E. globulus ssp. *globulus*	–	Geeveston TAS			200
E. globulus ssp. *globulus*	–	Henty River TAS			50
E. globulus ssp. *globulus*	–	Swansea TAS			100
E. globulus ssp. *globulus*	–	Leprena TAS			20
E. globulus ssp. *globulus*	–	Taranna TAS			120
E. globulus ssp. *globulus*	–	Pepper Hill TAS			560
E. globulus ssp. *globulus*	–	Rheban TAS			80
E. globulus ssp. *globulus*	–	Scamander TAS			50
E. globulus ssp. *globulus*	–	St. Helens TAS			50
E. globulus ssp. *globulus*	–	Flinders Is. TAS			50
E. globulus ssp. *globulus*	–	Channel TAS			100
E. globulus ssp. *globulus*	–	Yunnan			
E. globulus ssp. *maidenii*	–	Yunnan			
E. globulus ssp. *maidenii*	12126	Bimmil Hill NSW	37° 7′	149°53′	360
E. globulus ssp. *maidenii*	12132	SW Nelligen Bolaro NSW	35°41′	150° 4′	335
E. globulus ssp. *maidenii*	12321	Cann Valley VIC	37°18′	149°12′	290
E. globulus ssp. *maidenii*	12130	Mt Dromedary NSW	36°17′	150° 3′	305
E. globulus ssp. *maidenii*	12125	Tantawangalo Mtn NSW	36°48′	149°34′	381
E. globulus ssp. *bicostata*	9539	Stanley VIC	36°11′	146°40′	580
E. globulus ssp. *bicostata*	9246	Wee Jasper NSW	35°28′	148°10′	910
E. globulus ssp. *bicostata*	9541	NE of Mansfield VIC	37°03′	146°20′	850
E. globulus ssp. *bicostata*	11310	Mt Lonarch VIC	37°15′	143°22′	680
E. globulus ssp. *bicostata*	11742	5 km N of Bruthen VIC	37°40′	147°47′	310
E. nitens	12401	Federation Rd VIC	37°27′	147°57′	1100
E. nitens	15016	Barnewell Plains VIC	37°27′	147°57′	1100
E. nitens	15015	Marshalls Spur VIC	37°50′	146°21′	1165
E. nitens	13281	ENE of Armidale NSW	30°28′	152°15′	1277
E. nitens	12867	Bonang SF VIC	37°12′	148°42′	800
E. nitens	14450	Barrington Tops NSW	32°00′	151°30′	1500
E. nitens	14454	Mt Toorong Plateau VIC	37°47′	146°16′	900
E. nitens	14455	Brown Mtn NSW	36°38′	149°24′	1130
E. nitens	14437	Tallaganda SF NSW	35°54′	149°30′	1300
E. nitens	14449	Tallaganda SF NSW	35°49′	149°31′	1200
E. viminalis	12973	Tallaganda SF NSW	35°31′	149°33′	900
E. viminalis	12568	Forest Lands SF NSW	29° 9′	152° 6′	1100
E. viminalis	15017	Silver Ck, Morwell VIC	38°20′	146°14′	240
E. viminalis	11743	40 km NNW Bruthen VIC	37°26′	147°34′	900
E. viminalis	12556	9 km NE Mathinna TAS	41°24′	147°58′	340
E. viminalis	15018	Warburton VIC	37°44′	145°45′	220
E. viminalis	14523	Nullo Mt NE Rylstone NSW	32°43′	150°13′	900
E. viminalis	12564	Nundle SF Tamworth NSW	31°27′	151°15′	1250
E. viminalis	12555	15 km NW Swanport TAS	42°15′	147°51′	580
E. viminalis	12554	28 km NNW Maydena TAS	42°36′	146°28′	400
E. viminalis	12651	Erica VIC	37°46′	146°18′	325
E. viminalis	14525	Warung SF Coolah NSW	31°45′	149°58′	1080
E. viminalis	14201	14 km SE of Bendoc VIC	37°15′	148°58′	850
E. viminalis	14198	Cotter Flats ACT	35°38′	148°50′	1100
E. viminalis	14511	Barrington Tops NSW	31°58′	151°23′	1300

Table 2. Details of the seed sources of species planted at Haikou.

Species	CSIRO seedlot no.	Origin	Lat. (S)	Lon. (E)	Alt. (m)
E. amplifolia	13349	Paddy's Land SF Guyra NSW	30°04′	152°09′	880
E. badjensis	13286	23 km from Nimmatabel NSW	36°32′	149°15′	900
E. benthamii	14214	Near Wentworth Falls NSW	33°48′	150°24′	150
E. camphora	12448	Coree Flat ACT	35°17′	148°49′	1070
E. chapmaniana	9755	Kiew Northeast VIC	36°16′	147° 1′	140
E. cinerea	11711	Gunning Area NSW	35°41′	148°54′	490
E. cypellocarpa	9440	Fitzroy Falls NSW	34°39′	150°29′	-
E. cypellocarpa	12914	Jeeralang Nth VIC	38°25′	146°29′	520
E. cypellocarpa	12655	Bonang VIC	37°12′	148°42′	860
E. dalrympleana	12563	Nundle SF Tamworth NSW	31°27′	151°15′	1250
E. deanei	10340	SW of Thirlmere NSW	34°13′	150°31°	240
E. deanei	11688	Watagan Mtns NSW	33° 2′	151°25′	370
E. deanei	14521	Glen Innes NSW	29°48′	152° 7′	950
E. globulus ssp. *globulus*	-	Yunnan			
E. globulus ssp. *maidenii*	-	Yunnan			
E. grandis	8602	Unknown			
E. johnstonii	11825	Misery Plateau TAS	42°30′	147°35′	760
E. laevopinea	14840	S. New England NSW	31°30′	151° 6′	186
E. macarthurii	15057	ENE of Marulan NSW	34°39′	150°10′	600
E. mannifera	12159	Yass-Dalton Dist VIC	34°51′	148°54′	-
E. neglecta	9751	Buckland River VIC	36°42′	146°53′	760
E. nitens	14012	Brown Mountain NSW	36°38′	149°24′	1100
E. nova-anglica	13606	24 km SSW of Walcha NSW	31° 9′	151°31′	1045
E. parvifolia	12284	Badja R. Sth T'lands NSW	36° 4′	149°30′	1300
E. pauciflora	13831	Mt Coree ACT	35°19′	148°49′	1390
E. propinga	-	Unknown			
E. scoparia	12576	Nat. Bot. Gdns Canberra ACT	35°17′	149° 6′	620
E. smithii	15092	Wingello NSW	34°42′	150°10′	650
E. smithii	15090	Towamba NSW	37° 5′	149°47′	220
E. smithii	15059	Mt Dromedary NSW	36°18′	150° 1′	305
E. smithii	15091	NW of Narooma NSW	36°00′	150°00′	450
E. triflora	14207	Morton Nat. Park NSW	35° 6′	150° 9′	760

Table 3. Summarised results of analysis of variance for height, diameter, survival and crown diameter of the species/provenance trial at 18 months old at Jindian. *, ** and *** indicates significance at the 5, 1 and 0.1% level respectively; ns indicates not significant at the 5% level.

Souce of variation	d.f.	m.s.	F-ratio
Height			
Replicate	7	2.251	1.690ns
Species	6	4.539	1.669ns
Provenance within species	42	2.719	1.832**
Residual	332(4)	1.492	
Diameter at breast height			
Replicate	7	2.707	21.887***
Species	6	1.104	2.592*
Provenance within species	42	0.426	3.367***
Residual	334(2)	0.126	
Survival			
Replicate	7	5.232	30.681***
Species	6	2.430	8.408***
Provenance within species	42	0.289	1.697***
Residual	334(2)	0.170	
Crown diameter			
Replicate	7	1.243	38.066***
Species	6	0.926	1.493ns
Provenance within species	42	0.620	1.916***
Residual	334(2)	0.033	

Table 4. Means of measurements taken at 18 months after planting for the trial at Jindian.

Species	CSIRO seedlot no.	Surviving trees/plot	Height (m)	DBH (cm)	Crown diameter (m)
E. globulus					
ssp. *globulus*	Bruny Is. TAS	7.7	2.5	1.7	1.7
ssp. *globulus*	Geeveston TAS	7.9	2.8	2.4	1.8
ssp. *globulus*	Henty River TAS	7.5	2.4	2.0	1.6
ssp. *globulus*	Swansea TAS	8.3	2.3	1.9	1.7
ssp. *globulus*	Leprena TAS	7.7	2.5	2.1	1.7
ssp. *globulus*	Taranna TAS	7.5	2.6	2.2	1.7
ssp. *globulus*	Pepper Hill TAS	8.3	2.5	2.1	1.8
ssp. *globulus*	Rheban TAS	7.3	2.5	2.0	1.7
ssp. *globulus*	Scamander TAS	7.4	2.2	2.1	1.6
ssp. *globulus*	St Helens TAS	8.2	2.2	1.8	1.6
ssp. *globulus*	Flinders Is. TAS	7.4	2.3	1.8	1.7
ssp. *globulus*	Channel TAS	6.5	2.5	2.1	1.7
Mean		**7.6**	**2.4**	**2.0**	**1.7**
ssp. *maidenii*	12126	6.7	2.5	1.9	1.7
ssp. *maidenii*	12132	8.2	2.2	1.6	1.8
ssp. *maidenii*	12321	7.6	2.2	1.9	1.8
ssp. *maidenii*	12130	7.2	2.3	1.7	1.7
ssp. *maidenii*	12125	8.4	2.1	1.5	1.7
Mean		**7.6**	**2.3**	**1.7**	**1.7**
ssp. *bicostata*	9539	8.1	2.7	2.1	1.8
ssp. *bicostata*	9246	6.9	2.2	1.7	1.9
ssp. *bicostata*	9541	7.5	2.4	1.9	1.7
ssp. *bicostata*	11310	7.7	2.3	1.9	1.7
ssp. *bicostata*	11742	6.8	2.7	1.9	1.8
Mean		**7.4**	**2.5**	**2.0**	**1.8**
E. nitens	12401	6.9	2.3	1.7	1.6
E. nitens	15016	6.9	2.0	1.5	1.6
E. nitens	15015	7.3	1.9	1.4	1.6
E. nitens	13281	7.5	2.2	1.6	1.8
E. nitens	12867	8.1	2.6	1.8	1.6
E. nitens	14450	6.9	2.1	1.5	1.8
E. nitens	14454	7.3	2.3	1.8	1.7
E. nitens	14455	7.0	2.1	1.6	1.7
E. nitens	14437	6.7	1.9	1.4	1.7
E. nitens	14449	7.3	2.1	1.6	1.8
Mean		**7.3**	**2.1**	**1.6**	**1.7**
E. viminalis	12973	7.2	2.5	1.7	1.3
E. viminalis	12568	8.1	2.3	1.6	1.7
E. viminalis	15017	8.0	2.7	1.6	1.5
E. viminalis	11743	7.2	2.9	2.0	1.5
E. viminalis	12556	5.3	2.4	1.6	1.5
E. viminalis	15018	7.2	2.6	1.6	1.3
E. viminalis	14523	5.7	2.2	1.5	1.4
E. viminalis	12564	7.4	2.0	1.1	1.5
E. viminalis	12555	6.9	2.5	1.7	1.4
E. viminalis	12554	7.1	2.7	1.7	1.4
E. viminalis	12651	7.5	2.8	2.0	1.6
E. viminalis	14525	5.8	2.0	1.3	1.4
E. viminalis	14201	7.0	2.2	1.6	1.4
E. viminalis	14198	7.8	2.4	1.7	1.4
E. viminalis	14511	6.9	2.0	1.4	1.4
Mean		**7.0**	**2.4**	**1.6**	**1.5**
E. globulus	Yunnan	7.6	2.6	2.0	1.8
E. maidenii	Yunnan	8.0	1.9	1.4	1.6

Table 5. Analysis of variance of height, diameter at breast height and crown diameter 18 months after planting at Haikou. *** indicates significance at the 0.1% level.

Measurement	Source	d.f.	m.s.	F-ratio
Height	Replicate	8	8.013	67.892***
	Species	31	0.826	7.000***
	Residual	243(5)	0.118	
DBH	Replicate	8	28.525	114.647***
	Species	31	0.582	2.337***
	Residual	243(5)	0.249	
Crown diameter	Replicate	8	3.481	84.716***
	Species	31	0.429	10.446***
	Residual	243(5)	0.041	

Table 6. Mean values of measurements taken at 18 months after planting for the trial at Haikou.

Species	CSIRO seedlot no.	Surviving trees/plot	Height (m)	DBH (cm)	Crown diameter (m)
E. amplifolia	13349	9.0	1.7	1.9	1.3
E. badjensis	13286	8.2	1.9	2.2	1.3
E. benthamii	14214	8.3	2.0	2.5	1.5
E. camphora	12488	8.8	2.4	2.0	1.4
E. chapmaniana	9755	8.9	2.2	2.3	1.6
E. cinerea	11711	8.7	1.4	1.9	1.4
E. cypellocarpa	9440	8.9	1.7	2.2	1.4
E. cypellocarpa	12914	8.6	1.9	2.3	1.5
E. cypellocarpa	12655	8.2	1.9	2.0	1.5
E. dalrympleana	12563	8.2	1.3	2.0	1.2
E. deanei	10340	8.6	1.4	1.5	1.3
E. deanei	11688	8.7	1.4	2.1	1.2
E. deanei	14521	9.0	1.9	2.2	1.4
E. globulus	Yunnan	8.9	1.6	2.0	1.5
E. grandis	8602	8.9	2.0	2.2	1.5
E. johnstonii	11825	7.2	1.5	1.7	0.7
E. laevopinea	14840	7.9	1.5	1.8	1.1
E. macarthurii	15057	8.2	1.4	2.2	1.3
E. maidenii	Yunnan	8.9	1.9	2.0	1.5
E. mannifera	12159	8.3	1.6	1.6	1.3
E. neglecta	9751	9.0	1.3	1.7	1.3
E. nitens	14012	8.6	1.7	2.5	1.5
E. nova-anglica	13606	8.8	1.8	2.3	1.5
E. parvifolia	12284	7.0	1.4	1.7	1.1
E. pauciflora	13831	7.4	1.2	1.7	0.8
E. propinqua	unknown	7.7	1.7	2.0	1.2
E. scoparia	12576	8.2	1.7	2.1	1.3
E. smithii	15092	8.9	2.3	2.4	1.7
E. smithii	15090	8.3	1.8	2.0	1.3
E. smithii	15059	8.4	2.0	2.0	1.5
E. smithii	15091	8.6	2.1	2.2	1.5
E. triflora	14207	5.6	1.4	1.5	0.9

E. viminalis. Of the 15 provenances, four Victorian seedlots (11743, 12651, 15017 and 15018) and two Tasmanian seedlots (12554 and 12555) were the tallest. In contrast, three provenances from southern New South Wales (14525, 14511, 12564) were the lowest in height growth. The better-performing provenances were notable for their narrow leaves and dull-coloured foliage while the poorer provenances had broad and shiny green leaves.

Among the 10 provenances of *E. nitens*, seedlot 12867 from Bonang State Forest, Victoria, was the best and followed by seedlots from Toorong Plateau (14454) and Federation Road (12401), both also from Victoria. Trees from Bonang State Forest (12867) possessed quite sparse foliage and much narrower juvenile leaves with oil glands on serrated leaf margins. This provenance had produced intermediate leaves, thus possessing a very short juvenile leaf phase.

Species Trial at Haikou

There were marked differences between species in height, diameter at breast height and crown diameter (Table 5). Mean values for these parameters are given in Table 6. There were also highly significant differences amongst replicates.

Several species have shown promising growth. *Eucalyptus camphora, E. chapmaniana* and *E. smithii* (15092 from Wingello, NSW) were the fastest growing in height, whereas *E. benthamii, E. nitens, E. smithii* (15092), *E. chapmaniana* and *E. cypellocarpa* had greater diameters than other species. Species with greatest height and diameter growth tended to have wider crowns.

Most species survived well. The only exception was for *E. triflora* with an average of 5.56 surviving trees per plot. However, it should be noted that most species belonging to subgenus *Monocalyptus* (Pryor and Johnson 1971) were not stratified, failed to germinate satisfactorily in the nursery and did not get planted in the trial (e.g. *E. fraxinoides, E. dendromorpa, E. fastigata* and *E. elata*).

Discussion and Conclusions

Blue gums and other temperate eucalypts tested in Jindian and Haikou have been growing satisfactorily and appear to be adapted to the local environment. The early results have shown that the best performing provenances of all five species tested at Jindian originated from Victoria and Tasmania. This suggests that the natural populations of those species in Victoria and Tasmania could be considered for seed sources for plantation establishment in the Kunming area. Climatically this does not accord well with Booth's (see Chapter 4) exercise in climatic matching.

The good early performance of *E. globulus* ssp. *globulus* and ssp. *bicostata* is encouraging, but continued evaluation over a longer period is necessary. It is possible that ssp. *bicostata* would show greater drought tolerance given the empirical evidence of their comparative natural distributions. The good performance (to 18 months) of the local form of ssp. *globulus* was surprising, but this may reflect some local genetic selection. One serious local problem is spiral grain in local ssp. *globulus* and this, together with the search for improved growth performance, has led to a proposed tree improvement program (Raymond 1988) to be implemented in 1989.

Some species, for example *E. nitens, E. viminalis, E. camphora* and *E. smithii,* etc., are very promising and merit testing on a wider range of sites. Some lesser-known species such as *E. smithii, E. badjensis* and *E. scoparia* should be explored further for potential uses in oil production and urban forestry. Unfortunately, some potentially promising species such as *E. dunnii* were not included in the trial.

Eucalyptus camphora is worthy of special mention as it displayed good survival (but moderate growth) even on one replicate which was exposed to drying summer winds. This is unusual given its natural habitat (cold swampy areas), but this result mirrors similar results in other dry area trials where Australian species from very moist natural habitats have performed well (e.g. *E. occidentalis, E. camaldulensis,* and *Melaleuca quinquenervia*). The physiological reasons for this are not fully understood.

As a general rule in eucalypt introduction work it has been emphasised that 'the transfer of species in the subgenus *Monocalyptus* is unreliable' (Pryor 1976; Turnbull and Pryor 1978). Therefore, although three monocalypts (viz. *E.pauciflora, E. laevopinea* and *E. triflora*) have grown well in the present trial, long-term observation is needed. All *Monocalyptus* species that failed in the nursery should be planted in a new trial. World experience has shown that monocalypts are often difficult species to raise in field nurseries and great care has to be exercised. Further tests with more species from subgenus *Monocalyptus* are warranted.

There are two factors, low temperatures in winter and a long dry season, that affect the growth of eucalypts in this area. The results reported here were obtained at 18 months after planting and the area has not yet suffered very cold conditions during that time, and hence it would be unwise to predict future performance.

Acknowledgments

We wish to acknowledge the assistance of many colleagues from the following collaborating organisations who have assisted with species selection, establishment, maintenance, measurement, and assessment of the trial: Australian Centre for International Agricultural Research; Division of Forestry and Forest Products CSIRO; Forestry Academy of Yunnan Province, China; Jindian Experimental Forestry Farm and Haikou State Forestry Farm; Dr J. Turnbull, Mr D. Boland, Ms S. Searle, Dr C. Matheson, Dr. E. Williams, Mr Bai Jiayu, Mr Yang Chunhui, Mr Dong Jingxi and Ms Li Bojing.

Chapter 7

Tropical Eucalypt Trials on Hainan Island, People's Republic of China

Zhou Wenlong and Bai Jiayu

Abstract

A *Eucalyptus* species and provenance trial was established in tropical areas of the People's Republic of China. This trial differs from many previous trials in that it included a select range of potentially promising species and provenances. Eighteen months after planting, *E. camaldulensis* from Western Australia, *E. tereticornis* from North Queensland and *E. urophylla* from Indonesia had grown best on the poor soils of the test site. These species grew much better than *E. exserta* and *E. citriodora* which have been widely planted in South China. The trial will provide important information on better eucalypt species and provenances for planting in tropical areas of the PRC.

Introduction

Australian eucalypts were first introduced into China 70–80 years ago. It is estimated that over 200 *Eucalyptus* species have been tried in China. Currently over 400 000 ha of eucalypts have been planted in the tropics and subtropics of South China. The main species now grown are *E. exserta* and *E. citriodora* in tropical areas, and *E. globulus* in the more temperate regions (e.g. Yunnan Province).

Since 1985, the Research Institute of Tropical Forestry, Chinese Academy of Forestry (RITF), has cooperated with the CSIRO Division of Forestry and Forest Products to conduct a trial of tropical *Eucalyptus* species and provenances in Qionghai County, Hainan Island. This work has been supported by ACIAR and the trial consists of 11 species and includes 79 provenances. Results obtained at 18 months after planting are reported here.

The aim of the project is to test a wide range of newly acquired seed from tropical species and provenances in Australia. At present the most widely planted species on Hainan Island is *E. exserta*, a slow-growing species. It should be possible to find a seed source from Australia that would be more productive than *E. exserta* for use as fuelwood, poles, rough sawnwood purposes and possibly pulp in the future.

Materials and Methods

Trial Site

The trial site (see Fig. 1 in Chapter 1) is located at Shang Yong Forest Farm about 10 km from Qionghai (19°16′N, 110°24′E, altitude 20–40 m).

Yellow-red latosol with coarse chad developed from sediment of shallow seabed is distributed over large areas of Qionghai County, eastern Hainan Island. The fertility of the soil is poor (Table 1). In the surface layer of the soil (0–20 cm), humus content is 0.8–1.5%, total nitrogen is <0.05%, and available P and K are low. Generally, the soil contains a high percentage of gravel, and the percentage increases with soil depth. At 85–120 cm gravel content is 62%, which is cemented into a hard pan.

The climate is tropical with a strong monsoonal influence. Mean annual temperature is 24°C, annual

Table 1. Soil analyses for the eucalypt trial site in Qionghai County.

Depth (cm)	Humus (%)	Total N (%)	Total P (%)	Available P (mg/100g)	Available K (mg/100g)	pH	
						H_20	KCl
0–2	1.31	0.0406	0.025	0.914	0.830	6.0	4.8
2–31	1.41	0.0211	0.022	0.732	–	6.0	4.8
31–59	0.63	0.0166	0.025	0.288	0.841	6.0	4.8
59–96	0.74	0.0136	0.026	0.498	1.048	6.0	4.8
96–110	0.79	0.0050	0.029	0.264	2.528	6.0	4.8

Table 2. Meteorological data[a] for the eucalypt trial site in Qionghai County.

	Jan.	Feb.	Mar.	Apr.	May	June	July	Aug.	Sept.	Oct.	Nov.	Dec.	Annual
Mean temp (°C)[b]	18.0	19.1	22.3	25.1	27.5	28.1	28.3	27.5	26.6	24.6	21.7	19.2	24.0
Precipitation (mm)	50.1	47.5	72.8	112.0	176.9	240.1	184.7	296.6	428.6	328.0	174.0	70.6	2181.5
Relative humidity (%)	86	88	87	86	84	84	84	83	86	87	85	85	86
Evaporation	98.6	95.4	139.3	176.0	215.6	199.0	217.9	180.3	151.5	144.7	112.8	96.7	1828

[a] The data are averages of 27 years.
[b] Absolute minimum temperature is 5°C (recorded 12 June 1955); absolute maximum temperature is 39.8°C (recorded 4 May 1957).

Table 3. Details of the seed sources of *Eucalyptus* species and provenances included in the trial on Hainan Island.

Seedlot	Species	Location		Lat	Long	Alt (m)	No. parent trees
14860	*E. grandis*	Embrapa	Brazil	–	–	–	–
14849	*E. grandis*	NE Atherton	QLD	17°06′	145°36′	1050	22
14838	*E. grandis*	WNW Cardwell	QLD	18°14′	143°00′	620	7
14519	*E. grandis*	Mt George, Taree	NSW	31°50′	152°01′	230	25
13019	*E. grandis*	NW Coffs Harbour	NSW	30°13′	153°02′	135	10
14509	*E. grandis*	Urbenville	NSW	28°31′	152°30′	600	25
14431	*E. grandis*	Belthorpe S. F.	QLD	26°52′	152°42′	500	12
14420	*E. grandis*	12 km S Ravenshoe	QLD	17°42′	145°28′	860	20
14393	*E. grandis*	25–36 km SE Mareeba	QLD	17°06′	145°33′	900	11
14210	*E. grandis*	27 km SE Ravenshoe	QLD	17°50′	145°33′	720	5
13965	*E. grandis*	Seed Orchard South Africa		–	–	–	–
13431	*E. grandis*	Mt Lewis	QLD	16°36′	145°16′	840	7
14861	*E. granids*	Embrapa	Brazil	–	–	–	–
13365	*E. grandis*	Seed Orchard South Africa		–	–	–	–
13020	*E. grandis*	NNW Coffs Harbour	NSW	30°10′	153°01′	98	10
13663	*E. camaldulensis*	Wrotham Park	QLD	16°48′	144°10′	230	21
15062	*E. camaldulensis*	NE of Katherine	NT	14°23′	132°21′	200	8
14518	*E. camaldulensis*	Tennant Ck	NT	19°34′	134°13′	335	10
13941	*E. camaldulensis*	Victoria River	NT	16°20′	131°07′	100	5
15052	*E. camaldulensis*	Isdell River	WA	16°50′	125°32′	250	7
15050	*E. camaldulensis*	Gibb River	WA	16°30′	126°10′	400	7
14540	*E. camaldulensis*	Pentecost River	WA	15°48′	127°53′	10	10
13933	*E. camaldulensis*	N Fitzroy Crossing	WA	18°06′	125°42′	110	10
15049	*E. camaldulensis*	Bullock Creek	QLD	20°46′	143°55′	400	10
14918	*E. camaldulensis*	Laura	W QLD	15°34′	144°27′	90	15
14917	*E. camaldulensis*	NW of Mt Carbine	W QLD	16°22′	144°43′	400	13

(*Continued*)

Table 3. Details of the seed sources of *Eucalyptus* species and provenances included in the trial on Hainan Island.

Seedlot	Species	Location		Lat	Long	Alt (m)	No. parent trees
14847	*E. camaldulensis*	Emu Ck Petford	QLD	17°10′	145°15′	500	20
14016	*E. camaldulensis*	Gilbert River	N QLD	18°00′	143°00′	150	9
12187	*E. camaldulensis*	8 km W Irvinebank	QLD	17°24′	145°09′	680	16
15011	*E. saligna*	Kroombit Tops. Monto	QLD	24°51′	151°01′	730	45
15054	*E. saligna*	SE of Tamworth	NSW	31°31′	151°31′	1100	25
14527	*E. saligna*	Barrington Tops	NSW	32°00′	151°50′	450	26
14526	*E. saligna*	Glen Innes	NSW	29°47′	152°09′	1030	26
14524	*E. saligna*	Armidale	NSW	30°39′	152°08′	900	26
14508	*E. saligna*	Urbenville	NSW	28°34′	152°30′	600	25
14507	*E. saligna*	Chaelundi S.F.	NSW	30°13′	152°46′	640	25
14435	*E. saligna*	Kenilworth S.F.	QLD	26°38′	152°33′	600	26
14429	*E. saligna*	Blackdown Tableland	QLD	23°50′	149°05′	780	5
13340	*E. saligna*	NE of Warwick	QLD	27°58′	152°12′	850	3
13263	*E. saligna*	Consuelo T'lands	QLD	24°57′	148°03′	1090	11
13029	*E. saligna*	NE of Bulahdelah	NSW	32°22′	152°28′	80	1
13015	*E. saligna*	North Nelligen	NSW	35°33′	150°11′	30	8
13418	*E. tereticornis*	Sirinumu Sogeri Plat	PNG	9°30′	147°26′	580	20
13443	*E. tereticornis*	Kennedy River	QLD	15°26′	144°11′	60	10
14115	*E. tereticornis*	S of Helenvale	QLD	15°46′	145°14′	120	30
13442	*E. tereticornis*	N of Mareeba	QLD	16°55′	145°25′	380	7
14424	*E. tereticornis*	Ravenshoe	QLD	17°39′	145°21′	700	30
12965 E/326	*E. tereticornis*	SW Mt Garnet	QLD	18°30′	144°45′	800	25
13446	*E. tereticornis*	Nth of Cardwell	QLD	18°16′	146°00′	40	4
13994	*E. tereticornis*	Crediton S.F.	QLD	21°00′	148°30′	700	15
13544	*E. tereticornis*	40 km N Gladstone	QLD	23°44′	151°01′	10	10
13541	*E. tereticornis*	9 km SW Imbil	QLD	26°30′	152°37′	100	10
13350	*E. tereticornis*	S of Urbenville	NSW	28°36′	152°24′	400	10
13319	*E. tereticornis*	N of Woolgoolga	NSW	29°55′	153°12′	30	6
13307	*E. tereticornis*	Windsor	NSW	33°32′	150°50′	100	8
13304	*E. tereticornis*	Nerrigundah	NSW	36°13′	149°48′	80	9
13303	*E. tereticornis*	Sale	VIC	38°07′	147°04′	10	5
15089	*E. urophylla*	Mt Egon, Flores	INDO	8°38′	122°27′	500	14
14532	*E. urophylla*	Mt Lewotobi, Flores	INDO	8°31′	122°45′	398	31
13828	*E. urophylla*	Mt Mutis W Timor	INDO	10°35′	123°35′	1200	-
12898	*E. urophylla*	Mt Boleng Adonara	INDO	8°21′	123°15′	890	16
12362	*E. urophylla*	S Dili East Timor	INDO	8°37′	125°38′	1100	-
10140	*E. urophylla*	S of Hato Bulico	INDO	8°53′	125°32′	2100	7
12895	*E. urophylla*	Mt Mandiri, Flores	INDO	8°15′	122°58′	415	23
14852	*E. citriodora*	Mt Garnet	QLD	17°41′	145°07′	850	18
14851	*E. citriodora*	Herberton	QLD	17°23′	145°23′	1000	9
14850	*E. citriodora*	Irvinebank	QLD	17°26′	145°12′	900	34
14703	*E. citriodora*	W of Mt Carbine	QLD	16°18′	145°05′	940	10
13472	*E. citriodora*	ESE of Mt Molloy	QLD	16°42′	145°23′	600	12
14864	*E. exserta*	Herberton Area	QLD	17°25′	145°23′	950	16
13282	*E. exserta*	N of Marlborough	QLD	22°40′	149°54′	30	-
12411	*E. resinifera*	14.5 km S Ravenshoe	QLD	17°42′	145°28′	940	6
12418	*E. resinifera*	Mt Lewis	QLD	16°36′	145°17′	1100	7
13321	*E. propinqua*	W of Woolgoolga	NSW	30°04′	153°06′	200	4
12018	*E. propinqua*	Kangaroo Ck SF	NSW	30°07′	152°46′	335	21
13657	*E. paniculata*	SW Nowra	NSW	35°00′	150°30′	120	5
14130	*E. torelliana*	SSW Kuranda	QLD	16°53′	145°36′	420	17

rainfall 2182 mm, annual evaporation 1826 mm and average relative humidity 86% (Table 2). Typhoons occur frequently in the area and are particularly damaging to tree growth.

The trial site originally carried a poor stand of *E. exserta* together with about 40 shrub and grass species as understorey. The stand was 10 years old with trees averaging 11.8 m in height and 9.9 cm in diameter at breast height. There was an average of 1775 trees/ha and volume growth increments ranged from 4.05 to 8.7 m^3/ha/year.

Establishment of Trial

After cutting and removing the original trees and digging out the stumps, the land was ploughed twice, using tractors, to a depth of 30 cm. Planting holes (40 × 40 × 40 cm) were dug and fertiliser (4 kg burned soil) and 120 g compound fertiliser (N:P:K = 15:15:12) was applied to each hole and covered with soil.

Seed sown in mid March 1986 germinated 3–5 days after sowing. After 40–50 days, when seedlings were 4–5 cm tall with 8–10 true leaves, they were transplanted into containers. The seedlings were watered twice daily. Carbamid (0.5% solution) was applied every 7–10 days, i.e. 4–5 times during whole seedling stage. When 20–30 cm in height the seedlings were ready for planting.

The seedlings were planted in early June 1986. Because of high temperatures and strong sunshine, they were planted during a rainy period (or irrigated) to improve survival.

In August 1986 when the young trees were established, 25 g of compound fertiliser was applied to each tree. During late 1986, the soil was cultivated to reduce weed growth.

A complete randomised block design with four replications was used. There were 18 trees (3 × 6) in each square plot (9 × 9 m). The spacing between rows and within rows was 3.0 and 1.5 m respectively. Each block contained one plot of provenance (i.e. 79 plots) and with four blocks there were 316 plots in total. The species and provenances included are listed in Table 3.

Tree height and ground-level diameter were first measured in January 1987. In June 1987, tree height, ground-level diameter and breast-height (1.3 m) diameter were measured for the entire trial. Height to the green crown, crown diameter and diameter at mid height of trees were also measured in one of the blocks. In December 1987, height and breast-height diameter were measured for the third time, together with crown diameter and height to green crown within one block.

Data were analysed using multiple-range tests. Mean height and diameter at age 18 months was calculated for each of the 79 provenances to determine whether growth of seedlots differed significantly. Analyses of variance (F-tests) were also conducted within each species.

Results

Mean tree height and diameter for each provenance, after 18 months growth in the field, are shown in Table 4.

Analyses of variance were conducted among provenances within each species. There were significant differences for height and diameter for *E. grandis, E. camaldulensis, E. tereticornis, E. exserta* and *E. urophylla* and for the diameter of *E. citriodora* (Table 5). There are no significant differences for the other species.

The height and diameter of the best provenance of each species at 18 months, together with their standard deviations and coefficients of variation, are shown in Table 6. The 20 best provenances based on height in the Qionghai trial are listed in Table 7.

Discussion

In general *E. camaldulensis* grew better than any other species in the trial. Among the 14 provenances of *E. camaldulensis* tested, 8 were placed in the 10 best provenances for height growth, and 3 in the 10 best provenances for diameter growth. *Eucalyptus tereticornis* was the second best species in the trial, and two provenances were placed in the 10 best provenances for height growth, and one was in the 10 best provenances for diameter growth. The height growth of *E. urophylla* was not as good as that of the two red gums (*E. camaldulensis* and *E. tereticornis*). However, 5 out of the 7 *E. urophylla* provenances were among the 10 best provenances in diameter growth. Local *E. exserta* rated 28th in height growth.

Provenances of *E. camaldulensis* from the northern parts of Queensland and Western Australia grew well. Provenances of interest (15052, 12187, 14918, 13933, 14847, 14540, 15050, 14917) are all originally from north of 18°15′ latitude in Australia. The two best provenances of *E. tereticornis* were 13443 and 13544, both of which were from North Queensland. Good provenances of *E. urophylla* were 12898, 14532, 15089, 12895 and 12362, with the Timor Island provenances growing less well than those from other Indonesian islands.

Eucalyptus camaldulensis (15052) from Isdell River, northern Western Australia, has performed exceptionally well in the Qionghai trial. The average height of 18-month-old trees was 7.58 m and average diameter at breast height was 5.34 cm.

Table 4. Ranking for mean height and diameter of the eucalypt trial on Hainan Island showing partial results of Duncan's new multiple range test.

Species/provenance	Height (m)	Species/provenance	Diameter (cm)
15052 *camaldulensis*	7.58	15052 *camaldulensis*	5.34
13443 *tereticornis*	6.80	12898 *urophylla*	5.29
12187 *camaldulensis*	6.75	14918 *camaldulensis*	5.27
14918 *camaldulensis*	6.73	14532 *urophylla*	5.21
13933 *camaldulensis*	6.59	15089 *urophylla*	5.19
13544 *tereticornis*	6.54	12895 *urophylla*	5.10
14847 *camaldulensis*	6.53	12362 *urophylla*	4.98
14540 *camaldulensis*	6.57	14917 *camaldulensis*	4.84
15050 *camaldulensis*	6.47	14420 *grandis*	4.81
14917 *camaldulensis*	6.45	13443 *tereticornis*	4.80
15062 *camaldulensis*	6.33	15062 *camaldulensis*	4.78
13663 *camaldulensis*	6.32	13544 *tereticornis*	4.78
13282 *exserta*	6.28	14540 *camaldulensis*	4.75
14106 *camaldulensis*	6.22	15050 *camaldulensis*	4.64
15049 *camaldulensis*	6.21	12187 *camaldulensis*	4.63
14532 *urophylla*	6.10	14847 *camaldulensis*	4.61
15089 *urophylla*	6.05	13282 *exserta*	4.57
12898 *urophylla*	5.93	13663 *camaldulensis*	4.47
12895 *urophylla*	5.92	14424 *tereticornis*	4.47
14420 *grandis*	5.92	13933 *camaldulensis*	4.45
14703 *citriodora*	5.91	local *citriodora*	4.44
13472 *citriodora*	5.78	15049 *camaldulensis*	4.40
13319 *tereticornis*	5.73	13431 *grandis*	4.38
14424 *tereticornis*	5.66	14703 *citriodora*	4.37
local *citriodora*	5.58	13994 *tereticornis*	4.30
13941 *camaldulensis*	5.58	13418 *tereticornis*	4.28
13418 *tereticornis*	5.54	13828 *urophylla*	4.28
local *exserta*	5.42	13541 *tereticornis*	4.27
12362 *urophylla*	5.42	14106 *camaldulensis*	4.26
13431 *grandis*	5.39	13319 *tereticornis*	4.24
14518 *camaldulensis*	5.36	12965 *tereticornis*	4.24
14852 *citriodora*	5.36	13941 *camaldulensis*	4.18
13541 *tereticornis*	5.29	local *exserta*	4.17
14851 *citriodora*	5.25	13446 *tereticornis*	4.14
14864 *exserta*	5.18	13350 *tereticornis*	4.14
13446 *tereticornis*	5.15	13307 *tereticornis*	4.12
12965 *tereticornis*	5.10	14849 *grandis*	4.02
13442 *tereticornis*	5.08	13304 *tereticornis*	4.02
13304 *tereticornis*	5.04	13472 *citriodora*	3.95
13307 *tereticornis*	5.03	13321 *propinqua*	3.95
14849 *grandis*	5.00	12418 *resinifera*	3.91
13828 *urophylla*	4.99	12411 *resinifera*	3.77
13350 *tereticornis*	4.95	13442 *tereticornis*	3.77
13994 *tereticornis*	4.92	14518 *camaldulensis*	3.70
13321 *propinqua*	4.83	14210 *grandis*	3.69
14850 *citriodora*	4.76	14864 *exserta*	3.67
14210 *grandis*	4.74	14852 *citriodora*	3.67
14838 *grandis*	4.65	14861 *grandis*	3.67
12018 *propinqua*	4.52	14839 *grandis*	3.66
14861 *grandis*	4.51	14115 *tereticornis*	3.65
12418 *resinifera*	4.48	12018 *propinqua*	3.49
12411 *resinifera*	4.44	13019 *grandis*	3.44
13020 *grandis*	4.39	14851 *citriodora*	3.44
13019 *grandis*	4.38	13020 *grandis*	3.40

(*Continued next page*)

Table 4. (*Concluded*)

Species/provenance	Height (m)	Species/provenance	Diameter (cm)
14115 *tereticornis*	4.37	14431 *grandis*	3.38
14507 *saligna*	4.27	14509 *grandis*	3.34
14431 *grandis*	4.20	14850 *citriodora*	3.30
14393 *grandis*	4.08	14860 *grandis*	3.25
14860 *grandis*	4.08	13965 *grandis*	3.24
15011 *saligna*	4.02	15011 *saligna*	3.23
13029 *saligna*	3.98	14597 *saligna*	3.21
14526 *saligna*	3.97	13029 *saligna*	3.13
14509 *grandis*	3.89	14393 *grandis*	3.12
13303 *tereticornis*	3.89	13365 *grandis*	3.11
14519 *grandis*	3.88	14429 *saligna*	2.95
13965 *grandis*	3.84	14519 *grandis*	2.94
14508 *saligna*	3.75	14526 *saligna*	2.93
13365 *grandis*	3.74	10140 *urophylla*	2.82
14524 *saligna*	3.69	15424 *saligna*	2.73
13015 *saligna*	3.61	14508 *saligna*	2.73
14435 *saligna*	3.55	14130 *torelliana*	2.72
10140 *urophylla*	3.54	14425 *saligna*	2.70
13340 *saligna*	3.50	13015 *saligna*	2.53
14527 *saligna*	3.38	13303 *tereticornis*	2.47
15054 *saligna*	3.22	14527 *saligna*	2.46
14130 *torelliana*	3.20	13340 *saligna*	2.38
14429 *saligna*	3.14	15054 *saligna*	2.28
13657 *paniculata*	3.12	13657 *paniculata*	2.16
13263 *saligna*	2.84	13263 *saligna*	1.71

Table 5. Summarised results[a] of analyses of variance for height and diameter for 28-month-old *E. urophylla, E. exserta, E. grandis, E. camaldulensis, E. tereticornis* and *E. citriodora.* * and ** indicate significance at the 5 and 1% levels respectively; ns indicates no significance at the 5% level.

Source of variation	Degree of freedom	Mean squares	F-ratio
E. urophylla—height			
Provenance	6	2.41	15.49**
Replicate	3	0.09	0.41ns
Error	18	0.22	
E. urophylla—diameter			
Provenance	6	3.39	13.86**
Replicate	3	0.17	0.73ns
Error	18	0.24	
E. exserta—height			
Provenance	2	1.3	9.78*
Replicate	3	0.3	2.28ns
Error	6	0.13	
E. exserta—diameter			
Provenance	2	0.98	6.22*
Replicate	3	0.38	2.42ns
Error	6	0.15	
E. grandis—height			
Provenance	14	1.56	4.71**
Replicate	3	0.21	0.63ns
Error	42	0.33	
E. grandis—diameter			
Provenance	14	1.41	3.16**
Replicate	3	0.15	0.35ns
Error	42	0.44	
E. camaldulensis—height			
Provenance	13	1.08	5.82**
Replicate	3	0.18	1.00ns
Error	39	0.18	
E. camaldulensis—diameter			
Provenance	13	0.71	4.76**
Replicate	3	0.09	0.64ns
Error	39	0.15	
E. tereticornis—height			
Provenance	14	2.21	6.17**
Replicate	3	0.09	0.26ns
Error	42	0.25	
E. tereticornis—diameter			
Provenance	14	1.27	5.01**
Replicate	3	1.31	5.15**
Error	42	0.25	
E. citriodora—diameter			
Provenance	5	0.92	3.69*
Replicate	3	0.60	2.39ns
Error	15	0.25	

[a] Results are given only for those species and parameters displaying significant differences.

Table 6. A list of the best provenances for each species giving height, diameter at breast height (D.B.H.) and survival percentage at 18 months.

Provenance number	Species	Height H (m)	Height SD	Height CV (%)*	D.B.H. D (cm)	D.B.H. SD	D.B.H. CV (%)*	Survival (%)
15052	*E. camaldulensis*	7.6	0.8	10.4	5.3	0.8	15.1	99
13443	*E. tereticornis*	6.8	0.9	14.0	4.8	1.0	21.2	99
14532	*E. urophylla*	6.1	1.2	19.2	5.2	1.6	31.1	85
14703	*E. citriodora*	5.9	0.9	15.5	4.4	1.0	23.1	85
14507	*E. saligna*	4.3	0.9	21.2	3.2	1.0	31.1	79
14402	*E. grandis*	5.9	1.2	20.5	4.8	1.4	29.1	86
13282	*E. exserta*	6.3	1.2	18.5	4.6	1.2	26.0	82
13321	*E. propinqua*	4.8	0.8	16.8	3.9	1.1	27.9	89
13657	*E. paniculata*	3.1	1.0	32.3	2.2	1.1	49.5	81
14130	*E. torelliana*	3.2	0.6	17.4	2.7	0.8	28.1	94

*Coefficient of variation based on a plot basis.

Table 7. List of the 20 best eucalypt provenances (based on height) in the Qionghai trial.

No.	Species	Seedlot no.	Height (m)	Diameter (cm)	Location		
1	*E. camaldulensis*	15052	7.58	5.34	Isdell River WA	16°50′	125°32′
2	*E. tereticornis*	13443	6.80	4.80	Kennedy River Q	15°26′	144°11′
3	*E. camaldulensis*	12187	6.75	4.63	W. Irvinebank Q	17°24′	145°09′
4	*E. camaldulensis*	14918	6.73	5.27	Laura Q	15°34′	144°27′
5	*E. camaldulensis*	13933	6.59	4.45	N Fitzroy Crossing WA	18°06′	125°42′
6	*E. tereticornis*	13544	6.54	4.78	N Gladstone Q	23°44′	151°01′
7	*E. camaldulensis*	14847	6.53	4.61	Petford Q	17°10′	145°15′
8	*E. camaldulensis*	14540	6.51	4.75	Pentecost Riv. WA	15°48′	127°53′
9	*E. camaldulensis*	15050	6.47	4.64	Gibb River WA	16°30′	126°30′
10	*E. camaldulensis*	14917	6.45	4.84	NW Mt Carbine Q	16°22′	144°43′
11	*E. camaldulensis*	15062	6.33	4.78	NE Katherine NT	14°23′	132°21′
12	*E. camaldulensis*	13663	6.32	4.47	Wrotham Park Q	16°18′	144°10′
13	*E. exserta*	13282	6.28	4.57	N Marlborough Q	22°40′	149°54′
14	*E. camaldulensis*	14106	6.22	4.26	Gilbert River Q	18°00′	143°00′
15	*E. camaldulensis*	15049	6.21	4.40	Bullock Creek Q	20°46′	143°55′
16	*E. urophylla*	15432	6.10	5.21	Mt Lewotobi Indonesia	08°31′	122°45′
17	*E. urophylla*	15089	6.05	5.19	Mt Egon Fores Indonesia	08°38′	122°27′
18	*E. urophylla*	12898	5.93	5.29	Mt Boleng Indonesia	08°21′	123°15′
19	*E. urophylla*	12895	5.92	5.10	Mt Mandiri Indonesia	08°15′	122°58′
20	*E. grandis*	14420	5.92	4.81	S Ravenshoe Q	17°42′	145°33′

Eucalyptus tereticornis (13443) from Kennedy River, North Queensland, was second best. The average height of 18-month-old trees was 6.80 m and average diameter at breast height was 4.8 cm. The second best provenance for diameter growth was *E. urophylla* (12898) from Mt. Boleng in Indonesia. Its average height was 5.93 m and diameter at breast height was 5.29 cm. It grew well with straight stems.

Survival of *E. camaldulensis* was most promising, with all provenances having greater than 95% survival at age 18 months. *Eucalyptus tereticornis* came second with survival being greater than 90%. The survival of *E. propinqua* and *E. torelliana* was also above 90%, but their growth increments were not as great as for the other two species.

The heights of 18-month-old trees of *E. saligna, E. propinqua, E. resinifera, E. paniculata* and *E. torelliana* were all less than 5 m, and differed significantly from those of *E. camaldulensis* and *E. tereticornis.*

Conclusion

Although the trial was still young (18 months) when last assessed, it is tentatively concluded that *E. camaldulensis* and *E. tereticornis* from northern Australia are well-adapted, fast-growing species suitable for the poor site conditions in this tropical part (Quionghai County) of the PRC. Both have performed better than *E. exserta* which is commonly planted on Hainan Island. All three species belong to the red gum group of eucalypts (section *Exsertaria*). *Eucalyptus urophylla* also grew fast with very straight stems and is a species worth considering for plantations in southern China, perhaps in areas having a much shorter dry season. This is because *E. urophylla* is found naturally in Indonesia in areas having a shorter dry season than that experienced on Hainan Island. Nevertheless, of the species tested in section *Transversaria, E. urophylla* appears to tolerate the tropical dry periods much better than *E. grandis* and *E. saligna.*

The trial indicates that *Eucalyptus* species and provenances from the tropical areas of North Queensland and northern Western Australia are best suited for Qionghai County. Differences among provenances were significant for every species, but apart from a few badly performing provenances within each species, the differences are generally not important. The trials have demonstrated that in the red gum group both *E. tereticornis* and *E. camaldulensis* from tropical Australia can perform better at 18 months than local *E. exserta.*

The poor performance of *E. grandis* and *E. saligna* was somewhat surprising, but the poor soils of the area and the extended dry season are not conducive to their success in the Qionghai trial. Overall, some caution is necessary in interpreting all results as the trial has not yet experienced a major typhoon. This undoubtedly will affect future species recommendations.

1987 Acacia species trial (photo taken April 1988), Bai Shi Ling Forest Farm near, Qionghai, Hainan Island, Hainan Province, People's Republic of China.

Chapter 8

Tropical Australian Acacia Trials on Hainan Island, People's Republic of China

Yang Minquan, Bai Jiayu and Zeng Yutian

Abstract

Differences between species and provenances of acacias in trials on Hainan Island, China, are described. *Acacia crassicarpa* (two provenances, S13682, S13683 from Papua New Guinea) grew fastest while *A. mangium*, *A. auriculiformis*, *A. cincinnata*, and *A. aulacocarpa* also appear suitable for reforestation on poor soils on the east coast of Hainan. One serious limiting factor for some species of tropical acacias is their apparent susceptibility to typhoon damage because of their large dense crowns. Other species such as *A. cincinnata* and *A. aulacocarpa* grow rapidly and produce straight trees. Superior individual trees of these species offer great potential for vegetative propagation.

Introduction

Hainan is a large tropical island with an area of 34 290 km^2. It is the most southerly province in the People's Republic of China. The area is currently undergoing a rapid economic change and a number of industries are being attracted to the island. The main species currently being planted is *Eucalyptus exserta* (about 67 000 ha established), and there is a need to find well-adapted species to produce high-yielding forest plantations for a variety of uses (e.g. industrial use, poles, fuelwood and protection forests). On Hainan Island there are many areas where soil fertility is low and these areas may be more suitable for nitrogen-fixing leguminous species.

Since 1985, the Research Institute of Tropical Forestry, Chinese Academy of Forestry, has been establishing a series of acacia species and provenance trials with ACIAR support. Currently, there is one Chinese indigenous species planted (*A. confusa*) that has a good reputation for producing fuelwood and resisting typhoon damage. This species, together with the more recently introduced *A. auriculiformis*, is usually planted along roadsides and around villages and homes. A major limitation of both species is their slow growth and poor stem form, which limits their utilisation for poles or industrial use. The current trials are designed to explore the potential of lesser-known tropical Australian species of *Acacia*, thus widening the range of species and seedlots that could be used to fulfil timber needs in this region of the People's Republic of China. Several of those species also extend to Papua New Guinea.

Two series of *Acacia* species and provenance trials were planted in 1986 and 1987. Growth performance of the trials at 24 and 12 months after planting of the trials established in 1986 and 1987, respectively, is reported in this paper. Results obtained at nursery stages are also discussed.

Materials and Methods

Trial Site

The *Acacia* species/provenances trials were established at Bai Shi Ling forestry farm. The farm is located at an elevation of 60–80 m (19°00′N, 110°15′E) 30 km south of Qionghai on the east coast of Hainan (see Fig. 1 in Chapter 1).

The experimental site is on the coastal plains with

a gentle topography of low hills. It has a humid tropical climate with distinct dry and wet seasons, and is influenced by monsoon winds. Average temperature in the coldest month is above 15°C, extreme low temperature is 5°C. Mean annual rainfall is 2072 mm. The soils are laterites derived from granite and contain grits and a few small iron concretions. They are deep but are relatively infertile.

The trial site was formerly a *Eucalyptus exserta* plantation together with herbaceous plants and grasses, including *Lygodium microstachyum*, *Dicraopteris linearis*, *Strophanthus divaricatus*, *Eupatorium odoratum*, *Melastoma candidum*, *Heteropogon contortus*, *Imperata cylindrica* var. *major* and *Sida acuta*. The site suffered considerable disturbance prior to trial establishment when all trees were cut and the roots and stumps dug up and removed.

Seed Material

Seeds for the two trials (1986 and 1987 trial) came from the Australian Tree Seed Centre of the CSIRO Division of Forestry and Forest Products, with the exception of two local seedlots. Details of the seed sources for each trial are given in Tables 1 and 2.

Nursery Techniques

Most seeds were treated with boiling water before sowing in order to promote rapid and even germination. Swollen seeds were sown in plastic tubes and covered with a thin layer (0.5cm) of fine soil. Potting mix was 40% burnt soil, 40% subsoil and 20% acacia root nodule bacterium soil obtained from a nearby *Acacia confusa* and *A. auriculiformis* plantation. Superphosphate (2.5 kg) was added to each 100 kg of potting mix. Seedling beds were covered with a layer of straw to provide shade. The cover was removed after 5–10 days when the seeds had germinated.

The seedlings were watered twice daily during the first 3 months in the nursery. When the seedlings produced the first pair of phyllodes, manure was applied as a fertiliser. About every 15 days thereafter, fertiliser in the form of 5% urine or 0.1% urea was applied. Topsin-M® or Bavistin® were used to control powdery mildew (*Oidium* sp.).

Site Preparation and Planting

The planting site was ploughed once using a tractor fitted with a heavy disc harrow. Planting holes (40 × 40 × 30 cm) were then dug and 100 g of superphosphate was put in each hole. Trials were planted in April 1986 and in April 1987 in a randomised complete block with four replicates. Twenty-four (4 × 6) seedlings were planted per plot. *Acacia confusa* and local *A. auriculiformis* were used as controls. The experimental area was surrounded by two rows of *Acacia* species.

Measurement

At nursery stage, germination rate was counted for each seedlot (after 20 days) based on 100 seeds. Seedling height and stem diameter at the root collar were measured monthly up to 6 months.

Field data presented in this report are as follows: 1986 trial: height, diameter at ground level, crown width, number of stems per tree and wind resistance indices were recorded at 20 months after planting; 1987 trial: the above parameters except wind resistance were recorded at 8 months after planting. Survival was assessed at 1 month and at 12 months after planting for both trials.

A wind resistance index was calculated in October 1987 following a typhoon that landed in Qionghai County. The typhoon had wind speeds of grade 7 and gusts of grade 8. A wind resistance index for each species was calculated using the formula:

$$\text{index} = \frac{1x + 2x + 3x + 4x + 5x}{\overline{x}}$$

where number 1,2, ... are subjective scores of wind damage from 1 = no damage to 5 = tree blown on ground; x represents the number of observed trees in each category and $\overline{x}$ is the total number of observed trees.

Data Analysis

Plot means were calculated for height, diameter, crown width and number of stems per tree and analyses of variance were conducted. Survival was calculated for each plot and transformed (using $\sin^{-1}$) prior to analysis. Duncan's multiple range test was used to compare provenance means for height and diameter.

Results

Nursery Stage

There were clear differences between species in germination rate, seedling height and diameter at root collar, although statistical analyses were not conducted (Tables 3 and 4). In the 1986 trial, species having satisfactory (over 75%) germination were *A. oraria*, *A. melanoxylon*, *A. crassicarpa* and *A. cincinnata*. In the 1987 trial high germination rates were recorded for *A. leptocarpa*, *A. melanoxylon* and *A. crassicarpa*.

After 6 months in the 1986 trial the best height and diameter were obtained for *A. auriculiformis*. After 6 months in the 1987 trial seedlings of *A. mangium* and *A. crassicarpa* tended to have better height and diameter than other species. Seedlings of *A. aulacocarpa* from Julattan area in

Table 1. Details of the seed sources of *Acacia* species and provenances used in the 1986 trial.

CSIRO seedlot no.	Species	Collection locality		Lat (S)	Long (E)	Alt (m)	No. of parent trees
14969	*Acacia aulacocarpa*	31 km S Cooktown	QLD	15°41′	145°12′	125	10
13689	*Acacia aulacocarpa*	Oriomo River	PNG	8°48′	143°09′	20	5
13869	*Acacia auriculiformis*	Springvale Holding	QLD	15°48′	144°55′	150	3
13854	*Acacia auriculiformis*	Oenpelli	NT	12°20′	133°04′	50	200
13686	*Acacia auriculiformis*	Iokwa	PNG	8°41′	141°29′	35	10
13684	*Acacia auriculiformis*	Balamuk	PNG	8°54′	141°18′	18	17
13878	*Acacia cincinnata*	Julatten area	QLD	16°35′	145°25′	410	12
13864	*Acacia cincinnata*	Shoteel L.A.	QLD	16°57′	145°38′	440	5
13361	*Acacia cincinnata*	13 km SSE Mossman	QLD	16°37′	145°20′	457	9
13863	*Acacia crassicarpa*	Shoteel L.A.	QLD	16°57′	145°38′	440	5
13683	*Acacia crassicarpa*	Woroi Wimpim	PNG	8°49′	143°00′	20	15
13682	*Acacia crassicarpa*	Oriomo River	PNG	8°50′	143°10′	20	11
15063	*Acacia mangium*	7 km SSE Mossman	QLD	16°31′	145°24′	65	100
13622	*Acacia mangium*	Sidei, Indonesia		0°46′	133°34′	30	15
13504	*Acacia mangium*	Gadgarra S.F. Res.	QLD	17°18′	145°41′	720	1
14766	*Acacia melanoxylon*	NW of Samford	QLD	27°22′	152°47′	300	4
14176	*Acacia melanoxylon*	Atherton	QLD	17°17′	145°26′	1022	10
14961	*Acaica oraria*	39 km NW Cairns	QLD	16°41′	145°35′	5	8
86001	*Acacia auriculiformis*	Guangzhou, Guangdong		23°08′	113°19′	150	unknown
86002	*Acacia confusa*	Lufeng, Guangdong		23°00′	115°40′	100	unknown

Table 2. Details of the seed sources of *Acacia* species and provenances used in the 1987 trial.

CSIRO seedlot no.	Species	Collection locality		Lat (S)	Long (E)	Alt (m)	No. of parent trees
13877	*Acacia aulacocarpa*	Julatten	QLD	16°35′	145°25′	410	10
13865	*Acacia aulacocarpa*	Buckley L.A.	QLD	17°09′	145°37′	720	5
13878	*Acacia cincinnata*	Julatten	QLD	16°35′	145°25′	410	12
15283	*Acacia crassicarpa*	40 km from Cooktown	QLD	15°23′	145°02′	–	10
13680	*Acacia crassicarpa*	Wemenever	PNG	8°51′	141°26′	30	21
14966	*Acacia leptocarpa*	1–26 km S Musgrave	QLD	14°53′	143°31′	98	10
14139	*Acacia leptocarpa*	Mt Molloy	QLD	16°40′	145°18′	400	10
15316	*Acacia mangium*	NW of Ingham	QLD	18°35′	146°05′	50	20
13229	*Acacia mangium*	Claudie River	QLD	12°44′	143°13′	60	6
14766	*Acacia melanoxylon*	NW of Samford	QLD	27°22′	152°47′	300	4
87001	*Acacia auriculiformis*	Jianfeng, Hainan		18°32′	109°48′	78	1
87002	*Acacia auriculiformis*	Guangzhou, Guangdong		23°08′	113°19′	150	unknown
87003	*Acacia confusa*	Lufeng, Guangdong		23°00′	115°40′	100	unknown

Queensland (S13877) were also developing satisfactorily.

Field Performance

In the 1986 trial there were marked statistical differences between seedlots in all parameters measured except survival at 1 month after planting (Table 5). There were statistical differences in the 1987 trial between seedlots in height, diameter at ground level (d.g.l.), survival (at 1 and 12 months), crown width and number of stems per tree (Table 6).

Table 7 shows performances of the *Acacia* species and provenances for the parameters measured in the 1986 trial. Fastest height and diameter growth was recorded for *A. crassicarpa* and *A. auriculiformis*. *Acacia melanoxylon*, *A. oraria* and *A. confusa* were the slowest growing in height and diameter. Multiple range tests for height and diameter are given in Table 8.

The fastest-growing species for height and diameter in the 1987 trials were *A. mangium*, *A. crassicarpa* and *A. leptocarpa* (Table 9). Survival of these species was also high. Slow-growing species were *A. aulacocarpa*, *A. cincinnata*, *A. melanoxylon* and *A.confusa*. Multiple range tests for height and diameter are given in Table 8.

Most species were multistemmed, particuarly *A. aulacocarpa* and *A. confusa* which had on average more than three stems per tree. *Acacia leptocarpa* had the least number of stems per tree. It was noted that trees of *A. auriculiformis* from Springvale, Queensland, were mostly single-stemmed (see Table 7 for 1986 trial).

Most species survived well after 1 year but *A. cincinnata* and *A. melanoxylon* suffered greater losses. Virtually all had better survival than *A. confusa*.

Faster-growing species also had wide crowns. The values recorded for *A. crassicarpa* and *A. auriculiformis* were greater than those recorded for other species (Table 7).

Results for the wind-resistance index for each species are given in Table 7. There were indications that the least wind-resistant species and provenance was *A. mangium* S13622 from Sidei, Indonesia, and the next worst was the fastest-growing *A. crassicarpa* S13682 and S13683.

Provenance Differences

A. crassicarpa The five provenances were the fastest growing and had a high survival rate among the species. Their wind resistance was poor. The best-growing provenance came from Oriomo River, Papua New Guinea. At 20 months it grew to 6 m tall, with a trunk 5.8 cm in diameter at breast height (see Table 7).

A. mangium Of the five provenances in the trials the best provenance was S15063 from Mossman, Queensland. It reached 4.3 m in height with 4.1 cm at breast height in 20 months. The average number of stems per tree was 1.6. The slowest provenance was S13622 from Sidei, Indonesia, which reached 3.5 m in height with 2.5 cm d.g.l. Average number of stems per tree was two.

A. aulacocarpa Four provenances were included in the trials. The best provenance was S13689 from Oriomo River Province, Papua New Guinea. This provenance had fast growth with a straight stem and strong apical dominance. The other three provenances were from North Queensland. They were slow-growing and possessed a shrub form.

A. auriculiformis There were seven provenances including two control (local) provenances from Guangdong Province, China. The best provenance was S13686 from Iokwa, Papua New Guinea. It attained 5.3 m in height with 5.2 cm in d.g.l. which were 120 and 142% respectively better than the controls. The provenance with the best stem form was S13869 from Springvale Holding, Queensland. Most of the trees from this provenance were single-stemmed and straight.

Table 3. Germination percent (after 20 days), height and diameter at ground level (dgl) after 3 and 6 months of seedlings of the *Acacia* species and provenances included in the 1986 trial.

			3 months		6 months	
CSIRO seedlot no.	Species	Germination (%)	Height (cm)	D.G.L. (cm)	Height (cm)	D.G.L. (cm)
14961	*A. oraria*	80	10.4	0.21	79	0.67
14766	*A. melanoxylon*	68	25.0	0.33	124	0.76
14176	*A. melanoxylon*	89	17.3	0.27	105	0.63
13863	*A. crassicarpa*	74	10.0	0.28	73	0.74
13683	*A. crassicarpa*	87	11.0	0.29	97	0.88
13682	*A. crassicarpa*	82	10.7	0.28	87	0.77
13361	*A. cincinnata*	81	11.0	0.29	126	0.76
13878	*A. cincinnata*	76	13.3	0.28	135	0.83
13864	*A. cincinnata*	70	12.3	0.28	133	0.79
14969	*A. aulacocarpa*	58	8.3	0.22	108	0.65
13689	*A. aulacocarpa*	48	13.4	0.27	146	1.09
13622	*A. mangium*	42	14.6	0.25	120	0.84
15063	*A. mangium*	68	11.9	0.29	137	1.01
13504	*A. mangium*	19	17.7	0.33	140	1.11
13686	*A. auriculiformis*	32	21.3	0.32	188	1.24
13869	*A. auriculiformis*	54	14.9	0.32	151	0.99
13684	*A. auriculiformis*	32	23.4	0.33	169	1.20
13854	*A. auriculiformis*	66	13.7	0.28	145	1.08

Table 4. Germination percent (after 20 days), height and diameter at ground level (dgl) after 3 and 6 months of seedlings of the *Acacia* species and provenances included in the 1987 trial.

CSIRO seedlot no.	Species	Germination (%)	3 months Height (cm)	3 months D.G.L. (cm)	6 months Height (cm)	6 months D.G.L. (cm)
15316	*A. mangium*	62	18.9	0.36	93	0.65
13229	*A. mangium*	39	16.9	0.34	110	0.83
14139	*A. leptocarpa*	82	16.5	0.28	71	0.61
14966	*A. leptocarpa*	47	12.0	0.21	37	0.34
14766	*A. melanoxylon*	74	18.6	0.29	70	0.57
13878	*A. cincinnata*	23	11.0	0.28	61	0.64
13865	*A. aulacocarpa*	24	11.7	0.28	51	0.51
13877	*A. aulacocarpa*	46	18.7	0.25	105	0.70
13680	*A. crassicarpa*	83	23.7	0.38	102	0.68
15283	*A. crassicarpa*	39	17.4	0.24	100	0.58

Table 5. Analyses of variance, based on plot means, for height, diameter at ground level, survival, crown width and number of stems per tree for the 1986 *Acacia* trial.

Source of variation	DF	MS	F-ratio
Height (m)			
Provenance	19	4.61	8.90**
Replication	2	10.766	20.78**
Error	38	0.518	
Diameter ground level (cm)			
Provenance	19	10.388	13.28**
Replication	2	9.523	12.18**
Error	38	0.782	
Survival rate of planted of sin $^{-1}$ P value in 1986	(1 mth after planting)		
Provenance	19	46.621	1.49^{ns}
Replication	3	46.191	1.48^{ns}
Error	57	31.240	
Preserved survival rate of planted of sin $^{-1}$ P value in 1986	(1 yr after planting)		
Provenance	19	415.736	3.91**
Replication	2	1685.807	15.85**
Error	38	106.380	
Crown width (m)			
Provenance	19	0.749	3.24**
Replication	2	0.717	3.10^{ns}
Error	38	0.231	
Number of stems per tree			
Provenance	19	0.998	10.08**
Replication	2	1.653	16.70**
Error	38	0.099	

Note: * and ** show significant differences at the 5 and 1% level respectively; ^{ns} = not significant at the 5% level.

Table 6. Analyses of variance, based on plot means, for height, diameter ground level, survival, crown width and number of stems per tree for the 1987 *Acacia* trial.

Source of variation	DF	MS	F-ratio
Height			
Provenance	12	1.153	28.83**
Replication	3	0.005	0.13^{ns}
Error	36	0.040	
Diameter ground level			
Provenance	12	3.814	27.64**
Replication	3	0.113	0.82^{ns}
Error	36	0.138	
Survival rate of planted of $\sin^{-1}$ P value in 1987			
Provenance	12	210.163	2.10*
Replication	3	241.278	2.41^{ns}
Error	36	99.978	
Preserved survival rate of planted of $\sin^{-1}$ P value in 1987			
Provenance	12	303.713	4.93**
Replication	3	46.754	0.76^{ns}
Error	36	61.642	
Crown width			
Provenance	12	0.461	14.87**
Replication	3	0.662	2.00^{ns}
Error	36	0.031	
Number of stems per tree			
Provenance	12	2.302	22.79**
Replication	3	0.315	3.12*
Error	36	0.101	

*, ** indicate significant differences at the 5 and 1% level respectively;
^{ns} = not significant at the 5% level.

Table 7. Results of the 1986 trials after various time periods for height, diameter, survival, multiple stems per tree, crown width and wind resistance.

Seedlot no.	Species	Height[c] (m)	Diameter ground level[c] (cm)	Diameter breast height[c] (cm)	Survival[a] %	Survival[b] at 1 year %	Stems mean number[c]	Crown width (m)[c]	Index wind resistance[c]
13863	*Acacia crassicarpa*	4.7	7.4	4.6	99	88	1.7	2.4	1.00
13683	*Acacia crassicarpa*	5.7	8.0	5.6	99	96	1.4	2.7	1.67
13682	*Acacia crassicarpa*	6.0	7.8	5.8	99	93	1.7	2.8	1.79
13504	*Acacia mangium*	3.8	5.9	3.8	99	89	1.4	2.1	1.08
13622	*Acacia mangium*	3.5	4.3	2.5	99	88	2.0	2.1	1.89
15063	*Acacia mangium*	4.3	5.8	4.1	100	86	1.6	2.2	1.05
13869	*Acacia auriculiformis*	4.1	6.7	4.1	99	98	1.03	2.3	1.00
13854	*Acacia auriculiformis*	4.5	7.1	3.8	97	96	1.7	2.5	1.00
13686	*Acacia auriculiformis*	5.3	7.8	5.2	99	97	1.5	2.5	1.03
13684	*Acacia auriculiformis*	4.3	6.3	4.2	99	99	1.5	2.5	1.01
14969	*Acacia aulacocarpa*	2.7	3.0	1.6	100	93	2.2	2.2	1.07
13689	*Acacia aulacocarpa*	4.9	6.9	4.8	99	85	1.5	2.4	1.48
13864	*Acacia cincinnata*	3.7	4.9	3.0	97	62	1.8	2.0	1.00
13878	*Acacia cincinnata*	3.8	4.5	2.8	99	74	1.5	2.0	1.00
13361	*Acacia cincinnata*	3.7	4.9	2.9	100	67	1.7	2.7	1.00
14176	*Acacia melanoxylon*	2.2	2.5	0.6	97	56	1.7	1.3	1.07
14766	*Acacia melanoxylon*	2.0	3.3	1.4	97	56	1.3	1.1	1.00
14961	*Acacia oraria*	1.9	2.7	0.9	99	89	3.2	1.6	1.00
86001	*Acacia auriculiformis*	4.4	5.5	3.6	97	96	2.6	2.6	1.00
86002	*Acacia confusa*	1.6	2.0	0.7	97	74	3.1	1.5	1.00

[a] surveyed 1 month after planting; [b] surveyed 1 year after planting; [c] surveyed December 1987 (20 months).

Table 8. Duncan's multiple range tests for height and diameter (ground level) in the 1986 trial (age 2 years) and the 1987 trial (age 1 year).

1986 trial (2 years old)		1987 trial (1 year old)	
Height(m)	Diameter (cm)	Height (m)	Diameter (cm)
13682 Cras 6	13683 Cras 8.0	14966 Lep 2.6	15316 Man 4.8
13683 Cras 5.7	13682 Cras 7.8	13680 Cras 2.6	13229 Man 4.7
13686 Aur 5.3	13686 Aur 7.8	13229 Man 2.5	13680 Cras 4.5
13689 Aul 4.9	13863 Cras 7.4	15316 Man 2.4	14966 Lep 4.4
13863 Cras 4.7	13854 Aur 7.1	14139 Lep 2.2	87001 Aur 4.1
13854 Aur 4.5	13689 Aul 6.9	15283 Cras 2.0	15283 Cras 4.0
86001 Aur 4.4	13869 Aur 6.7	87001 Aur 2.0	14139 Lep 3.9
13684 Aur 4.3	13685 Aur 6.3	87002 Aur 1.7	87002 Aur 3.8
15063 Man 4.3	13504 Man 5.9	13878 Cin 1.6	13878 Cin 3.2
13869 Aur 4.1	15063 Man 5.8	13877 Aul 1.5	13877 Aul 2.9
13878 Cin 3.8	86001 Aur 5.5	87003 Con 1.2	13865 Aul 2.7
13504 Man 3.8	13864 Cin 4.9	13865 Aul 1.2	87003 Con 2.2
13864 Cin 3.7	13361 Cin 4.9	14766 Mel 1.1	14766 Mel 1.8
13361 Cin 3.7	13878 Cin 4.5		
13622 Man 3.5	13622 Man 4.3		
14969 Aul 2.7	14766 Mel 3.3		
14176 Mel 2.2	14969 Aul 3.0		
14766 Mel 2.0	14961 Ora 2.7		
14961 Ora 1.9	14176 Mel 2.5		
86002 Con 1.6	86002 Con 2.0		

Table 9. Results of 1987 *Acacia* species/provenance trial. Trial was planted in April 1987 and different characters were measured at different times during the first year of growth.

Seedlot no.	Species	Height[c] (m)	Diameter[c] ground level (cm)	Survival percentage[a] %	Preserved survival[b,d] rate %	Stems mean no.[c]	Crown width[c] (m)
15316	*Acacia mangium*	2.4	4.8	94	100	2.6	1.7
13229	*Acacia mangium*	2.5	4.7	88	94	2.5	1.7
13865	*Acacia aulacocarpa*	1.2	2.7	75	91	3.1	1.0
13877	*Acacia aulacocarpa*	1.5	2.9	85	91	3.5	1.2
15283	*Acacia crassicarpa*	2.0	4.0	92	99	2.0	1.5
13680	*Acacia crassicarpa*	2.5	4.5	96	95	2.3	1.9
13878	*Acacia cincinnata*	1.6	3.2	76	83	2.2	1.3
14966	*Acacia leptocarpa*	2.6	4.4	88	100	1.3	1.6
14139	*Acacia leptocarpa*	2.2	3.9	93	99	1.4	1.4
14766	*Acacia melanoxylon*	1.1	1.8	95	89	1.3	0.8
87001	*Acacia auriculiformis*	2.0	4.1	90	91	2.2	1.6
87002	*Acacia auriculiformis*	1.7	3.8	96	95	2.7	1.4
87003	*Acacia confusa*	1.2	2.2	89	79	3.6	1.0

[a] Surveyed 1 month after planting.
[b] Surveyed 1 year after planting.
[c] Surveyed December 1987 (at 8 months old).
[d] Higher survival rate at 8 months over at 1 month indicates some replacements were made.

A. cincinnata This species produced some individuals with very straight stems in all trials. Growth of the three provenances was satisfactory.

A. leptocarpa There were two provenances in the trial, and both were fast-growing and single-stemmed but not very straight.

A. melanoxylon The two provenances were slow-growing and had poor stem form. They appear not to be suitable to the trial site.

A. oraria A special feature of this species is the high survival of trees and the uniform growth. It is admirably suited for ornamental planting on Hainan Island.

Discussion and Conclusion

The experiments show that *A. crassicarpa*, *A. mangium*, *A. auriculiformis* and *A. cincinnata* have good potential for larger-scale planting on Hainan. These species are fast-growing trees, with moderately straight stems and tolerate tropical, low-fertility acidic soils. One surprising aspect was the slow growth of *A. crassicarpa* in the nursery in the 1986 trials but its fast growth when out-planted. Of particular interest was the number of individuals of *A. aulacocarpa* and *A. cincinnata* in the trials that had straight stems and small-diameter branches. The potential for vegetative propagation of the phenotypes could be explored further.

The species/provenance trials have been very important in demonstrating the potential for new species and provenances to this area of China. In particular large provenance differences were observed for *A. aulacocarpa* and *A. mangium*. The success of *A. crassicarpa* was most encouraging. Results suggest that it would be useful to extend these trials to other parts of tropical China.

Hainan Island is often threatened by typhoons. The wind-resistance index is a very important parameter for evaluating species/provenance trials in this area. The fast-growing, large-leaved acacias may be particularly prone to wind throw (e.g. *A. crassicarpa*).

These early results are considered to be important as they provide basic information which may be used to determine suitable provenances for future breeding work.

Chapter 9

Acacia mearnsii Provenance Trials in the People's Republic of China

Gao Chuanbi

Abstract

Provenance trials of *Acacia mearnsii* were established in the central and southern districts of subtropical zones in the People's Republic of China. Imported seed from Australia, South Africa and Brazil were compared with a number of local provenances. Some of the trials suffered heavy loss due to drought or disease. Available early results have shown marked differences amongst provenances in many growth characteristics. Several newly introduced seedlots have performed better than the local seedlots in height and diameter. Surprisingly all local provenances had first flowering at 18 months after planting but none of the newly introduced provenances did so at the same age.

Introduction

Acacia mearnsii is a fast-growing, multipurpose tree species. Its bark is a superior source of condensed tannin extracts. Its wood is used for mine props, furniture manufacture and fuelwood. The species is especially suited for eroded hillsides because the trees grow fast and develop symbiotic root nodules capable of fixing atmospheric nitrogen, thus improving soil fertility.

The species was introduced into China in the early 1930s, and has been planted in the central and southern districts of the subtropical zone (10°N, 110°E). The planting areas range from sea level to about 1500 m. The climatic and soil conditions vary greatly among the planting areas. Therefore, selection of suitable provenances is essential to achieve the best results.

The seed of *A. mearnsii* used in the early plantings in China was commercial seed, of unknown Australian origin, imported from several countries including Indonesia, Japan, Kenya, Algeria, Netherlands, France and Australia. The seed being used in the current plantings has been collected from these earlier plantations. Special problems in the development of *A. mearnsii* plantations in China are the availability of large quantities of improved seed and frost-resistant provenances for planting in the cooler areas where land is available.

In order to select suitable seed sources for planting in China, provenance trials of this species have been established as a collaborative ACIAR-supported project between CSIRO Division of Forestry and Forest Products and the Research Institute of Subtropical Forestry, Chinese Academy of Forestry. The trials were established in 1986-87 in the main black wattle growing areas in Fujian, Jiangxi, Zhejiang and Guangxi provinces. The main aim of this chapter is to report the results of a nursery experiment at Changtai County, Fujian Province, and a provenance trial established at Ganzhou, Jiangxi Province.

Materials and Methods

Trial Locations

Six trials were established in the main black wattle-growing regions as follows:

(1) Chenxiang Tree Farm, Changtai County, Fujian Province (24°49′N, 117°52′E, 19 m);
(2) Experimental Field of Fujian Forestry College, Nanping City, Fujian Province (26°39′N, 118°10′E, 127 m);

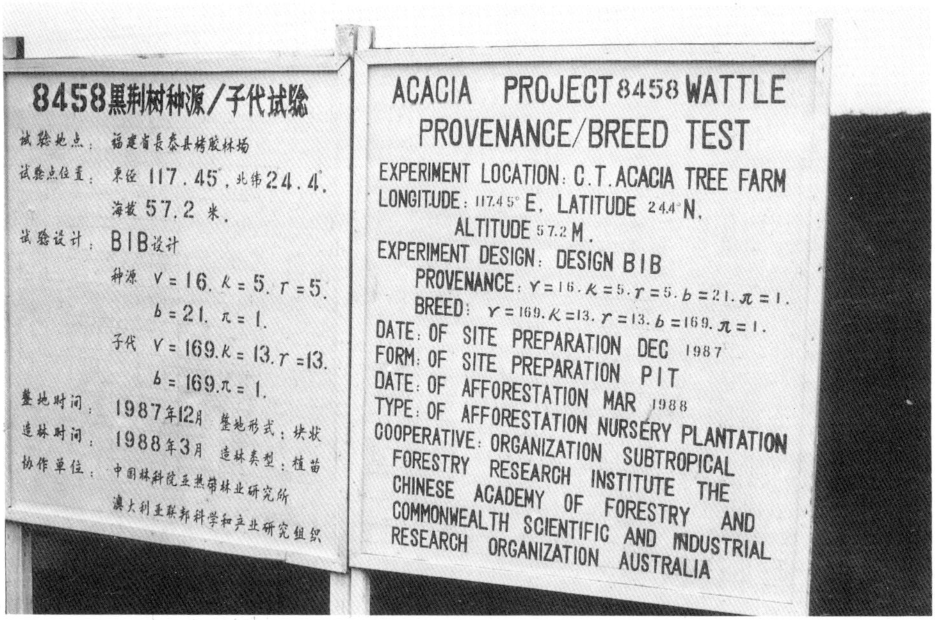

Acacia mearnsii provenance/progeny test and seedling seed orchard established in March 1988 near Zhangzhou, Fujian Province, People's Republic of China. Some of the staff (top) associated with the establishment of the trial, and (bottom) display sign at the planting site.

(3) Hubian Horticultural Farm, Ganzhou City, Jiangxi Province (25°21′N,114°50′E,123 m);
(4) Kongtian Tree Farm, Anyuan County, Jiangxi Province (25°59′N,115°20′E,266 m);
(5) Forestry Institute of Hechi District, near Nandan, Guangxi Autonomous Region (24°49′N,107°41′E,697 m); and
(6) Subtropical Crop Institute, Wenzhou City, Zhejiang Province (28°01′N,120°40′E,50 m).

In the above locations, annual mean temperature is between 17 and 21°C. Absolute minimum temperature ranges from -2.5 to -7.9°C. Mean annual rainfall ranges from 1100 to 1700 mm. Booth (see Chapter 4) has provided a climatic match between Ganzhou (the main trial site in this paper) and similar areas in Australia (see Fig 2 in Chapter 4). Location of trial sites is indicated in Fig. 1 of Chapter 1.

Seed Material

Twenty-four provenances (two from South Africa) were imported from Australia but only 18 were used in the trials. In addition, one provenance from Brazil and six from the main black wattle-growing regions in China were included for comparison. Details of the seed sources used in the trials are given in Table 1.

Site Preparation

The planting sites were ploughed and planting holes were dug (60 × 40 × 40 cm). Before planting, 12.5 kg of organic soil, 100 g phosphorus and 50 g of nitrogen fertiliser were placed into each planting hole.

Trial Design and Layout

Nursery Experiment

A nursery experiment was conducted at Chenxiang using randomised complete block design with three replicates consisting of 100 seedlings each.

Field Plantings

Seedlings were raised in black polythene tubes (20 × 12 cm) with seeds being soaked in 95°C water prior to planting. The trial design was a 5 × 5 balanced incomplete block with six replicates and 25 (5 × 5) trees per plot. Spacing was 2 × 2 m. Two weeding operations were conducted during the first year after planting, followed by a dressing of nitrogen just after the first weeding.

Measurement and Analysis

At nursery stage (6 months after germination) seedling height, diameter at root collar and biomass were assessed.

Table 1. Details of the seed sources of *Acacia mearnsii* used in the provenance trial in China.

Seedlot no.	Location	No. of parent trees	Lat S*	Long E	Alt (m)
14394	Candelo, NSW	13	36°45′	149°40′	80
14395	Lake George, NSW	7	35°15′	149°20′	700
14397	Bodalla, NSW	11	36°08′	150°05′	75
14398	N Batemans Bay, NSW	10	35°42′	147°15′	40
14416	Dargo, VIC	3	37°28′	147°15′	200
14725	NE of Bungendore, NSW	12	35°12′	149°32′	760
14769	Googong Rsvr, NSW	12	35°29′	149°16′	670
14770	Polacks Flat Ck, NSW	6	36°39′	149°35′	260
14771	S of Cooma, NSW	9	36°28′	149°01′	940
14922	NW of Braidwood, NSW	12	35°15′	–	–
14923	S of Bombala, NSW	13	37°09′	149°20′	500
14924	Merimbula, NSW	7	36°55′	149°54′	–
14925	Blackhill Reserve, VIC	6	37°12′	144°28′	500
14926	Omeo Highway, VIC	9	37°10′	147°45′	300
14927	Sth. Gippsland, VIC	7	37°44′	146°51′	100
14928	Cann R & Orbost, VIC	5	37°34′	148°28′	100
15087	Harding Natal, S Africa	Unknown	30°35′	129°51′	932
15088	Natal, S Africa	Unknown	29°32′	30°28′	838
C	Ganzhou CHINA	–	25°21′	114°50′	100
C21	Wenzhou, CHINA	–	28°01′	120°40′	50
C22	Guangnan, CHINA	–	23°50′	105°10′	1540
C23	Ganzhou, CHINA	–	25°50′	114°50′	123
C24	Sichuan, CHINA	–	31°56′	107°14′	690
C25	Yunnan, CHINA	–	24°20′	103°20′	1600
20	Brazil	Unknown	–	–	–

*Latitude N for Chinese seedlots.

In the field trials, measurements were made of height and diameter at breast height at 18 months after planting. Dieback and first flowering were also recorded. Analyses of variance were performed for nursery results at Chenxiang Tree Farm, and field data recorded at Ganzhou.

Results

Nursery Experiment at Chenxiang Tree Farm

The results obtained from the nursery experiment are summarised in Table 2. There were marked differences between provenances in height, diameter at root collar and biomass of seedlings. The best provenances at the nursery stage were two South African provenances (S15087, S15088) and two Australian ones (S14725 from Bungendore and S14395 from Lake George, both from New South Wales). The worst provenances included seed from Polacks Flat Creek, NSW (S14770), Bombala, NSW, (S14923) and local seed from Wenzhou (C21). Seed from Brazil also performed poorly.

Field Trial at Hubian Horticultural Farm, Ganzhou

Some of the trials suffered heavy losses as a result of drought and disease after they were planted in the field. The results presented here are thus limited to those obtained from the trials at Hubian Horticultural Farm, Ganzhou.

Provenances of *A. mearnsii* varied considerably in height and diameter (Table 3). There were significant differences amongst provenances. Several provenances attained 3 m in height and over 2 cm in diameter at breast height. These included two local seedlots (C21, C24), six Australian (S14397 Bodalla, S14398 Batemans Bay, S14725 Bungendore, S14771 Cooma, S14922 Braidwood, S14925 Blackhill Reserve), two South African (S15087, S15088) and the one Brazilian.

One local provenance (C22) from Guangnan was the poorest in height and the second poorest in diameter. Two Australian provenances from New South Wales (S14769 Googong, S14924 Merimbula) were also growing slowly in both height and diameter.

Table 2. Mean height, diameter at root collar and biomass of seedlings of *A. mearnsii* at 6 months of age in a nursery experiment at Chenxiang Forestry Farm.

Seedlot no.	Provenance location	Height (cm)	Diameter (cm)	Biomass
14394	Candelo, NSW	16.9	0.3	1.1
14395	Lake George, NSW	22.9	0.3	1.5
14397	Bodalla, NSW	16.4	0.2	1.0
14398	N. Batemans Bay, NSW	19.5	0.3	1.9
14416	Dargo, VIC	21.3	0.3	1.3
14725	NE Bungendore, NSW	21.8	0.3	1.6
14769	Googong Rsvr, NSW	11.8	0.2	1.1
14770	Polacks Flat Ck, NSW	8.1	0.2	0.8
14771	S. Cooma, NSW	11.5	0.2	0.9
14922	NW Braidwood, NSW	16.1	0.2	1.0
14923	S. Bombala, NSW	9.8	0.2	0.9
14924	Merimbula, NSW	20.7	0.3	1.4
14925	Blackhill Reserve, VIC	17.3	0.2	1.2
14926	Omeo Highway, VIC	18.7	0.2	1.1
14927	S. Gippsland, VIC	18.0	0.2	1.4
14928	Cann R & Orbost, VIC	13.0	0.2	1.0
15087	Harding Natal, S. Africa	22.6	0.4	2.8
15088	Natal, S. Africa	25.7	0.3	2.0
C	Ganzhou, China	12.3	0.2	0.9
C21	Wenzhou, China	11.5	0.2	0.9
C22	Guangnan, China	14.7	0.2	1.2
C23	Ganzhou, China	17.1	0.2	1.1
C24	Sichuan, China	20.5	0.2	0.8
C25	Yunnan, China	17.7	0.2	1.6
20	Brazil	13.2	0.2	1.3
	General Mean	16.8	0.2	1.3
	Least significant difference (.05)	4.3	0.06	0.5
	F-test (.001)	8.9***	4.3***	3.2***

Table 3. Mean height and diameter at breast height at 18 months old of *A. mearnsii* provenance trials at Ganzhou.

Seedlot no.	Provenance location	Mean height (m)	Mean DBH (cm)
14394	Candelo, NSW	2.9	1.9
14395	Lake George, NSW	2.7	1.7
14397	Bodalla, NSW	3.2	2.0
14398	N. Batemans Bay, NSW	3.2	1.9
14416	Dargo, VIC	2.7	1.6
14725	NE Bungendore, NSW	3.2	2.2
14769	Googong Rsvr, NSW	2.5	1.6
14770	Polacks Flat Ck, NSW	2.9	1.3
14771	S. Cooma, NSW	3.2	2.2
14922	NW Braidwood, NW	3.3	2.1
14923	S. Bombala, NSW	2.7	1.6
14924	Merimbula, NSW	2.4	1.3
14925	Blackhill Reserve, VIC	3.1	2.2
14926	Omeo Highway, VIC	2.9	2.1
14927	S. Gippsland, VIC	2.9	1.6
14928	Cann R & Orbost, VIC	2.8	1.7
15087	Harding Natal, S. Africa	3.1	2.3
15088	Natal, S. Africa	3.1	2.4
C	Ganzhou, China	2.7	1.9
C21	Wenzhou, China	3.0	2.1
C22	Guangnan, China	2.3	1.4
C23	Ganzhou, China	2.7	1.9
C24	Sichuan, China	3.3	2.5
C25	Yunnan, China	2.8	2.1
20	Brazil	3.3	2.3
	General Mean	2.9	1.9
	Least significant difference (0.5)	0.4	0.5
	F-test (.001)	3.94***	3.10***

Provenances varied in their first flowering. All local provenances flowered at 18 months old but none of the newly introduced provenances did at the same age. There were also differences between the local provenances in percentage of flowering trees. Highest percentage (62.5%) was recorded for Ganzhou provenance (C23) as compared to 2.8% for Wenzhou provenance (C21).

Some provenances in these trials suffered dieback either caused by drought or a disease identified as *Colleotrichum* sp. However, it was noted that the provenance from Omeo Highway, Victoria (S14926), had no loss, and that the provenance from Polacks Flat Creek, NSW (S14770) suffered only slight damage.

Discussion and Conclusion

Available results to date of the *A. mearnsii* provenance trials have shown marked differences between provenances in many growth characteristics at the nursery and field trials. Some provenances have performed consistently well since the nursery stage (i.e. two South African provenances and one Australian provenance from Bungendore, NSW). However, some provenances that were establishing poorly in the nursery grew very fast in the field (i.e. seed from Brazil and a local seed from Wenzhou). The early flowering of the local Chinese provenances was a most surprising result and no adequate explanation has been advanced to explain this behaviour.

Another five provenance trials have been established but so far no data have been analysed. Because our results are preliminary, it will be some time before complete data on best provenances are available. In order to reduce time to produce improved seed, a design for a seedling seed orchard was proposed by Raymond (1987) and implemented at Zangzhou in Fujian Province in March 1988.

Chapter 10

Growth and Survival of Australian Tree Species in Field Trials in Kenya

P.B. Milimo

Abstract

Field trials with Australian tree species, mainly eucalypts and acacias, were established in semi-arid to humid areas in Kenya. Early results showed that *Eucalyptus saligna* and *E. grandis* had best growth and survival in the humid area, whereas *Acacia crassicarpa* did well in semi-humid to semi-arid areas. The trials planted in semi-arid areas failed, and possible factors responsible for the failure are discussed.

Introduction

In many parts of Kenya insufficient fuelwood has long been a problem and has become more serious with the high population growth rate (4.1% per annum). However, development, with its high energy demands and a continuously growing standard of living, has ignored a simple ecological tenet — that many natural resources are finite, and that excessive exploitation hastens their exhaustion (Janick et al. 1981). In semi-arid Kenya, once luxuriant forest land is now desert, and seemingly limitless virgin humid forests are also suffering significant degradation. After centuries of cutting and depletion, Kenya has commenced reforestation work with selected species.

Recognising the fuelwood crisis and the need to search for new productive exotic species, the Australian Centre for International Agricultural Research (ACIAR) is funding a research project in Kenya on 'Australian Hardwoods for Fuelwood and Agroforestry.' The Kenya Forestry Research Institute (KEFRI) is the implementing agency.

The objective of the project is to determine the potentials of the lesser-known Australian tree species for provision of fuelwood, roundwood and other uses. Most of the species tested are in the genera *Eucalyptus* and *Acacia*. Three sites were selected for field trial establishment during 1986–88: Gede, Turbo and Loruk. Early results from field trials are presented below.

Materials and Methods

Trial Sites

These were established at Gede, Turbo and Loruk (see Fig. 1 in Chapter 1) and the sites previously carried exotic tree plantations. Gede is classified as 45–50% semihumid to semi-arid, Turbo as 80% humid and Loruk as 25–40% semi-arid (Teel 1985).

Gede is located on the coast (3°18'S, 40°01'E) at 40 m above sea level, and receives a mean annual rainfall of 988 mm. The mean maximum temperature is 32.3°C, mean minimum 21.7°C, and average annual potential evaporation of 155–2200 mm. The vegetation is composed of a low-deciduous forest composed mainly of dry woodland and bushland species.

Turbo (0°37'N, 35°05'E) is 1800 m above sea level and receives a mean annual rainfall of 1315 mm with 4 months of severe drought. The mean annual temperature is 17.9°C, a mean maximum of 28.3°C, and a mean mimimum of 14°C. The site originally carried an *Acacia mearnsii* plantation for production of tannin. This was replaced with *Pinus patula* in the early 1970s for pulpwood production. In 1985 the *P. patula* stand was clearfelled.

Loruk (1°2′N, 36°3′E) is located within the semi-arid zone, 1000 m above sea level. The site receives about 450 mm rainfall annually and an average annual potential evaporation of 1650–2300 mm. The vegetation is generally classified as bushland dominated by *Salvadora persica, Acacia tortilis,* etc.

Trial Establishment

Sites were cleared prior to planting, but not ploughed, which necessitated pit planting. Because of potential animal damage to trees, the Turbo site was fenced. Plots at Loruk were not fenced, hence high mortalities resulted from goat, camel and donkey browsing. Fencing was not necessary at Gede.

Seeds for the trials were received from CSIRO's Tree Seed Centre, Australia. At all sites planting commenced after the start of the rainy season (April–May), and beating up (replacement of dead plants) done within 3 months of planting. Plants were not watered after they were planted in the field. At Gede, plots were clean-weeded whenever it was necessary. At Turbo, they were spot-weeded.

Trial Design and Layout

The trials were established in a completely randomised block design with four replicates. Each treatment plot comprised 25 plants, at 2.5 × 2.5 m spacing. However, for seedlots with insufficient seedlings, unreplicated plots were established.

Assessment

Assessment of height and survival was done at 3 and 6 months after establishment. Thereafter, assessment was done annually.

Results

Turbo

Mean height at 6 months and at 18 months for the May 1986 plantings is presented in Table 1. Height growth at 6 and 18 months differed significantly at 5% confidence level (results not presented here). *Eucalyptus saligna* (14527) had the best height growth at age 6 (0.96 m) and 18 months (6.6 m) after out-planting. At 6 months, the poorest height growth (0.53m) was observed in *E. laevopinea* (14840) and at 18 months, (4 m) in *E. urophylla* (14534), *E. laevopinea* (14840) and *E. robusta* (14128).

Mean percent survival rates are presented in Table 1. At 6 months after planting, most species had high survival rates (>90%) except *E. laevopinea* (77%). At 18 months after planting, *E. saligna* (15054) and

Table 1. Mean height growth and percent survival at Turbo for 1986 planting at 6 and 18 months.

Seedlot no.	Species	Mean height (m)		Mean survival (%)	
		6 months	18 months	6 months	18 months
13965	*Eucalyptus grandis*	0.91 ± 0.09	5.87 ± 0.81	92 ± 9.24	89.33 ± 9.71
13024	″ ″	0.85 ± 0.10	4.15 ± 0.9	92 ± 5.66	66.38 ± 18.88
13021	″ ″	0.83 ± 0.23	5.26 ± 1.30	93 ± 8.25	91.67 ± 9.71
13020	″ ″	0.88 ± 0.11	5.65 ± 0.59	93 ± 3.27	95.65 ± 7.51
15054	″ *saligna*	0.86 ± 0.17	5.56 ± 1.09	100 ± 0.00	50.00 ± 0.0
14527	″ ″	0.96 ± 0.15	6.63 ± 1.63	94 ± 6.93	89.33 ± 9.71
14508	″ ″	0.88 ± 0.07	5.88 ± 0.73	89 ± 6.00	70.67 ± 29.87
14421	″ *resinifera*	0.78 ± 0.10	4.60 ± 1.08	84 ± 21.66	79.00 ± 22.07
14916	″ *pellita*	0.77 ± 0.13	4.75 ± 0.97	93 ± 6.83	79.00 ± 22.07
14534	″ *urophylla*	0.70 ± 0.15	3.95 ± 1.12	93 ± 3.83	89.33 ± 4.04
14128	″ *robusta*	0.73 ± 0.04	4.05 ± 0.21	88 ± 11.31	88.67 ± 6.51
9424	″ ″	0.79 ± 0.04	4.83 ± 0.49	96 ± 5.66	81.00 ± 16.82
14840	″ *laevopinea*	0.53 ± 0.23	4.00 ± 2.13	77.67 ± 33.68	95.67 ± 51
13037	″ *andrewsii* ssp. *andrewsii*	0.63 ± 0.05	4.70 ± 0.74	94.22 ± 6.68	93.67 ± 11.00
Unreplicated single plots					
13983	*Eucalyptus resinifera*	0.65	–	–	–
14532	″ *urophylla*	0.97	5.44	–	–
12895	″ ″	0.85	5.04	–	–
14913	″ *pryrocarpa*	0.82	3.90	–	–
14433	″ *pilularis*	0.30	1.44	–	–
10345	″ *oreades*	0.63	4.14	–	–

E. grandis (13024) had poorest survival, i.e. 50 and 66% respectively.

Results obtained after 3 months for the May 1987 plantings at Turbo are presented in Table 2. The best height was 53 cm for *E. punctata* (13265) and the poorest was 12 cm for *Acacia flavescens* (S14175). The best survival was 96% for *A. auriculiformis* (15477) and *E. paniculata* (S13657). The poorest survival was 50% for *A. flavescens* (14175).

Gede

The plantings of eucalypts and acacias were established as two separate experiments and results obtained at 3 months old are given in Table 3. Mean height growth varied among eucalypts, with *E. tereticornis* (14108) attaining the best height growth and survival. Amongst the acacias, *A. crassicarpa* has shown promising growth and survival at a young age.

Table 2. Mean height and percent survival at Turbo for 1987 planting at 3 months.

Seedlot no.	Species	Height (m)	Survival (%)
13687	*Acacia aulacocarpa*	0.16 ± 0.03	79
14591	,, ,,	0.31 ± 0.01	89
14969	,, ,,	0.13 ± 0.03	66
15483	,, *auriculiformis*	0.30 ± 0.02	92
15477	,, ,,	0.25 ± 0.06	96
14175	,, *flavescens*	0.12 ± 0.04	50
12991	,, *mangium*	0.17 ± 0.01	89
15063	,, ,,	0.20 ± 0.02	71
13139	*Casuarina glauca*	0.32 ± 0.04	89
13508	,, *cunninghamiana*	0.41 ± 0.02	79
13265	*Eucalyptus punctata*	0.53 ± 0.06	94
13657	,, *paniculata*	0.37 ± 0.02	96
15145	,, *propinqua*	0.19 ± 0.04	81
13570	,, *siderophloia*	0.35 ± 0.02	94

Table 3. Mean height and percent survival at Gede for 1987 planting at 3 months.

Seedlot no.	Species	Mean height (m)	Mean survival (%)
Experiment 1			
14861	*Eucalyptus grandis*	1.21 ± 0.32	16.5 ± 6.4
14849	,, ,,	1.29 ± 0.37	22.0 ± 15.6
14436	,, ,,	1.18 ± 0.76	29.3 ± 12.7
13431	,, ,,	1.38 ± 0.28	55.7 ± 38.7
15011	,, *saligna*	1.93 ± 0.75	22.0 ± 9.0
14524	,, ,,	0.48 ± 0.00	22.0 ± 0.0
14508	,, ,,	–	–
14435	,, ,,	–	–
14534	,, *urophylla*	1.75 ± 0.68	91.8 ± 16.5
12895	,, ,,	1.22 ± 0.49	58.0 ± 32.1
14532	,, ,,	1.72 ± 0.42	72.3 ± 14.3
13398	,, *tereticornis*	1.21 ± 0.51	80.3 ± 24.8
14444	,, ,,	1.75 ± 0.32	97.3 ± 5.5
14108	,, ,,	2.19 ± 0.18	100.0 ± 0.0
Experiment 2			
14969	*Acacia aulacocarpa*	0.48 ± 0.03	69.8 ± 10.5
14591	,, ,,	0.99 ± 0.07	63.8 ± 26.5
13862	,, *auriculiformis*	1.41 ± 0.21	97.3 ± 5.5
13686	,, ,,	1.27 ± 0.36	69.8 ± 5.5
15063	,, *mangium*	0.38 ± 0.04	11.0 ± 0.0
13622	,, ,,	0.51 ± 0.16	22.0 ± 15.6
12991	,, ,,	0.34 ± 0.06	16.5 ± 7.8
13680	,, *crassicarpa*	1.72 ± 0.30	94.5 ± 11.0
14175	,, *flavescens*	0.20 ± 0.00	11.0 ± 0.0

Table 4. Selection of species for planting at Loruk in 1986 and 1987.

1986		1987	
Seedlot	Species	Seedlot	Species
15062	*Eucalyptus camaldulensis*	15062	*Eucalyptus camaldulensis*
14045	*E. camaldulensis*	14045	*E. camaldulensis*
15050	*E. camaldulensis*	15050	*E. camaldulensis*
13663	*E. camaldulensis*	13663	*E. camaldulensis*
14847	*E. camaldulensis*	14847	*E. camaldulensis*
11633	*E. ochrophloia*	11730	*E. ochrophloia*
11465	*E. bigalerita*	11473	*E. bigalerita*
13713	*E. argophloia*	13713	*E. argophloia*
13678	*E. orgadophila*	13678	*E. orgadophila*
13265	*E. punctata*	13265	*E. punctata*
14660	*Acacia holosericea*	14660	*Acacia holosericea*
14632	*A. holosericea*	14632	*A. holosericea*
14631	*A. ampliceps*	14631	*A. ampliceps*
14650	*A. ampliceps*	14650	*A. ampliceps*
14683	*A. cowleana*	14683	*A. cowleana*
14655	*A. cowleana*	14655	*A. cowleana*
14622	*A. shirleyi*	14622	*A. shirleyi*
14753	*A. shirleyi*	14753	*A. shirleyi*
13483	*A. cambagei*	13483	A. cambagei
14904	*Melaleuca argentea*	14904	*Melalueca argentea*
14099	*M. pauperiflora*	14099	*M. pauperiflora*
14027	*M. uncinata*	14027	*M. uncinata*
14879	*M. nervosa*	14879	*M. nervosa*
13749	*M. glomerata*	13749	*M. glomerata*

Loruk

Seedlings for planting were raised in 1986 and 1987 (see Table 4). In 1986, seedlings died before out-planting due to lack of water in the nursery. In 1987, seedlings were successfully raised at Muguga and transported to a temporary nursery at the site prior to planting, but a combination of browse damage, termite attack and drought led to unsuccessful establishment.

Discussion

There is an urgent need to identify multipurpose woody perennials suitable for reforestation or integration into farming systems in semi-arid Kenya. Although this chapter is mostly based on 18-month-old plots, trends of performance are encouraging. With sufficient caution, some practical decisions could be made about nursery treatments, site preparations, planting, and genetic selection.

The most promising results to date include the good growth and survival of *E. saligna* and *E. grandis* at Turbo and the fast early height growth of *Acacia crassicarpa* at Gede. The surprisingly fast early growth of *A. crassicarpa* mirrors the good results achieved with this species in the ACIAR trials in Thailand (see Chapter 11) and China (see Chapter 8).

Since 2 February 1985, more than 181 seedlots have been received from Australia for planting. These belong to the genera *Eucalyptus* (91), *Acacia* (76), *Melaleuca* (11), *Casuarina* (2), and *Grevillea* (1). Only 57 species of these have been successfully established in the field and these include *Eucalyptus* (38), *Acacia* (17) and *Casuarina* (2). Some of the factors responsible for the poor performance are: (a) very poor germination, (b) high rates of seedling mortality in the nursery, (c) harsh environmental conditions, (d) high susceptibility of most species to termite attack and, (e) browse damage.

About 50% of the seedlot received has been characterised by low germination. Although all species had poor germination, the problem was more severe in *Grevillea* and *Melaleuca*. For the 1988 planting at Turbo the *Acacia* seeds failed to germinate in the nursery. For the 1988 plantings at Loruk and Kibwezi (Gede seedlings), the *Eucalyptus* seeds failed to germinate. This poor germination could be due to the low seed viability, dormancy or poor nursery techniques. Most acacias have hard seeds and germinate more quickly following hot water scarification (Crocker 1916).

According to Zummer-Linder (1983), hot-water treatment promotes faster germination of the smaller seeds. Studies with boiling water treatment in West Africa indicated that it is ineffective on some African acacias. As for Australian acacias, boiling water treatment is suspected of destroying 30% of the seed, or the treatment is simply not effective. Pure water boils at 100°C at sea level and, because the boiling point is depressed by impurities and high elevation, it is possible that the boiling water treatment is simply not as effective at high elevations in Kenya. This could partly explain the better germination results observed at Gede (at sea level) compared with either Muguga or Turbo.

Performance of field plots at Gede and Turbo, based on height growth, was not unexpected as there already exist some data on Australian species and provenances (*E. grandis* and *E. saligna*) from Elburgon and Turbo. Stressful environmental conditions are an important factor contributing to poor field performance at Loruk. Therefore, it is not appropriate, under arid and semi-arid conditions, to screen species and their provenances for adaptability without first conducting research into how to condition nursery stock to resist out-planting stresses.

Seedlings of high physiological quality are those that will flourish despite the relatively harsh environment into which they are transplanted (Duryea and Landis 1984). Producing such seedlings consistently and economically should be the nursery's prime objective. Therefore, nursery personnel should thoroughly familiarise themselves with the physiology of the plant species they are working with.

Termite attack on out-planted seedlings is a serious problem in semi-arid areas. Among those that have shown high susceptibility are eucalypts, casuarinas and some acacias. For the 1988 planting, investigations with slow-release, inert plastic granules of carbosulfan insecticide (Incitec Ltd Australia) are planned.

Grevillea pteridifolia in the RFD/ACIAR field trials at Ratchaburi, Thailand. The seed source was from near Cooktown, Qld, and tree form is bushy compared with the columnar form from Dimbulah, Qld, seed source (*left*, Khun Bunyarit Puriyakorn, *right*, Khun Sathit Sawintara).

Chapter 11

Growth and Survival of Australian Tree Species in Field Trials in Thailand

K. Pinyopusarerk

Abstract

Field trials of Australian tree species of the genera *Acacia, Eucalyptus* and *Melaleuca* were planted at different sites across Thailand during the period 1985 to 1987. Early results obtained for the trials planted in 1985 and 1986 have shown marked differences between species in growth and survival. Several acacias (e.g. *Acacia crassicarpa, A. auriculiformis, A. torulosa* and *A. julifera*) and eucalyptus (e.g. *Eucalyptus camaldulensis, E. tereticornis, E. citriodora* and *E. urophylla*) were amongst the fastest-growing while most melaleucas and casuarinas were slow-growing. Some species (e.g. *A. oraria, Albizia procera* and most melaleucas) grew slowly but survived well. A little-known species (*Grevillea pteridifolia*) has grown well with a dense crown, and has maintained a healthy appearance throughout the year. Provenance variation has been noted for some species. Northern provenances of *A. crassicarpa* and *A. aulacocarpa* grew faster than southern provenances. Some species were also found to differ in tree form between different sites (e.g. trees of *A. polystachya* and *A. holosericea*, normally multistemmed with heavy branching at dry sites, were single-stemmed and had light branching patterns at two wet sites).

Introduction

Thailand has a total area of 514 000 km^2. Early this century, 70% of the country's area was covered with forests (Feeny 1984). The forests have been seriously and rapidly depleted through widespread tree-cutting together with subsequent land clearing for agricultural expansion. By 1985 Landsat Satellite Imagery revealed that the forests were reduced to only 29% (Thailand Royal Forest Department 1986). Supply of wood produced from natural forests is consistently declining while demand continues to increase. Thus there is a pressing need to locate tree species to supplement production from the decreasing native forests. Such species should be fast-growing and capable of surviving in the severely deforested areas.

Many Australian tree species are fast-growing and capable of tolerating harsh conditions (e.g. drought and soils of low fertility — Boland and Turnbull 1981). In Thailand, some Australian species (e.g. *Eucalyptus camaldulensis* and *Acacia auriculiformis*) have been used as plantation species with great success. Many other species remain unexplored and may be of great value for the national tree-planting programs. An ACIAR-supported project on Australian hardwoods for fuelwood and agroforestry was set up with the Royal Forest Department of Thailand (RFD). The prime objective was to test, under field trials, a wide but selected range of Australian species in many genera including *Acacia, Eucalyptus, Casuarina* and *Melaleuca*. Three series of field trials were planted during 1985 to 1987. This chapter outlines growth performance at 24 and 12 months of those trials planted in 1985 and 1986 respectively.

Materials and Methods

Location of Trials

The trials were established over seven sites selected to represent a range of climatic and geographic conditions in Thailand: rationale for selection of the trial sites discussed in Chapter 1.

The seven sites are:

(1) Ratchaburi Forest Experiment Station, Ratchaburi (central-west);
(2) Sai Thong Forest Experiment Station, Prachuap Khiri Khan (south);
(3) Huai Tha Forest Experiment Station, Si Sa Ket (far northeast);
(4) Sakaerat Thai-Japan Project, Nakhon Ratchasima (northeast);
(5) Khao Soi Dao Seed Orchard, Chanthaburi (central-east);
(6) Huai Bong Forest Experiment Station, Chiang Mai (north — high altitude); and
(7) Ban Hong Plantation, Lamphun (north—low altitude).

Sites 1-6 were planted in both 1985 and 1986 whereas site 7 was planted in 1986 only (see Fig. 1 in Chapter 1 for location of trial sites). Detailed climatic data for five of the seven sites are given in Chapter 4.

Plant Material

Seed for all plantings was supplied by the CSIRO Australian Tree Seed Centre in Canberra. Details of the seed sources for the species used are given in Table 1. At each planting, local material (native species or local exotics) was included for comparison with the newly introduced Australian seedlots.

All six sites planted in 1985 were given a similar set of seedlots. The number of seedlots represented at each site varied but most seedlots were planted at four to six sites. In the 1986 plantings the trial sites were divided broadly into two types (i.e. wet site, Sai Thong, Chanthaburi and Sakaerat, and dry site, Ratchaburi, Si Sa Ket, Huai Bong and Ban Hong). Seed was then allocated according to each of the site types. However, there were some eucalypts planted at both site types (e.g. *E. camaldulensis, E. tereticornis, E. raveretiana, E. punctata* and *E. houseana*).

Plants were raised at each planting site. No special inoculation with microorganisms (*Rhizobium*/*Frankia*) was made.

Site Preparation

Following clearing and burning, the planting sites were disc-ploughed twice in cross directions before the rainy season. Weedicide (Roundup at 1:100 in water) was sprayed over each planting spot 2-3 weeks prior to planting. The planting sites at Sai Thong and Huai Bong were fenced to exclude cattle. Planting holes were dug to a depth of 25 cm and a width of 25 cm.

Design and Layout

A randomised complete block design with three replicates was used. Each replicate consisted of 25 trees arranged in a plot of 5 × 5 trees. Spacing was 2 × 2 m.

Table 1. Origin data for seedlots used in the field trials in Thailand.

CSIRO seedlot no.	Species		No. of parent trees in collection	Location	Lat S	Long E	Alt (m)
1985 Plantings							
13877	*Acacia aulacocarpa*	(ACAAUL)	10	Julatten Area QLD	16°35′	145°25′	410
13866	*aulacocarpa*	(ACAAUL)	6	Garioch QLD	16°40′	145°18′	400
13689	*aulacocarpa*	(ACAAUL)	5	Oriomo River PNG	8°48′	143° 9′	20
13688	*aulacocarpa*	(ACAAUL)	6	Keru PNG	8°32′	141°45′	40
13861	*auriculiformis*	(ACAAUR)	4	Springvale Holding QLD	15°50′	144°55′	500
13854	*auriculiformis*	(ACAAUR)	200	Oenpelli NT	12°20′	133° 4′	50
13686	*auriculiformis*	(ACAAUR)	10	Iokwa PNG	8°41′	141°29′	35
13684	*auriculiformis*	(ACAAUR)	17	Balamuk PNG	8°54′	141°18′	18
13864	*cincinnata*	(ACACIN)	5	Shoteel QLD	16°57′	145°38′	440
13863	*crassicarpa*	(ACACRA)	5	Shoteel LA QLD	16°57′	145°38′	440
13683	*crassicarpa*	(ACACRA)	15	Woroi Wipim PNG	8°49′	143° 0′	20
13681	*crassicarpa*	(ACACRA)	10	Mata PNG	8°40′	141°45′	30
13680	*crassicarpa*	(ACACRA)	21	Wemenever PNG	8°51′	141°26′	30
14623	*difficilis*	(ACADIF)	41	Daly Waters NT	16°21′	133°22′	235
14175	*flavescens*	(ACAFLA)	9	Mt Molloy QLD	16°40′	145°18′	400
14660	*holosericea*	(ACAHOL)	26	Turkey Creek WA	17° 4′	128°12′	400
13691	*leptocarpa*	(ACALEP)	4	Woroi Wipim PNG	8°52′	143° 3′	30
13653	*leptocarpa*	(ACALEP)	1	Starcke Holding QLD	14°16′	144°26′	2
13846	*mangium*	(ACAMAN)	75	7 Km SSE of Mossman QLD	16°31′	145°24′	60
13621	*mangium*	(ACAMAN)	9	Piru, Ceram INDONESIA	3° 4′	128°12′	150
14176	*melanoxylon*	(ACAMEL)	10	Atherton QLD	17°17′	145°26′	1,022
13871	*polystachya*	(ACAPOL)	4	Bridle LA QLD	16°58′	145°37′	480
14622	*shirleyi*	(ACASHI)	10	Daly Waters NT	16°19′	133°23′	225
13876	*Allocasuarina littoralis*	(ALLLIT)	5	Gordon and Chili Cks QLD	12°42′	143°20′	80

CSIRO seedlot no.	Species		No. of parent trees in collection	Location	Lat S	Long E	Alt (m)
13519	*Casuarina cunninghamiana*	(CASCUN)	10	9 Km N Rollingstone QLD	19° 1′	146°20′	20
13514	*cunninghamiana*	(CASCUN)	5	11 Km SE of Petford QLD	17°25′	144°59′	560
13148	*cunninghamiana*	(CASCUN)	5	5 Km E of Cobargo NSW	36°24′	149°56′	100
13990	*equisetifolia* subsp. *incana*	(CASEQU)	20	N of Stradbroke Is QLD	27°24′	153°26′	0
14537	*Eucalyptus camaldulensis*	(EUCCAM)	10	Isdell River WA	16°56′	125°35′	315
14106	*camaldulensis*	(EUCCAM)	9	Gilbert River QLD	18° 0′	143° 0′	150
12013	*pellita*	(EUCPEL)	unknown	5 Km S of Helenvale QLD	15°45′	145°15′	152
14130	*torelliana*	(EUCTOR)	17	SSW of Kuranda QLD	16°53′	145°36′	420
14485	*Melaleuca bracteata*	(MELBRA)	unknown	N of Alice Springs NT	23°36′	133°52′	840
14166	*dealbata*	(MELDEA)	unknown	Weipa QLD	12°39′	141°49′	5
11935	*dealbata*	(MELDEA)	unknown	NT	12°35′	131°18′	20
14170	*symphyocarpa*	(MELSYM)	unknown	Weipa QLD	12°40′	141°53′	10
14152	*viridiflora*	(MELVIR)	10	Weipa QLD	12°31′	141°48′	10
Local seedlot							
L8511	*Eucalyptus citriodora*	(EUCCIT)	unknown	Ratchaburi	-	-	-
L8512	*Peltophorum dasyrachis*	(PELDAS)	unknown	Ratchaburi	-	-	-
L8513	*Pterocarpus indicus*	(PTEIND)	unknown	Ratchaburi	-	-	-
L8514	*Azadirachta indica*	(AZAIND)	unknown	Ratchaburi	-	-	-
L8515	*Cassia siamea*	(CSASIA)	unknown	Ratchaburi	-	-	-
L8516	*Casuarina junghuhniana*	(CASJUN)	unknown	Ratchaburi	-	-	-
L8521	*Alstonia macrophylla*	(ALSMAC)	unknown	Prachuap Khiri Khan	-	-	-
L8531	*Peltophorum pterocarpum*	(PELPTE)	unknown	Si Sa Ket	-	-	-
L8542	*Melia azedarach*	(MLAAZE)	unknown	Nakhon Ratchasima	-	-	-
L8552	*azedarach*	(MLAAZE)	unknown	Chanthaburi	-	-	-
L8553	*Peltophorum dasyrachis*	(PELDAS)	unknown	Chanthaburi	-	-	-
L8555	*Cassia siamea*	(CSASIA)	unknown	Chanthaburi	-	-	-
L8556	*Acacia auriculiformis*	(ACAAUR)	unknown	Chanthaburi	-	-	-
L8561	*Casuarina equisetifolia*	(CASEQU)	unknown	Prachuap Khiri Khan	-	-	-
L8562	*Acacia auriculiformis*	(ACAAUR)	unknown	Chiang Mai	-	-	-
L8563	*Pinus kesiya*	(PINKES)	unknown	Chiang Mai	-	-	-
1986 Plantings							
14958	*Acacia bidwillii*	(ACABID)	5	ENE Georgetown QLD	18°12′	143°57′	385
14965	*brassii*	(ACABRA)	43	28-30 Km N Coen QLD	13°44′	143° 7′	165
14981	*falciformis*	(ACAFAL)	100	15 Km NE Ravenshoe QLD	17°31′	145°26′	1,050
13872	*flavescens*	(ACAFLA)	1	Claudie River QLD	12°45′	143°13′	100
14968	*flavescens*	(ACAFLA)	10	40-43 Km NW Cooktown QLD	15°19′	145° 2′	240
15100	*harpophylla*	(ACAHAR)	25	Pasha,via Rd to Mt. Coolon C. 70 Km NW Moranbah	21°44′	147°36′	300
14657	*hemignosta*	(ACAHEM)	10	98 Km N Halls Ck WA	17°30′	127°56′	395
14977	*hylonoma*	(ACAHYL)	1	14 Km NE Gordonvale QLD	17° 1′	145°50′	110
14885	*julifera* ssp. *gilbertensis*	(ACAJUL)	6	61.5 Km NW Chillagoe QLD	16°47′	144° 8′	280
14974	*julifera* ssp. *julifera*	(ACAJUL)	20	3 Km SW Balfes Ck QLD	20°13′	145°53′	330
14886	*oraria*	(ACAORA)	12	E Lakeland Downs QLD	15°46′	144°58′	180
14961	*oraria*	(ACAORA)	8	39 Km NW Cairns QLD	16°41′	145°35′	5
14542	*platycarpa*	(ACAPLA)	10	SE Katherine NT	14°35′	132°30′	190
14960	*platycarpa*	(ACAPLA)	30	61 Km NW Chillagoe QLD	16°47′	144° 8′	280
14967	*rothii*	(ACAROT)	10	51.4 Km NE Laura QLD	15°18′	144°39′	62
14553	*simsii*	(ACASIM)	10	SSW Port Douglas QLD	16°31′	145°27′	10
14576	*simsii*	(ACASIM)	10	NE Mareeba QLD	16°52′	145°35′	355
14183	*torulosa*	(ACATOR)	3	NW Chillagoe QLD	16°36′	144° 7′	275
14888	*torulosa*	(ACATOR)	15	29 Km NW Laura QLD	15°27′	144°13′	110
14180	*Adenanthera abrosperma*	(ADEABR)	10	WMW Wrotham Park QLD	16°30′	143°21′	108
14557	*abrosperma*	(ADEABR)	6	NW Chillagoe QLD	16°59′	144°18′	220
14959	*Albizia procera*	(ALBPRO)	12	14 Km NE Cairns QLD	16°50′	145°41′	10
14962	*procera*	(ALBPRO)	5	12 Km S Port Douglas QLD	16°33′	145°29′	15
14190	*Alphitonia excelsa*	(ALPEXC)	5	Dingo QLD	23°11′	149°17′	212
14976	*Atalaya hemiglauca*	(ATAHEM)	10	34 Km W Georgetown QLD	18°17′	143°14′	220
14504	*Callitris intratropica*	(CALINT)	6	Murgonella NT	11°33′	132°55′	7
14188	*Cassia brewsteri*	(CSABRE)	25	Blackwater QLD	23°35′	149° 3′	195
14556	*Desmodium umbellatum*	(DESUMB)	10	SE Almaden QLD	17°20′	144°41′	520
11465	*Eucalyptus bigalerita*	(EUCBIG)	3	58 Km SW Katherine NT	14°53′	131°54′	90
13397	*brassiana*	(EUCBRA)	27	Woroi to Wipim PNG	8°51′	143° 2′	30
13692	*camaldulensis*	(EUCCAM)	25	Gilbert River QLD	18°12′	142°53′	150
14338	*camaldulensis*	(EUCCAM)	129	Region E Petford QLD	17°17′	145° 3′	500

(Continued).

Table 1. *(Concluded).*

CSIRO seedlot no.	Species		No. of parent trees in collection	Location	Lat S	Long E	Alt (m)
14852	*citriodora*	(EUCCIT)	18	Mt Garnet QLD	17°41′	145° 7′	850
10691	*cloeziana*	(EUCCLO)	12	Veteran LA NE Gympie QLD	26° 7′	152°42′	135
13461	*deglupta*	(EUCDEC)	3	Seed Orchard, Philippines	-	-	-
13329	*dunnii*	(EUCDUN)	10	NW Kyogle NSW	28°24′	152°41′	400
14864	*exserta*	(EUCEXS)	16	Herberton Area QLD	17°25′	145°23′	950
14431	*grandis*	(EUCGRA)	25	Belthorpe SF QLD	26°52′	152°42′	500
14700	*grandis*	(EUCGRA)	11	NE Atherton QLD	17°11′	145°36′	780
9091	*houseana*	(EUCHOU)	1	Prince Regent River WA	15°50′	125°30′	45
13973	*microcorys*	(EUCMIC)	10	Fraser Island QLD	25°29′	153° 2′	65
14442	*paniculata*	(EUCPAN)	11	Coffs Harbour NSW	29°41′	152°56′	90
10863	*punctata* var. *longirostrata*	(EUCPUN)	5	Barakula SF Chinchilla QLD	26°22′	150°26′	350
10857	*pyrocarpa*	(EUCPYR)	5	Barcoongere SF NSW	29°57′	153°10′	180
13546	*raveretiana*	(EUCRAV)	6	R'hampton Racecourse QLD	23°23′	150°30′	30
13166	*resinifera*	(EUCRES)	7	Mt Lewis Timb Res 66 QLD	16°36′	145°17′	1,100
15011	*saligna*	(EUCSAL)	45	Kroombit Tops, Monto QLD	24°51′	151° 1′	730
13598	*suffulgens*	(EUCSUF)	10	45.9 Km E Rolleston QLD	24°39′	149° 3′	400
14108	*tereticornis*	(EUCTER)	8	Kennedy River QLD	15°26′	144°11′	60
14212	*tereticornis*	(EUCTER)	25	5-12 Km S Helenvale QLD	15°45′	145°15′	500
14532	*urophylla*	(EUCURO)	31	Mt Lewotobi Indonesia	8°31′	122°45′	398
14534	*urophylla*	(EUCURO)	30	Mt Egon Indonesia	8°38′	122°27′	500
14143	*Grevillea parallela*	(GREPAR)	11	Weipa QLD	12°33′	141°52′	10
14980	*pinnatifida*	(GREPIN)	10	Julatten Area QLD	16°34′	145°22′	415
14905	*pteridifolia*	(GREPTE)	10	49 Km NW Cooktown QLD	15°17′	145°59′	280
14502	*Leptospermum flavescens*	(LEPFLA)	7	SW Atherton QLD	17°20′	145°25′	910
14554	*flavescens*	(LEPFLA)	10	SW Atherton QLD	17°15′	145°23′	1,255
14900	*longifolium*	(LEPLON)	10	33.5 Km NW Laura QLD	15°26′	144°11′	90
14873	*Melaleuca acacioides*	(MELACA)	15	SSE Laura QLD	15°37′	144°28′	90
14899	*argentea*	(MELARG)	2	SE Musgrave QLD	15° 2′	143°39′	55
14904	*argentea*	(MELARG)	10	W. Wrotham Park QLD	16°41′	143°54′	135
14903	*bracteata*	(MELBRA)	15	W. Lakeland Downs QLD	15°50′	144°54′	180
14982	*bracteata*	(MELBRA)	27	Basalt Gully, Mareeba QLD	17° 0′	145°25′	335
14878	*cajuputi*	(MELCAJ)	10	N. Mossman QLD	16°16′	145°23′	12
14147	*leucadendra*	(MELLEU)	10	Weipa QLD	12°31′	141°48′	10
14871	*saligna*	(MELSAL)	10	SSE Laura QLD	15°37′	144°28′	95
14148	*stenostachya*	(MELSTE)	5	Batavia Downs QLD	12°42′	142°42′	90
14500	*Melia azedarach* var. *australasica*	(MLAAZE)	10	Atherton QLD	17°17′	145°27′	752
14501	*azedarach* var. *australasica*	(MLAAZE)	10	SW Mt Garnet QLD	18° 5′	144°52′	780
14889	*Neofabrica myrtifolia*	(NEOMYR)	30	S Laura QLD	15°49′	144°16′	360
14896	*myrtifolia*	(NEOMYR)	4	C Weymouth QLD	12°38′	143°25′	10
14639	*Petalostigma nummularium*	(PETNUM)	30	SW of Hooker Creek NT	18°46′	130°13′	420
14189	*pubescens*	(PETPUB)	5	N Dingo QLD	23°11′	149°17′	192
14880	*Terminalia muelleri*	(TERMUE)	8	N Cairns QLD	16°47′	145°40′	3
14551	*orenicola*	(TERORE)	11	Mossman QLD	16°28′	145°27′	4
14874	*Xanthostemon umbrosus*	(XANUMB)	20	Cattle Ck 8.9 Km SSE Laura QLD	15°37′	144°28′	75
Local seedlot							
L8611	*Cassia siamea*	(CSASIA)	unknown	Ratchaburi	-	-	-
L8612	*Azadirachta indica*	(AZAIND)	unknown	Ratchaburi	-	-	-
L8613	*Peltophorum dasyrachis*	(PELDAS)	unknown	Ratchaburi	-	-	-
L8614	*Dalbergia sissoo*	(DALSIS)	unknown	Ratchaburi	-	-	-
L8615	*Adenanthera pavonin*	(ADEPAV)	unknown	Ratchaburi	-	-	-
L8621	*Alstonia macrophylla*	(ALSMAC)	unknown	Prachuap Khiri Khan	-	-	-
L8622	*Acacia auriculiformis*	(ACAAUR)	unknown	Prachuap Khiri Khan	-	-	-
L8623	*Casuarina equisetifolia*	(CASEQU)	unknown	Unknown	-	-	-
L8624	*Tabebuia rosea*	(TABROS)	unknown	Unknown	-	-	-
L8632	*Cassia siamea*	(CSASIA)	unknown	Si Sa Ket	-	-	-
L8633	*Eucalyptus camaldulensis*	(EUCCAM)	unknown	Si Sa Ket	-	-	-
L8634	*deglupta*	(EUCDEG)	unknown	Si Sa Ket	-	-	-
L8641	*Pterocarpus macrocarpus*	(PTEMAC)	unknown	Nakhon Ratchasima	-	-	-
L8651	*Parkia javanica*	(PAKJAV)	unknown	Chanthaburi	-	-	-
L8652	*Peltophorum dasyrachis*	(PELDAS)	unknown	Chanthaburi	-	-	-
L8653	*Acacia catechu*	(ACACAT)	unknown	Chanthaburi	-	-	-
L8671	*Azadirachta indica*	(AZAIND)	unknown	Lamphun	-	-	-
L8672	*Dipterocarpus alatus*	(DIPALA)	unknown	Lamphun	-	-	-
L8673	*Acacia auriculiformis*	(ACAAUR)	unknown	Lamphun	-	-	-

Planting

Planting took place between June and August in each year, the date depending on the commencement of the rainy season at each planting site. Seedlings were approximately 6 months old when out-planted.

Fertilising

Following cultivation 50 g of complete fertiliser (15:15:15) was applied to each plant 1 month after planting. Another 50 g of complete fertiliser was also applied in the second year at the beginning of the rainy season.

Weed Control

Weed competition in the experimental areas was kept to a minimum by frequent application of slash-weeding or chemical spraying. Frequency of weed control was based on an as-required basis.

Assessments

All trees in each trial were first measured for height, diameter at ground level and survival at the age of 6 months after planting, and then at 6-month intervals until the trees attained the age of 24 months. Additional measurements of the diameter at breast height were carried out at 24 months. Two-monthly observations of phenological development (flowering, seeding and shoot elongation patterns) and damage to the trees were also carried out. These observations were made on the basis of overall appearance of each plot. Results of these observations will appear in separate reports.

Data Analysis

Separate analyses of variance were carried out for each trial for height, diameter at breast height, diameter at ground level and survival using the 'GENSTAT' statistical package. The data presented in this chapter are those recorded at 24 months for the 1985 plantings and at 12 months for the 1986 plantings. Arcsine transformation was applied to the survival data before analysis. Duncan's new multiple range test procedure (Duncan 1955) was used to test the significance of the differences between treatment means. A few treatments (seedlots) had only two replicates due to poor germination. In these cases missing values were computed by the 'GENSTAT' program.

Results

1985 Plantings — 24 months

Height

There were marked differences (amongst species) in height at all planting sites (Table 2).

Species showing fastest growth at 24 months after planting were *Eucalyptus camaldulensis*, *E. pellita*, *Acacia crassicarpa* and *A. auriculiformis*. These species were generally amongst the top group at each planting site. Some species were growing fast at a particular site (i.e. *A. difficilis* at Ratchaburi, *A. leptocarpa* (from Papua New Guinea) at Si Sa Ket, *A. cincinnata* at Sakaerat, and *E. torelliana* at Chanthaburi).

Poor height growth was recorded for most casuarinas, melaleucas and some acacias (*A. melanoxylon, A. polystachya, A. shirleyi, A. flavescens* and *A. mangium*), whereas *A. holosericea* and *A. aulacocarpa* (northern seedlots 13688, 13689) were around overall average.

There were only a few occasions where height growth of local species or local exotics was comparable to that of the newly introduced Australian species. At Ratchaburi, *Casuarina junghuhniana* (local clone) and *E. citriodora* were amongst the fastest-growing, whereas at Chanthaburi *Melia azedarach* was second only to *E. camaldulensis,* although the *Melia* had spent an extra 12 months in the nursery before the out-planting.

It was noted that trees of more northern provenances of *A. crassicarpa* and *A. aulacocarpa* were growing faster than their southern counterparts, although the differences were not always significant.

Diameter at Breast Height

There were significant differences between species in diameter at breast height (Table 3). Species having greatest diameter were those which grew tallest (i.e. *E. camaldulensis, E. pellita, A. crassicarpa* and *A. auriculiformis*). Similarly, smallest diameter was recorded for most casuarinas, melaleuca and some acacias (e.g. *A. polystachya*, *A. flavescens* and *A. aulacocarpa* from more southerly latitudes).

Survival

Survival differed significantly between species although there was no clear pattern for species ranking at each planting site (Table 4). In general, about two-thirds of all seedlots had better than 80% survival, with *E. camaldulensis* and *A. auriculiformis* having survived well at most sites. High mortality was found in *A. melanoxylon, A. cincinnata, A. shirleyi, Allocasuarina littoralis* and *Casuarina equisetifolia*. At Ratchaburi, *A. melanoxylon* and *Allocasuarina littoralis* were completely dead in the second dry season.

1986 Plantings — 12 Months

Height

All trials planted in 1986 showed marked

Table 2. Ranking for mean height (m) at 24 months of field trials planted in 1985 in Thailand. For each planting site vertical lines group treatments that are not significantly different (P = 0.05).

Ratchaburi			Sai Thong			Si Sa Ket		
Seedlot no.		Height	Seedlot no.		Height	Seedlot no.		Height
EUCCAM	14537	7.64	ACACRA	13683	10.82	EUCCAM	14537	8.39
EUCCAM	14106	7.63	ACACRA	13680	10.73	EUCCAM	14106	8.09
CASJUN	L8516	7.32	ACAAUR	13684	9.41	ACACRA	13683	7.38
EUCCIT	L8511	7.03	EUCCAM	14106	9.39	ACALEP	13691	7.35
EUCPEL	12013	6.73	ACACRA	13681	9.20	ACACRA	13681	6.79
ACADIF	14623	6.08	ACAAUR	13854	9.00	ACACRA	13863	6.77
ACACRA	13683	6.00	EUCPEL	12013	8.93	ACACRA	13680	6.74
ACACRA	13680	5.83	ACAAUR	13686	8.84	ACAAUR	13861	6.61
CSASIA	L8515	5.80	EUCCAM	14537	8.83	ACALEP	13653	6.58
ACACRA	13681	5.71	ACAAUR	13861	8.58	ACAAUR	13854	6.58
ACAAUR	13686	5.67	ACALEP	13691	8.52	ACAAUR	13684	6.41
ACALEP	13691	5.53	ACAAUL	13688	8.02	ACAAUL	13689	6.06
ACAHOL	14660	5.52	ACACRA	13863	7.99	ACAAUL	13688	5.70
CASCUN	13514	5.23	ACALEP	13653	7.96	EUCPEL	12013	5.66
ACALEP	13653	5.21	ACAAUL	13689	7.42	ACAMAN	13846	5.21
ACAAUR	13684	5.20	ACADIF	14623	6.94	ACAMAN	13621	4.53
CASCUN	13519	5.00	ACAHOL	14660	6.86	ACAFLA	14175	4.21
EUCTOR	14130	4.90	ACAFLA	14175	6.24	CASCUN	13519	4.20
ACAMAN	13846	4.73	ACACIN	13864	5.71	ACAHOL	14660	4.16
ACASHI	14622	4.67	ACAAUL	13877	5.03	EUCTOR	14130	4.14
ACAAUR	13861	4.65	ACAMAN	13621	4.97	ACAAUL	13866	3.99
AZAIND	L8514	4.59	ACAMAN	13846	4.83	ACASHI	14622	3.92
ACAAUL	13688	4.39	EUCTOR	14130	4.75	MELDEA	11935	3.63
ACAFLA	14175	4.36	ALLLIT	13876	4.44	ACACIN	13864	3.53
ACAAUR	13854	4.28	CASCUN	13519	4.34	MELSYM	14170	3.53
ACAAUL	13689	4.24	ACAAUL	13866	4.13	MELDEA	14166	3.38
ACAMAN	13621	4.20	ACAPOL	13871	3.92	CASCUN	13514	3.30
ACACRA	13863	4.16	CASCUN	13514	3.50	ACAAUL	13877	3.17
CASEQU	13990	4.06	MELDEA	11935	3.18	ALLLIT	13876	3.11
PELDAS	L8512	3.87	ALSMAC	L8521	2.84	MELVIR	14152	2.88
CASCUN	13148	3.83	CASCUN	13148	2.68	PELPTE	L8531	2.64
PTEIND	L8513	3.57	CASEQU	13990	2.63	ACAPOL	13871	2.61
ACAAUL	13866	3.48	ACAMEL	14176	2.59	CASEQU	13990	2.52
ACAAUL	13877	3.35				CASCUN	13148	2.21
ACACIN	13864	3.25				ACAMEL	14176	1.58
ACAPOL	13871	2.56				MELBRA	14485	0.97
MELSYM	14170	2.49						
MELDEA	11935	2.18						
MELBRA	14485	2.16						
MELDEA	14166	2.09						

Sakaerat Seedlot no.		Height	Chanthaburi Seedlot no.		Height	Huai Bong Seedlot no.		Height
ACACRA	13683	6.58	EUCCAM	14537	5.69	EUCCAM	14106	3.97
EUCCAM	14537	6.32	EUCCAM	14106	5.55	ACAAUR	13861	3.45
ACACRA	13681	6.10	MLAAZE	L8552	5.24	EUCCAM	14537	3.36
EUCCAM	14106	5.75	EUCTOR	14130	4.15	EUCPEL	12013	3.27
ACAAUR	13684	5.38	ACAAUR	13854	3.91	ACAAUR	13854	3.15
ACAAUR	13854	5.27	EUCPEL	12013	3.82	ACAAUR	L8562	3.12
ACAAUR	13861	5.23	ACAAUR	13861	3.79	ACAAUR	13684	2.93
ACAAUL	13689	4.63	ACAAUR	13684	3.64	ACACRA	13681	2.72
ACACRA	13863	4.53	ACAAUL	13688	3.63	ACAAUL	13689	2.54
ACACIN	13864	4.40	ACAHOL	14660	3.36	ACAHOL	14660	2.50
ACAHOL	14660	4.33	ACALEP	13653	3.06	EUCTOR	14130	2.37
ACALEP	13653	3.92	ACAAUR	L8556	2.98	ACAAUL	13688	2.00
ACAAUL	13688	3.75	ACAAUL	13689	2.88	ACAAUL	13866	1.96
CASCUN	13519	3.34	ACACRA	13683	2.85	CASEQU	L8561	1.91
EUCPEL	12013	3.32	ACALEP	13691	2.49	ACALEP	13653	1.89
EUCTOR	14130	3.10	PELDAS	L8553	2.49	CASCUN	13519	1.76
MLAAZE	L8542	3.07	ACACRA	13680	2.48	ACALEP	13691	1.67
ACAAUL	13877	3.04	ACAAUL	13866	2.05	CASCUN	13148	1.58
ACAMAN	13846	2.83	ACAMAN	13621	1.87	PINKES	L8563	1.51
CASCUN	13514	2.73	CASCUN	13519	1.85	MELSYM	14170	1.42
ACAFLA	14175	2.71	CASEQU	13990	1.71	ALLLIT	13876	1.36
ACAAUL	13866	2.51	ACACIN	13864	1.66	ACAMEL	14176	1.34
MELSYM	14170	2.35	ACAAUL	13877	1.63	ACAAUL	13877	1.24
ALLLIT	13876	2.26	CASCUN	13514	1.43	CASCUN	13514	1.23
CASCUN	13148	2.24	ACAMAN	13846	1.43	MELDEA	14166	1.19
CASEQU	13990	2.06	ACAPOL	13871	1.41	MELVIR	14152	1.13
ACASHI	14622	2.05	ACAFLA	14175	1.37	ACAMAN	13621	1.02
MELDEA	11935	1.91	CSASLA	L8555	1.34	MELDEA	11935	1.00
ACAPOL	13871	1.77	CASCUN	13148	1.32	ACAPOL	13871	0.96
MELDEA	14166	1.74	ALLLIT	13876	0.84			
ACAMAN	13621	1.66	MELBRA	14485	0.80			
ACAMEL	14176	1.61	MELDEA	11935	0.76			
MELVIR	14152	0.90	MELVIR	14152	0.61			
MELBRA	14485	0.63						

Table 3. Ranking for mean diameter at breast height (cm) at 24 months of field trials planted in 1985 in Thailand. For each planting site vertical lines group treatments that are not significantly different (P = 0.05)

Ratchaburi Seedlot no.		D.B.H.	Sai Thong Seedlot no.		D.B.H.	Si Sa Ket Seedlot no.		D.B.H.
EUCCAM	14537	5.96	ACACRA	13683	10.30	EUCCAM	14537	7.70
EUCCAM	14106	5.94	ACACRA	13680	9.92	ECACRA	13681	7.10
EUCPEL	12013	5.70	ACAAUR	13684	8.61	EUCCAM	14106	7.08
ACACRA	13683	5.43	ACACRA	13681	8.55	ACACRA	13863	6.88
EUCCIT	L8511	5.22	ACAAUR	13686	7.94	ACACRA	13683	6.79
CSASIA	L8515	5.00	EUCPEL	12013	7.75	ACACRA	13680	6.71
ACACRA	13681	4.95	ACACRA	13863	7.51	ACALEP	13653	6.49
ACACRA	13680	4.95	ACAAUL	13689	7.47	ACAAUR	13854	6.26
AZAIND	L8514	4.91	ACAAUR	13854	7.39	ACALEP	13691	6.23
ACALEP	13691	4.83	ACAAUL	13688	7.28	EUCPEL	12013	6.17
CASJUN	L8516	4.79	ACALEP	13691	7.28	ACAAUR	13684	6.10
ACAAUR	13686	4.76	EUCCAM	14106	7.25	ACAAUR	13861	5.92
ACAMAN	13846	4.76	ACAAUR	13861	6.94	ACAAUL	13689	4.98
ACAAUR	13684	4.51	ACALEP	13653	6.94	ACAAUL	13688	4.84
ACALEP	13653	4.34	EUCCAM	14537	6.47	EUCTOR	14130	4.84
ACADIF	14623	4.28	MELBRA	14485	5.48	ACAMAN	13846	4.64
EUCTOR	14130	4.22	ACADIF	14623	5.44	ACAFLA	14175	4.43
ACAAUL	13689	3.82	ACAHOL	14660	5.31	MELDEA	11935	4.05
CASCUN	13514	3.79	ACAFLA	14175	5.16	ACAMAN	13621	3.86
ACAMAN	13621	3.76	ACAMAN	13846	5.00	MELDEA	14166	3.73
ACAFLA	14175	3.67	EUCTOR	14130	4.93	ACAHOL	14660	3.65
ACAAUL	13688	3.63	ACACIN	13864	4.05	ACASHI	14622	3.44
ACAAUR	13861	3.58	ACAMAN	13621	3.77	ACAAUL	13866	3.25
ACACRA	13863	3.57	MELDEA	11935	3.76	MELSYM	14170	3.18
CASCUN	13519	3.55	ACAAUL	13877	3.51	MELVIR	14152	3.10
ACAHOL	14660	3.53	ACAAUL	13866	3.42	ACACIM	13864	2.88
PELDAS	L8512	3.42	ACAPOL	13871	3.05	CASCUN	13519	2.85
ACASHI	14622	3.18	ALSMAC	L8521	2.98	PELPTE	L8531	2.68
PTEIND	L8513	3.13	ALLLIT	13876	2.81	ALLLIT	13876	2.46
ACAAUR	13854	3.10	CASCUN	13519	2.80	ACCAUL	13877	2.39
CASCUN	13148	2.69	ACAMEL	14176	1.98	CASCUN	13514	2.27
ACAAUL	13866	2.41	CASCUN	13148	1.68	ACAPOL	13871	2.09
ACAAUL	13877	2.25	CASEQU	13990	1.56	CASEQU	13990	1.55
CASEQU	13990	2.21				CASCUN	13148	1.34
ACACIN	13864	2.15						
ACAPOL	13871	1.39						
MELDEA	11935	1.37						
MELSYM	14170	1.24						
MELDEA	14166	1.24						
MELBRA	14485	0.78						

Sakaerat Seedlot no.		D.B.H.	Chanthaburi Seedlot no.		D.B.H.	Huai Bong Seedlot no.		D.B.H.
ACACRA	13683	5.59	MLAAZE	L8552	4.94	EUCPEL	12013	3.56
ACAAUR	13684	5.39	EUCTOR	14130	4.53	EUCCAM	14106	3.56
EUCCAM	14537	5.05	EUCCAM	14537	4.49	ACAAUR	13861	3.38
ACACRA	13681	5.02	EUCCAM	14106	4.22	EUCCAM	14537	3.27
ACAAUR	13854	5.00	EUCPEL	12013	3.52	ACAAUR	13684	2.95
ACAAUR	13861	4.63	ACAAUR	13684	3.35	ACAAUR	13854	2.70
EUCCAM	14106	4.58	ACAAUR	13854	3.32	ACACRA	13681	2.57
ACACIM	13864	4.27	ACAAUL	13688	2.96	EUCTOR	14130	2.36
ACAAUL	13689	3.87	ACAAUR	13861	2.92	ACAAUR	L8562	2.28
ACACRA	13863	3.66	PELDAS	L8553	2.61	ACAAUL	13689	2.00
EUCPEL	12013	3.43	ACAAUL	13689	2.47	ACAHOL	14660	1.60
ACAHOL	14660	3.24	ACACRA	13683	2.24	PINKES	L8563	1.33
EUCTOR	14130	3.10	ACAAUR	L8556	2.19	ACALEP	13653	1.12
ACAAUL	13688	2.95	ACAHOL	14660	2.08	ACAAUL	13688	1.12
ACALEP	13653	2.94	ACALEP	13653	2.05	CASEQU	L8561	1.06
ACAMAN	13846	2.48	ACACRA	13680	1.81	ACALEP	13691	0.81
MLAAZE	L8542	2.42	ACALEP	13691	1.63	ACAAUL	13886	0.76
ACAFLA	14175	2.10	CSASIA	L8555	1.39	ACAMAN	13621	0.67
ACAAUL	13877	1.93	ACAMAN	13621	1.30	CASCUN	13148	0.65
CASCUN	13519	1.73	ACAAUL	13866	1.07	MELDEA	14166	0.64
AZAIND	L8541	1.63	ACACIN	13864	1.00	CASCUN	13519	0.60
ACAAUL	13866	1.47	ACAPOL	13871	0.98	MELDEA	11935	0.55
CASCUN	13514	1.44	ACAMAN	13846	0.97	CASCUN	13514	0.54
MELDEA	14166	1.34	CASCUN	13519	0.86	ALLLIT	13876	0.52
MELDEA	11935	1.32	CASEQU	13990	0.84	ACAMEL	14176	0.50
ALLLIT	13876	1.25	ACAFLA	14175	0.81	MELVIR	14152	0.49
ACAMAN	13621	1.24	ACAAUL	13877	0.78	ACAAUL	13877	0.45
MELSYM	14170	1.10	CASCUN	13148	0.62	MELSYM	14170	0.39
CASCUN	13148	1.09	CASCUN	13514	0.60	ACAPOL	13871	0.28
ACAPOL	13871	0.85	MELDEA	11935	0.50			
ACAMEL	14176	0.84	ALLLIT	13876	0.18			
CASEQU	13990	0.82	MELBRA	14485	0.01			
ACASHI	14622	0.77						
MELVIR	14152	0.71						

Table 4. Ranking for mean survival (arcsine transformed) at 24 months of field trials planted in 1985 in Thailand. For each planting site vertical lines group treatments that are not significantly different (P = 0.05).

Ratchaburi			Sai Thong			Si Sa Ket		
Seedlot no.		Survival	Seedlot no.		Survival	Seedlot no.		Survival
EUCCAM	14106	90.00	ACAAUR	13854	90.00	ACAAUR	13684	90.00
EUCCAM	14537	90.00	EUCCAM	14537	87.90	ACAAUR	13854	90.00
PELDAS	L8512	90.00	ACAAUR	13684	86.10	ACAAUR	13861	90.00
CSASIA	L8515	90.00	ACACRA	13680	85.70	ACAFLA	14175	90.00
ACALEP	13653	86.20	ACAAUR	13861	84.50	PELPTE	L8531	90.00
ACACRA	13680	86.20	EUCPEL	12013	82.20	MELDEA	11935	90.00
ACAAUL	13866	86.20	ACAAUL	13688	80.70	MELSYM	14170	90.00
ACAPOL	13871	86.20	EUCCAM	14106	79.50	CASCUN	13514	90.00
EUCTOR	14130	86.20	ACAAUL	13866	77.80	CASCUN	13519	90.00
CASJUN	L8516	86.20	ACAMAN	13846	75.70	EUCCAM	14106	90.00
ACAAUR	13854	84.50	CASCUN	13519	75.60	EUCCAM	14537	90.00
PTEIND	L8513	83.20	EUCTOR	14130	75.50	ACAMAN	13621	86.20
ACACRA	13681	82.30	ACACRA	13863	73.10	ACAAUL	13689	86.20
AZAIND	L8514	82.30	ACADIF	14623	73.00	ACALEP	13691	86.20
CASCUN	13514	80.70	ALSMAC	L8521	69.40	ACAAUL	13866	86.20
CASCUN	13519	79.40	ACALEP	13653	68.70	EUCPEL	12013	86.20
ACAAUR	13861	78.30	ACAPOL	13871	67.70	EUCTOR	14130	86.20
ACAAUR	13684	75.70	ACACRA	13681	66.80	ACACRA	13680	84.50
ACAHOL	14660	75.50	ACACRA	13683	64.00	ACACRA	13863	84.50
MELDEA	11935	75.20	CASEQU	13990	63.80	MELVIR	14152	84.50
ACACRA	13683	75.20	ACAHOL	14660	63.40	MELDEA	14166	84.50
MELDEA	14166	72.30	MELDEA	11935	62.90	ACAPOL	13871	83.20
ACAAUR	13686	71.80	ACALEP	13691	62.50	ACAAUL	13688	82.30
ACAAUL	13689	71.70	MELBRA	14485	58.30	ACAMAN	13846	82.30
EUCPEL	12013	68.90	ACAAUL	13689	57.70	ACACRA	13683	82.30
MELBRA	14485	67.10	ACAAUR	13686	57.20	ACAAUL	13877	82.30
ACAAUL	13688	66.60	ALLLIT	13876	56.10	ACASHI	14622	80.70
EUCCIT	L8511	65.60	CASCUN	13148	54.80	ACALEP	13653	78.50
ACAFLA	14175	65.30	ACAMAN	13621	54.10	ACACRA	13681	77.30
ACAMAN	13846	62.00	CASCUN	13514	52.30	ACACIN	13864	75.60
CASCUN	13148	61.50	ACACIN	13864	51.40	ALLLIT	13876	72.80
MELSYM	14170	60.00	ACAAUL	13877	47.40	MELBRA	14485	62.50
ACACRA	13863	55.00	ACAFLA	14175	45.10	CASCUN	13148	60.00
ACAMAN	13621	54.60	ACAMEL	14176	43.90	CASEQU	13990	59.00
ACALEP	13691	50.20				ACAHOL	14660	43.70
ACADIF	14623	49.40				ACAMEL	14176	19.50
ACAAUL	13877	48.70						
ACACIN	13864	36.90						
CASEQU	13990	28.70						
ACASHI	14622	27.40						

Sakaerat Seedlot no.		Survival	Chanthaburi Seedlot no.		Survival	Huai Bong Seedlot no.		Survival
ACAAUR	13861	90.00	ACAAUR	13861	90.00	ACAAUR	13861	90.00
ACAAUR	13854	90.00	MLAAZE	L8552	90.00	EUCPEL	12013	90.00
ACAAUR	13684	90.00	ACAAUR	L8556	90.00	PINKES	L8563	86.10
CASCUN	13514	82.10	ACAAUR	13854	86.10	ACAAUR	13854	86.10
ACAAUL	13688	82.10	EUCCAM	14106	86.10	EUCCAM	14106	79.00
CASCUN	13519	80.70	CSASIA	L8555	79.30	CASEQU	L8561	78.40
EUCTOR	14130	78.50	EUCCAM	14537	76.80	ACAAUR	13684	78.20
EUCCAM	14106	78.50	PELDAS	L8553	75.20	EUCCAM	14537	78.20
EUCCAM	14537	78.30	EUCTOR	14130	72.60	EUCTOR	14130	75.30
ACAAUL	13877	77.80	ACALEP	13653	71.50	ACAAUR	L8562	75.20
ALLLIT	13876	76.70	ACAHOL	14660	68.60	ACAAUL	13866	73.90
ACACRA	13863	76.70	ACAPOL	13871	68.40	CASCUN	13519	71.10
ACACRA	13683	75.70	EUCPEL	12013	67.90	MELDEA	14166	70.90
ACAMAN	13846	75.40	ACALEP	13691	66.60	MELSYM	14170	69.60
MELDEA	14166	75.20	CASCUN	13519	65.80	MELBRA	14485	68.90
ACAAUL	13689	73.50	CASCUN	13148	57.40	ACAAUL	13689	68.60
ACALEP	13653	73.50	CASCUN	13514	55.50	ACAAUL	13688	68.50
MLAAZE	L8542	70.90	ACAAUL	13688	53.30	MELVIR	14152	63.70
MELDEA	11935	70.40	ACAAUL	13866	53.30	MELDEA	11935	62.70
ACACRA	13681	70.00	ACACRA	13683	52.40	CASCUN	13514	62.60
MELBRA	14485	69.30	ACAAUL	13689	51.60	ALLLIT	13876	60.00
ACAPOL	13871	69.30	ACAMAN	13621	51.40	ACAPOL	13871	59.80
ACAHOL	14660	67.90	MELDEA	11935	45.30	ACALEP	13691	58.90
ACACIN	13864	67.70	ACACIN	13864	44.20	CASCUN	13148	56.80
MELSYM	14170	65.20	ACAFLA	14175	43.80	ACAAUL	13877	52.50
CASEQU	13990	59.90	ACAMAN	13846	43.70	ACALEP	13653	46.90
ACAAUL	13866	58.50	ACACRA	13680	41.40	ACACRA	13681	45.30
ACAMAN	13621	58.20	ALLLIT	13876	34.10	ACAMEL	14176	45.30
MELVIR	14152	55.00	ACAAUL	13877	33.20	ACAMAN	13621	29.70
EUCPEL	12013	55.00	MELBRA	14485	32.20	ACAHOL	14660	23.10
CASCUN	13148	54.50	ACAAUR	13684	32.20			
ACAFLA	14175	50.60	CASEQU	13990	25.00			
ACAMEL	14176	50.00	MELVIR	14152	24.20			
ACASHI	14622	38.20	ACAMEL	14176	5.40			

Table 5. Ranking for mean height (m) at 12 months of field trials planted in 1986 in Thailand. For each planting site vertical lines group treatments that are not significantly different (P = 0.05)

Ratchaburi Seedlot no.	Height	Si Sa Ket Seedlot no.	Height	Huai Bong Seedlot no.	Height	Ban Hong Seedlot no.	Height
EUCCAM 14338	4.74	EUCTER 14108	4.26	EUCCIT 14852	2.18	EUCTER 14108	2.49
EUCEXS 14864	4.58	EUCCAM L8633	4.25	EUCCAM 14338	2.02	EUCCAM 14338	2.31
EUCTER 14108	4.39	EUCCAM 14338	4.21	EUCTER 14108	1.97	EUCCAM 13692	2.21
EUCBRA 13397	4.25	ACATOR 14888	3.79	EUCCAM 13692	1.87	EUCEXS 14864	2.14
EUCCIT 14852	4.23	EUCBRA 13397	3.75	EUCTER 14212	1.65	ACATOR 14183	1.93
EUCCAM 13692	4.20	ACATOR 14183	3.68	EUCBRA 13397	1.48	ACATOR 14888	1.89
ACATOR 14888	4.09	EUCCAM 13692	3.63	EUCEXS 14864	1.47	EUCCIT 14852	1.87
EUCTER 14212	4.02	ACAJUL 14974	3.35	EUCPUN 10863	1.35	EUCTER 14212	1.84
ACAJUL 14885	3.37	EUCCIT 14852	3.32	MLAAZE 14500	1.12	ACAJUL 14885	1.79
EUCRAV 13546	3.27	ACAJUL 14885	3.26	ACAJUL 14885	1.11	EUCPUN 10863	1.78
ACATOR 14183	3.24	EUCTER 14212	2.97	ACASIM 14553	1.06	ACASIM 14576	1.70
ACAJUL 14974	3.05	EUCEXS 14864	2.90	ACAPLA 14960	1.05	EUCBRA 13397	1.63
ACAPLA 14960	3.01	EUCRAV 13546	2.81	ACAJUL 14974	1.01	ACAHAR 15100	1.52
ACAPLA 14542	2.93	ACAPLA 14960	2.78	ACASIM 14576	0.99	GREPTE 14905	1.46
GREPTE 14905	2.85	ACABRA 14965	2.56	EUCRAV 13546	0.89	EUCRAV 13546	1.45
EUCBIG 11465	2.84	GREPTE 14905	2.52	ACATOR 14183	0.89	AZAIND L8671	1.26
ACABRA 14965	2.80	CSASIA L8632	2.44	GREPTE 14905	0.89	EUCBIG 11465	1.16
EUCPUN 10863	2.70	ACAPLA 14542	2.33	ACAFAL 14981	0.83	MELARG 14899	1.14
MLAAZE 14500	2.63	MLAAZE 14500	2.30	LEPFLA 14502	0.86	MELARG 14904	1.13
EUCSUF 13598	2.47	ACASIM 14576	2.16	ACABRA 14965	0.73	LEPLON 14900	1.10
ACAROT 14967	2.35	EUCBIG 11465	2.10	LEPLON 14900	0.68	EUCHOU 9091	1.08
ACASIM 14553	2.34	ACASIM 14553	2.10	MELARG 14904	0.62	MELSAL 14871	1.05
AZAIND L8612	2.34	EUCPUN 10863	2.01	LEPFLA 14554	0.56	ACAPIA 14960	0.98
CSASIA L8611	2.27	EUCDEG L8634	1.99	ACATOR 14888	0.55	MLAAZE 14500	0.91
MELSAL 14871	2.23	MELSAL 14871	1.62	PETPUB 14189	0.54	EUCSUF 13598	0.83
ACASIM 14576	2.10	EUCHOU 9091	1.58	EUCBIG 11465	0.50	LEPFLA 14502	0.79
DALSIS L8614	2.09	MLAAZE 14501	1.51	MELSAL 14871	0.50	ACAFAL 14981	0.79
MELARG 14889	2.00	ADEABR 14557	1.36	EUCHOU 9091	0.49	DIPALA L8672	0.78
MELARG 14904	1.98	MELARG 14904	1.33	CALINT 14504	0.45	ACAAUR L8673	0.73
EUCCLO 10691	1.91	GREPAR 14143	1.32	MELARG 14899	0.45	MELSTE 14148	0.70
MLAAZE 14501	1.84	LEPLON 14900	1.32	GREPIN 14980	0.45	MELBRA 14903	0.62
ADEPAV L8615	1.82	ACAFAL 14981	1.27	EUCSUF 13598	0.34	NEOMYR 14889	0.43
EUCHOU 9091	1.79	NEOMYR 14896	1.26	PETNUM 14639	0.23	ACABRA 14965	0.39
GREPAR 14143	1.79	MELARG 14899	1.25	MELSTE 14148	0.21	MELBRA 14982	0.36
PELDAS L8613	1.78	EUCSUF 13598	1.10	ACAPLA 14542	0.19		
ATAHEM 14976	1.69	MELSTE 14148	1.08	NEOMYR 14889	0.19		
PETPUB 14189	1.60	PETPUB 14189	1.05	MELACA 14873	0.15		
GREPIN 14980	1.60	DESUMB 14556	1.03	GREPAR 14143	0.14		
LEPLON 14900	1.52	ADEABR 14180	0.92	CSABRE 14188	0.10		
MELSTE 14148	1.45	PETNUM 14639	0.84	ACAHAR 15100	0.09		
NEOMYR 14896	1.37	GREPIN 14980	0.83				
MELBRA 14982	1.36	NEOMYR 14889	0.82				
ACAHEM 14657	1.33	CSABRE 14188	0.59				
ACAFAL 14981	1.31	ACAHAR 15100	0.43				
ACABID 14958	1.28	MELACA 14873	0.41				
PETNUM 14639	1.22						
LEPFLA 14502	1.14						
DESUMB 14556	0.96						
NEOMYR 14889	0.95						
XANUMB 14874	0.81						
MELBRA 14903	0.80						
MELACA 14873	0.76						
ACAHAR 15100	0.62						
CSABRE 14188	0.43						

Sai Thong Seedlot no.	Height	Sakaerat Seedlot no.	Height	Chanthaburi Seedlot no.	Height
EUCCAM 13692	6.48	EUCCAM 14338	3.19	EUCCAM 14338	4.08
EUCTER 14212	6.48	EUCURO 14532	2.99	EUCTER 14108	3.78
EUCURO 14532	6.22	EUCCAM 13692	2.90	EUCCAM 13692	3.59
EUCCAM 14338	6.07	EUCTER 14108	2.66	EUCRAV 13546	3.36
EUCURO 14534	5.70	EUCTER 14212	2.58	EUCTER 14212	3.36
EUCCRA 14431	5.41	EUCURO 14534	2.44	EUCPUN 10863	3.23
EUCRAV 13546	5.10	EUCRAV 13546	2.37	EUCCRA 14431	2.94
EUCTER 14108	4.90	EUCCIT 14852	2.13	EUCBRA 13397	2.21
EUCCRA 14700	4.83	EUCPUN 10863	1.74	EUCDEG 13461	2.07
EUCSAL 15011	4.59	ACAFLA 13872	1.67	EUCEXS 14864	2.03
EUCPUN 10863	4.52	EUCPAN 14442	1.58	EUCDUN 13329	1.93
EUCPYR 10857	4.38	ACAORA 14961	1.57	ACACAT L8653	1.85
EUCDEG 13461	4.31	EUCCRA 14700	1.49	EUCCLO 10691	1.64
EUCDUN 13329	3.95	EUCCRA 14431	1.41	ACAFLA 13872	1.64
ACAAUR L8622	3.81	MELLEU 14147	1.34	ACAORA 14961	1.59
ACAFLA 13872	3.51	ACAHYL 14977	1.33	EUCPAN 14442	1.56
ACAFLA 14968	3.39	ALBPRO 14962	1.23	PELDAS L8652	1.47
EUCHOU 9091	3.36	ACAORA 14886	1.12	ACAORA 14886	1.28
EUCPAN 14442	3.24	EUCMIC 13973	1.10	PAKJAV L8651	1.19
MELLEU 14147	3.22	EUCCLO 10691	1.08	EUCHOU 9091	1.18
EUCMIC 13973	3.21	MELCAJ 14878	1.04	ACAFLA 14968	1.18
MELCAJ 14878	2.73	EUCHOU 9091	1.01	ALBPRO 14959	1.08
ALBPRO 14962	2.47	ALBPRO 14959	0.88	MELCAJ 14878	0.90
TABROS L8624	2.34	PTEMAC L8641	0.79	MELLEU 14147	0.53
ACAORA 14961	2.13	EUCDEG 13461	0.78		
ALSMAC L8621	1.97	ACAFLA 14968	0.55		
ALBPRO 14959	1.79				
ACAORA 14886	1.71				
CASEQU L8623	0.80				

Table 6. Ranking for mean diameter at ground level (cm) at 12 months of field trials planted in 1986 in Thailand. For each planting site vertical lines group treatments that are not significantly different (P = 0.05).

Ratchaburi		Si Sa Ket		Huai Bong		Ban Hong	
Seedlot no.	D.G.L.	Seedlot no.	D.G.L.	Seedlot no.	D.G.L.	Seedlot no.	D.G.L.
EUCCAM 14338	6.26	EUCCAM L8633	5.75	EUCCAM 14338	2.75	EUCTER 14212	2.99
EUCTER 14212	5.04	EUCCAM 13692	5.39	EUCCIT 14852	2.65	EUCCAM 14338	2.76
EUCCAM 13692	4.78	EUCCAM 14338	5.37	EUCTER 14212	2.56	EUCTER 14108	2.67
MLAAZE 14500	4.70	ACATOR 14888	5.27	EUCCAM 13692	2.55	EUCPUN 10863	2.62
EUCTER 14108	4.65	GREPTE 14905	5.27	EUCPUN 10863	2.52	EUCCAM 13692	2.55
ACAJUL 14885	4.56	EUCTER 14212	5.00	EUCTER 14108	2.47	ACAAUR L8673	2.46
ACAPLA 14542	4.47	EUCTER 14108	4.92	MLAAZE 14500	2.29	GREPTE 14905	2.36
AZAIND L8612	4.45	ACAPLA 14960	4.78	GREPTE 14905	1.95	EUCRAV 13546	2.30
ACAPLA 14960	4.42	CSASIA L8632	4.77	EUCBRA 13397	1.93	EUCEXS 14864	2.23
ACABRA 14965	4.36	ACATOR 14183	4.69	EUCEXS 14864	1.81	ACAJUL 14885	2.02
EUCEXS 14864	4.30	ACABRA 14965	4.64	EUCRAV 13546	1.62	ACASIM 14576	1.99
GREPTE 14905	4.30	EUCRAV 13546	4.55	ACAPLA 14960	1.58	ACATOR 14183	1.91
EUCRAV 13546	4.27	ACAJUL 14885	4.37	ACAJUL 14885	1.34	EUCBRA 13397	1.86
CSASIA L8611	4.26	EUCBRA 13397	4.33	ACASIM 14553	1.25	ACATOR 14888	1.84
ACATOR 14888	4.24	MLAAZE 14500	4.08	EUCHOU 9091	1.21	ACABRA 14965	1.82
EUCBRA 13397	4.23	EUCDEG L8634	4.06	EUCBIG 11465	1.21	EUCBIG 11465	1.71
ACAJUL 14974	3.91	ACAJUL 14974	3.90	ACASIM 14576	1.21	EUCHOU 9091	1.70
EUCBIG 11465	3.77	ACASIM 14576	3.76	ACATOR 14183	1.20	MLAAZE 14500	1.70
EUCPUN 10863	3.72	ACASIM 14553	3.55	ACAJUL 14974	1.20	MELARG 14899	1.67
EUCCIT 14852	3.72	EUCPUN 10863	3.54	ACABRA 14965	1.17	EUCCIT 14852	1.58
ACATOR 14183	3.66	EUCCIT 14852	3.46	LEPFLA 14502	1.12	AZAIND L8671	1.58
MLAAZE 14501	3.42	EUCBIG 11465	3.13	CALINT 14504	1.04	ACAPLA 14960	1.42
PELDAS L8613	3.41	ACAPLA 14542	3.02	LEPLON 14900	0.99	LEPLON 14900	1.37
GREPAR 14143	2.98	EUCEXS 14864	2.90	ACAFAL 14981	0.98	MELSAL 14871	1.31
ACASIM 14553	2.97	EUCHOU 9091	2.83	ACATOR 14888	0.89	MELARG 14904	1.21
EUCHOU 9091	2.88	GREPAR 14143	2.73	MELARG 14899	0.86	MELBRA 14982	1.04
EUCSUF 13598	2.79	DESUMB 14556	2.48	PETPUB 14189	0.79	EUCSUF 13598	1.01
MELARG 14899	2.77	MLAAZE 14501	2.46	MELARG 14904	0.78	MELSTE 14148	0.90
EUCCLO 10691	2.76	MELSAL 14871	2.37	GREPIN 14980	0.78	LEPFLA 14502	0.76
MELARG 14904	2.71	LEPLON 14900	2.02	LEPFLA 14554	0.69	DIPALA L8672	0.75
DALSIS L8614	2.62	MELARG 14899	2.00	MELSAL 14871	0.61	ACAFAL 14981	0.71
ADEPAV L8615	2.60	MELARG 14904	1.90	EUCSUF 13598	0.47	MELBRA 14903	0.69
ACASIM 14576	2.57	MELSTE 14148	1.69	GREPAR 14143	0.44	ACAHAR 15100	0.57
ACAROT 14967	2.53	ACAFAL 14981	1.61	ACAPLA 14542	0.41	NEOMYR 14889	0.25
MELSAL 14871	2.47	PETPUB 14189	1.61	PETNUM 14639	0.41		
LEPLON 14900	2.34	NEOMYR 14896	1.59	MELSTE 14148	0.36		
DESUMB 14556	2.22	ADEABR 14557	1.58	CSABRE 14188	0.33		
ATAHEM 14976	2.11	EUCSUF 13598	1.32	ACAHAR 15100	0.33		
ACABID 14958	1.97	ADEABR 14180	1.17	MELACA 14873	0.17		
MELSTE 14148	1.96	PETNUM 14639	1.17	NEOMYR 14889	0.16		
MELBRA 14982	1.95	CSABRE 14188	1.13				
PETPUB 14189	1.89	GREPIN 14980	1.08				
ACAFAL 14981	1.84	ACAHAR 15100	0.72				
GREPIN 14980	1.68	NEOMYR 14889	0.71				
NEOMYR 14896	1.63	MELACA 14873	0.63				
PETNUM 14639	1.56						
ACAHEM 14657	1.42						
LEPFLA 14502	1.36						
XANUMB 14874	1.33						
CSABRE 14188	1.15						
MELBRA 14903	1.13						
ACAHAR 15100	1.05						
NEOMYR 14889	0.72						
MELACA 14873	0.66						

Sai Thong Seedlot no.	D.G.L.	Sakaerat Seedlot no.	D.G.L.	Chanthaburi Seedlot no.	D.G.L.
EUCURO 14534	7.37	EUCURO 14532	4.26	EUCCAM 14338	4.89
EUCURO 14532	6.97	EUCURO 14534	3.93	EUCTER 14212	4.83
EUCCAM 13692	6.89	EUCCAM 14338	3.66	EUCRAV 13546	4.70
EUCTER 14212	6.82	EUCTER 14212	3.41	EUCCAM 13692	4.50
EUCRAV 13546	6.54	EUCCAM 13692	3.25	EUCPUN 10863	4.31
EUCCAM 14338	6.44	EUCRAV 13546	3.10	EUCTER 14108	4.13
EUCDEG 13461	6.20	EUCTER 14108	2.65	EUCCRA 14431	4.05
EUCCRA 14431	6.15	EUCPUN 10863	2.45	PELDAS L8652	3.83
EUCTER 14108	6.10	ACAORA 14961	2.43	EUCDEG 13461	3.35
EUCSAL 15011	5.97	ALBPRO 14962	2.25	ACAORA 14961	2.90
EUCCRA 14700	5.71	ACAFLA 13872	2.02	ACACAT L8653	2.56
EUCPUN 10863	5.54	EUCCRA 14431	1.90	EUCDUN 13329	2.49
ACAFLA 13872	5.41	EUCCIT 14852	1.86	ACAORA 14886	2.40
ACAFLA 14968	5.38	EUCCRA 14700	1.83	EUCBRA 13397	2.36
MELLEU 14147	5.19	EUCPAN 14442	1.75	EUCCLO 10691	2.26
ACAAUR L8622	5.12	MELLEU 14147	1.71	ALBPRO 14959	2.23
EUCPYR 10857	5.11	ACAORA 14886	1.64	PAKJAV L8651	2.18
EUCDUN 13329	5.08	ALBPRO 14959	1.60	EUCEXS 14864	2.05
EUCHOU 9091	4.92	EUCHOU 9091	1.56	EUCPAN 14442	2.04
ALBPRO 14962	4.83	PTEMAC L8641	1.37	ACAFLA 13872	2.03
MELCAJ 14878	4.13	EUCCLO 10691	1.17	ACAFLA 14968	1.96
EUCPAN 14442	3.84	EUCMIC 13973	1.16	EUCHOU 9091	1.95
ACAORA 14961	3.84	MELCAJ 14878	1.15	MELCAJ 14878	1.51
EUCMIC 13973	3.72	ACAHYL 14977	1.15	MELLEU 14147	1.21
ACAORA 14886	3.66	EUCDEG 13461	1.06		
ALBPRO 14959	3.55	ACAFLA 14968	0.76		
ALSMAC L8621	3.25				
TABROS L8624	2.03				
CASEQU L8623	1.66				

Table 7. Ranking for mean survival (arcsine % transformed) at 12 months of field trials planted in 1986 in Thailand. For each planting site vertical lines group treatments that are not significantly different (P = 0.05).

Ratchaburi

Seedlot	no.	Survival
ACABRA	14965	90.00
EUCTER	14108	90.00
MELARG	14899	90.00
AZAIND	L8612	90.00
CSASIA	L8611	90.00
EUCTER	14212	90.00
MELARG	14904	90.00
MLAAZE	14500	86.20
GREPTE	14905	86.20
EUCCAM	14338	86.20
PETPUB	14189	86.20
EUCBIG	11465	86.20
MLAAZE	14501	86.20
ATAHEM	14976	85.50
PELDAS	L8613	84.50
MELSAL	14871	83.20
MELBRA	14982	83.20
LEPLON	14900	82.30
EUCRAV	13546	82.30
EUCCAM	13692	82.30
EUCSUF	13598	80.70
EUCEXS	14864	78.50
EUCBRA	13397	78.50
PETNUM	14639	77.80
NEOMYR	14896	76.70
ADEPAV	L8615	76.40
GREPAR	14143	75.40
NEOMYR	14889	75.20
EUCPUN	10863	74.70
ACAROT	14967	74.40
GREPIN	14980	73.60
MELSTE	14148	73.60
DESUMB	14556	72.80
MELBRA	14903	71.80
ACASIM	14576	71.80
EUCCIT	14852	70.50
DALSIS	L8614	70.20
EUCHOU	9091	69.90
ACATOR	14888	69.40
ACAJUL	14885	68.70
ACABID	14958	66.00
XANUMB	14874	65.40
ACAPLA	14960	65.00
MELACA	14873	60.90
ACASIM	14553	54.90
ACATOR	14183	53.80
ACAJUL	14974	51.90
ACAPLA	14542	47.50
ACAHAR	15100	44.90
ACAHEM	14657	40.90
ACAFAL	14981	36.10
CSABRE	14188	28.90
EUCCLO	10691	26.90
LEPFLA	14502	23.20
LEPFLA	14554	10.70

Si Sa Ket

Seedlot	no.	Survival
CSASIA	L8632	90.00
EUCCAM	L8633	90.00
EUCRAV	13546	90.00
EUCCAM	13692	90.00
EUCTER	14108	90.00
EUCTER	14212	90.00
EUCCAM	14338	90.00
EUCBIG	11465	86.20
GREPAR	14143	86.20
PETPUB	14189	86.20
MLAAZE	14500	86.20
PETNUM	14639	86.20
MELSAL	14871	86.20
MELARG	14899	86.20
EUCBRA	13397	84.50
ACAJUL	14885	84.50
EUCEXS	14864	83.20
EUCDEG	L8634	82.30
ACASIM	14553	82.30
ACASIM	14576	82.10
EUCCIT	14852	82.10
MELSTE	14148	80.70
LEPLON	14900	80.70
ACAPLA	14960	80.70
NEOMYR	14889	79.40
GREPTE	14905	79.40
ACABRA	14965	78.30
ACATOR	14888	77.80
DESUMB	14556	77.10
EUCSUF	13598	75.70
ACATOR	14183	75.50
EUCHOU	9091	73.90
MLAAZE	14501	71.50
ACAJUL	14974	70.20
MELARG	14904	69.40
EUCPUN	10863	66.90
ADEABR	14557	66.80
NEOMYR	14896	65.20
GREPIN	14980	63.30
ACAPLA	14542	62.00
MELACA	14873	61.60
ADEABR	14180	59.20
CSABRE	14188	56.10
ACAHAR	15100	41.80
ACAFAL	14981	30.80

Huai Bong

Seedlot	no.	Survival
EUCTER	14108	90.00
EUCCAM	14338	90.00
EUCCAM	13692	86.20
MLAAZE	14500	86.20
ACASIM	14553	84.50
EUCBRA	13397	83.20
CALINT	14504	82.30
EUCBIG	11465	79.40
MELSTE	14148	78.30
ACASIM	14576	76.80
ACAJUL	14885	75.70
EUCTER	14212	75.60
LEPLON	14900	74.40
PETPUB	14189	73.90
EUCHOU	9091	72.60
EUCEXS	14864	72.30
GREPTE	14905	71.50
EUCRAV	13546	69.60
EUCCIT	14852	69.60
ACAJUL	14974	69.40
EUCSUF	13598	66.70
MELARG	14899	65.00
ACATOR	14888	64.60
ACATOR	14183	63.20
ACAPLA	14960	62.90
ACABRA	14965	62.00
EUCPUN	10863	56.40
GREPIN	14980	56.40
NEOMYR	14889	53.20
MELACA	14873	52.90
MELARG	14904	52.70
MELSAL	14871	50.10
ACAFAL	14981	50.00
PETNUM	14639	49.20
LEPFLA	14502	47.00
ACAPLA	14542	43.50
LEPFLA	14554	41.20
GREPAR	14143	41.00
CSABRE	14188	29.00
ACAHAR	15100	24.40

Ban Hong

Seedlot	no.	Survival
EUCHOU	9091	90.00
EUCCAM	14338	90.00
EUCTER	14108	82.30
EUCBRA	13397	80.70
EUCCAM	13692	80.70
EUCBIG	11465	79.40
EUCEXS	14864	79.40
AZAIND	L8671	77.80
EUCTER	14212	76.80
MELARG	14899	76.40
EUCSUF	13598	75.50
MELSAL	14871	74.70
ACABRA	14965	73.10
EUCCIT	14852	72.50
EUCRAV	13546	72.30
MELARG	14904	71.80
MELBRA	14982	71.80
ACASIM	14576	70.90
ACAAUR	L8673	70.80
LEPLON	14900	68.90
MELSTE	14148	68.50
MELBRA	14903	67.60
GREPTE	14905	65.50
ACAJUL	14885	64.60
DIPALA	L8672	63.20
ACATOR	14183	62.10
ACATOR	14888	61.20
MLAAZE	14500	60.00
EUCPUN	10863	55.10
NEOMYR	14889	51.70
ACAPLA	14960	25.40
ACAHAR	15100	18.60
ACAFAL	14981	18.30
LEPFLA	14502	13.60

Sai Thong Seedlot no.	Survival	Sakaerat Seedlot no.	Survival	Chanthaburi Seedlot no.	Survival
EUCCAM 14338	90.00	ALBPRO 14959	90.00	EUCRAV 13546	90.00
ALBPRO 14959	90.00	EUCTER 14108	90.00	EUCCAM 13692	90.00
ALBPRO 14962	90.00	EUCCAM 14338	90.00	EUCTER 14212	90.00
EUCRAV 13546	90.00	ACAORA 14886	90.00	EUCCAM 14338	90.00
EUCTER 14212	86.20	ACAORA 14961	86.20	ACAORA 14961	90.00
ACAORA 14961	86.20	EUCRAV 13546	82.30	EUCTER 14108	84.50
EUCURO 14534	86.20	ALBPRO 14962	82.10	ACACAT L8653	82.30
ACAAUR L8622	85.20	PTEMAC L8641	80.70	EUCDEG 13461	80.70
EUCURO 14532	84.50	EUCURO 14532	78.30	EUCHOU 9091	77.70
EUCTER 14108	84.50	EUCTER 14212	77.80	EUCCRA 14431	76.80
EUCCAM 13692	82.30	EUCCAM 13692	76.80	PAKJAV L8651	75.40
CASEQU L8623	80.80	EUCURO 14534	76.80	EUCDUN 13329	75.20
TABROS L8624	80.80	EUCPAN 14442	73.90	EUCBRA 13397	74.40
EUCHOU 9091	80.70	EUCHOU 9091	72.30	EUCPUN 10863	73.50
MELLEU 14147	80.70	EUCDEG 13461	71.20	ACAORA 14886	71.50
ACAORA 14886	79.40	EUCPUN 10863	69.70	ACAFLA 13872	65.50
EUCPUN 10863	76.80	ACAHYL 14977	67.60	EUCPAN 14442	64.50
EUCDEG 13461	75.20	EUCCIT 14852	66.30	PELDAS L8652	54.50
MELCAJ 14878	73.90	EUCCRA 14700	65.40	ACAFLA 14968	48.70
EUCCRA 14700	71.20	EUCCRA 14431	62.40	ALBPRO 14959	48.60
ALSMAC L8621	70.50	ACAFLA 14968	59.30	EUCEXS 14864	46.10
EUCMIC 13973	70.50	ACAFLA 13872	56.50	EUCCLO 10691	30.80
EUCPYR 10857	68.90	EUCMIC 13973	49.40	MELLEU 14147	29.20
EUCCRA 14431	68.50	EUCCLO 10691	48.50	MELCAJ 14878	24.80
ACAFLA 13872	66.60	MELLEU 14147	45.80		
EUCSAL 15011	64.40	MELCAJ 14878	39.80		
EUCPAN 14442	60.50				
EUCDUN 13329	58.00				
ACAFLA 14968	47.90				

differences in tree height between species with many eucalypts growing very fast (Table 5).

At the dry-site (Ratchaburi, Si Sa Ket, Huai Bong and Ban Hong) *E. camaldulensis, E. tereticornis, E. citriodora, E. exserta* and *E. brassiana* had excellent height while the best acacias were *A.torulosa* and *A. julifera*. Slow-growing species at these sites were *Cassia brewsteri, A. harpophylla, M. acacioides, M. bracteata, Petalostigma pubescens* and *Neofrabrica myrtifolia*. *Eucalytpus suffulgens* and *E. houseana* were slow-growing compared to other eucalypts.

Three *Grevillea* species (*G. pteridifolia, G. parallela* and *G. pinnatifida*) were included in these plantings and it was clear that *G. pteridifolia* was the tallest.

At the wet-site type (Sai Thong, Sakaerat and Chanthaburi) many eucalypts also grew fast (i.e. *E. camaldulensis, E. tereticornis, E. urophylla, E. raveretiana, E. grandis* and *E. punctata*). Slower-growing eucalypts were *E. houseana, E. microcorys* and *E. cloeziana*. Other species which were generally ranked in the slowest-growing groups included *A. flavescens, A. oraria, Albizia procera* and *M. cajuputi*.

It was also noted that trees of *E. camaldulensis, E. tereticornis* and *E. urophylla* from more northerly latitudes were taller than those from more southerly areas, although the differences were not statistically significant.

Diameter at Ground Level

Certain groups of species appeared to have greatest diameter at all planting sites (Table 6). At the dry sites, *E. camaldulensis* and *E. tereticornis* were clearly the best. Other species having relatively better diameter included *E. punctata* at Huai Bong and Ban Hong, *A. torulosa* at Si Sa Ket, *Melia azedarach* (from Atherton) at Ratchaburi and Huai Bong and *G. pteridifolia* at most sites. *Acacia julifera* and *A. platycarpa* also had a large diameter at Ratchaburi.

Trees of several species had very small diameter at most dry sites (i.e. *M. acacioides, A. harpophylla, N. myrtifolia, C. brewsteri)*.

For the wet sites, *E. urophylla, E. camaldulensis, E. tereticornis* and *E. raveretiana* had the largest diameters. These were followed by *E. grandis, E. punctata* and *A. oraria*.

Species possessing the smallest diameters differed between sites within the wet-site type. At Sai Thong, three local seedlots were the poorest (i.e. *Casuarina equisetifolia, Tabebuia rosea* and *Alstonia macrophylla*). At Sakaerat, *A. flavescens* (14968), *E. deglupta, M. cajuputi* and *A. hylonoma* were amongst the poorest, whereas at Chanthaburi poorest species included *M. leucodendra, M. cajuputi, E. houseana* and *A. flavescens* (two provenances).

Survival

There were statistical differences in survival between species at all sites, but, similar to the 1985 plantings, there was no clear pattern of species rankings (Table 7). Although species at most planting sites had high survival (i.e. two-thirds of the species had more than 80% survival), some species suffered high losses (i.e. *A. harpophylla, A. flavescens* and *Leptospermum flavescens,* both seedlots) at the dry-site types. At wet sites, *M. cajuputi, M. leucodendra* and *E. cloeziana* had highest mortality.

Some species had slow height and diameter growth but survived extremely well (e.g. *Albizia procera* and *Acacia oraria*).

Discussion

The results to date have shown marked differences in growth and survival between the Australian species being tested across Thailand. A number of species have performed well enough at this stage to justify further evaluation while many species did poorly and even failed in the trials.

The most promising species at 24 months after planting were *E. camaldulensis. E. pellita, A. crassicarpa* and *A. auriculiformis*, with the biomass of *A. crassicarpa* (as judged by heights and diameters and general observation) being especially promising. *Eucalyptus camaldulensis* and *A. auriculiformis* have been planted with acceptable performance elsewhere in Thailand, and the results for these two species in the present trials, while not surprising, vindicate the high priority accorded to them in current tree-planting programs in Thailand.

Acacia crassicarpa was first introduced to Thailand under the current program and appears to have a big future because of its fast growth and wide adaptability to various site conditions. Tree form of *A. crassicarpa* was generally poor with stems being notably sinuous although trees were mostly single-stemmed. It was also noted that Papua New Guinea provenances grew faster than the North Queensland provenance, and further more extensive provenance trials with the species are fully justified.

Provenance differences were also observed for *A. aulacocarpa*. Seedlots from Papua New Guinea grew much faster than their North Queensland counterparts. Some individuals from Papua New Guinea (Oriomo River and Keru) possessed excellent stem form and strong apical dominance. These trees have been propagated vegetatively by airlayering along with some other good-stem-formed acacias to investigate if their superior form is persistent.

Among the poorest species in these series of trials were *A. melanoxylon* and *Allocasuarina littoralis* which were all dead at one site (Ratchaburi) and suffered heavy losses at other sites. Melaleucas were all slow-growing but survived well.

The results obtained for the 1986 plantings revealed that growth of many eucalypts was similar to that of *E. camaldulensis. Eucalyptus tereticornis* was most outstanding while other interesting eucalypts included *E. urophylla, E. raveretiana, E. citriodora* and *E. exserta*. The results reflect the wide adaptability of this genus to all site conditions which makes it so successful as a plantation species in many countries. At one site (Chanthaburi) *E. raveretiana* was severely attacked by stem girdlers.

Some eucalypts (e.g. *E. houseana, E. cloeziana, E. dunnii* and *E. microcorys*) were slow-growing compared to those species mentioned above. However, the results were only obtained at 12 months after planting and growth rate may well change with time. Some species (e.g. *E. microcorys*) are believed to be slow starters and should grow faster later.

Other genera in the 1986 plantings were slower-growing than *Eucalyptus*. Many of these were planted outside their native habitats for the first time. One particularly interesting species is *Grevillea pteridifolia* which has performed relatively well, has a dense crown and maintains a healthy appearance throughout the year.

There is evidence of considerable variation in tree form of some species between different trial sites in Thailand. This is particularly perplexing in the 1985 plantings; trees of *A. polystachya* and *A. holosericea*, which are normally multistemmed or heavy branching from near ground level, were mostly found to be light branching with a visible main stem axis, at Sakaerat and Chanthaburi. The cause of such variation is not known. Nutritional problems have been suggested as one of the possible environmental determinants of stem form (Evans et al. 1987). It is also possible that heavy weed (i.e. imperata grass, *Imperata cylindrica*) at those two sites could have caused the light branching of the trees, though weed competition has always been kept to a minimum.

Acknowledgments

Thanks are due to staff members of the Royal Forest Department, Division of Silviculture, for trial establishment and measurement. I wish to thank Doug Boland for his valued comments, and Vitoon Luangviriyasaeng and Wiroj Ratanaporncharoen for the statistical analyses. I would also like to thank Miss Suthiwan Preecha for her patience in typing the manuscript.

Collecting fuelwood is very much a job for the whole family in rural areas of Zimbabwe.

Chapter 12

Growth and Survival of Australian Tree Species in Field Trials in Zimbabwe

D.P. Gwaze

Abstract

Six species trials located on five sites in Zimbabwe were assessed for height, diameter at ground level and survival 1.5–2 years after planting. Three of the trials were established at the beginning of 1985, with *Eucalyptus camaldulensis* being used as a check species. *Eucalyptus camaldulensis, Casuarina cunninghamiana, C. glauca, Grevillea glauca* and *Acacia holosericea* were showing most promise in these trials. The other three trials were established between December 1985 and January 1986. Promising species in two of these trials include *E. camaldulensis, A. auriculiformis, A. crassicarpa. A. cowleana, A. torulosa, A. podalyrifolia* and *A. leptocarpa*. On the driest site that had alkaline soils, the central American species *Senna atomaria, Leucaena shannonii, Parkinsonia aculeata* and *Leucaena diversifolia* were superior to the Australian species planted.

Introduction

Approximately 60% of Zimbabwe's 8 million people live in communal areas where wood is the prime source of energy for cooking and heating. There are shortages of fuelwood, poles, and fodder. Soil degradation caused largely by deforestation and overstocking is also a problem.

In order to improve the living standards of the communal people afforestation is playing a major role. Eucalypts, mainly *Eucalyptus grandis, E. camaldulensis* and *E. tereticornis* are being planted for poles and fuelwood. There is a need to identify species for fodder, for control of soil erosion and for improving the soil fertility. There is also a need to identify alternative species for poles and fuelwood adapted to very dry areas (<400 mm/annum) where productivity of the eucalypts is low.

Australia has a wide, diverse woody flora that is adapted to difficult environments. Many species have the important attributes of being able to fix nitrogen and to coppice. However, little of this flora has been tested for community forestry. The ACIAR-supported project was initiated in 1984 with the objective of evaluating these little-exploited, potentially valuable species for use in communal areas. Emphasis was placed on acacias and casuarinas because they fix nitrogen, they are good fuelwood, and some of them coppice and can be used as fodder. Some Central American species were included in the trials because they have the same important properties as the Australian acacias and casuarinas.

Results of six species trials 1.5–2 years after planting are reported. Three were established between January and March 1985 at Domboshawa (trial MV03), Grasslands (MV02) and Makoholi (MV01). The following planting season trials were established between December 1985 and January 1986 at Middle Sabi (MV08), Makoholi (MV06) and Kadoma (MV05).

Materials and Methods

Locations and other site details for each trial are shown in Table 1 and in Fig. 1, Chapter 1. The sites were chosen to be representative of communal areas, and their location on agricultural stations facilitated their monitoring and security. *Eucalyptus*

Table 1. Details of trial sites in Zimbabwe.

Trial site	Long	Lat	Alt (m)	Mean annual rainfall (mm)	Soil	Vegetation
Cotton Research Institute, Kadoma (MV05)	29°55′E	18°18′S	1180	780	Reddish brown clay loam soils	*Acacia, Terminalia* and combretum scrub vegetation
Domboshawa (MV03)	31°09′E	17°36′S	1500	1 000	Sandy soils	*Brachystegia* woodlands
Grasslands Marondera (MV02)	31°31′E	18°11′S	1640	1 000	Sandy soils	*Brachystegia* mixed with *Julbernadia* woodland
Middle Sabi (MV08)	32°18′E	20°21′S	440	400	Leached sandy soils with alkaline properties	*Colophospermum* mopane with *Acacia* species
Makoholi (MV01, MV06)	30°46′E	19°45′S	1200	600	Sandy soils with heavily textured sub-soils	*Brachystegia* mixed with *Julbernardia globiflora* woodland

camaldulensis was planted at all sites as the check species.

Details of seed origin are given in Table 2. All nurseries were located as close to the planting sites as possible to facilitate transport of seedlings to planting sites. Seed was sown directly into black polythene tubes (flat dimensions of 10 cm × 20 cm) between July and August. The seedlings were then thinned to leave one seedling per tube. The tubes were lifted every fortnight to prune the taproots.

Site preparation for all trials included manual stumping, then tractor ploughing and tractor disc-harrowing. Trials were planted between November and March depending on the beginning of the rainy season at each site. All trials were fenced using barbed wire to keep out domestic animals, especially cattle. Fences generally consisted of four-strand wire with *E. grandis* creosote-treated poles. Weeding was carried out when necessary. No fertiliser was applied in any of the trials.

Each trial had a randomised complete block design with each plot containing 36 plants per seedlot (first three trials, MV01, MV02, MV03) and 25 plants per seedlot (second three trials, MV05, MV06, MV08). The number of replicates varied from two to three, depending on nursery recoveries. Spacing was 2 m × 2 m (first three trials) and 2.5 m × 2.5 m (other three trials).

For the first three trials all trees were measured for height after 1.5 years, and for height and basal diameter (10 cm above ground level) after 2 years. For the second three trials the trees were measured for height after 1 year, and for height and basal diameter after 1.5 years.

The trials were first completely balanced by omitting any seedlots that had a plot in which all trees were dead. Two-way analysis of variance was performed on the height, basal diameter and on the survival percentages (transformed arcsine) recorded at 2 years for the first three trials, and at 1.5 years for the second three trials. Duncan's multiple range test was used to test the significance of the differences between treatment means of the top 10 leading seedlots at each site.

Results

With few exceptions there were significant differences ($P<0.05$) among species in terms of mean height, mean basal diameter and survival (Table 3). Only basal diameter at the Domboshawa trial (MV03) and survival at the Grasslands trial (MV02) were not significantly different. Significant differences between blocks were observed in mean height in the Kadoma trial and the Makoholi trial (MV06), in mean basal diameter in the Domboshawa, Kadoma and Makoholi (MV06) trials, and in survival at the Makoholi trial (MV06). Rankings of mean height, mean basal diameter and survival are given in Tables 4–9.

Ranking of most promising species in the first set of trials was generally consistent across sites although there were some large differences in the absolute growth. The first set of trials had more or less the same seedlots planted on each of the three sites and at all the sites, *Eucalyptus camaldulensis, Casuarina cunninghamiana* and *C. glauca* were most promising. Growth of these species was best at Grasslands (3.5–3.9 m tall at 2 years) and poorest at Domboshawa (2.2–2.3m). At Grasslands *Acacia holosericea* (3.5 m at 2 years) was also promising, as was *Grevillea glauca* (2.6 m at 2 years) at Makoholi. In all the trials the check species (*E. camaldulensis*) performed better than all the

Table 2. Details of seed sources used in species trials in Zimbabwe. x indicates the site at which each seedlot was planted.

CSIRO Seedlot	Species	Provenance	Lat	Long	Alt (ml)	Planting site		
						MV01	MV02	MV03
1985 planting								
1393	*E. camaldulensis*	Pentecost River, WA	15°48′	127°43′	60	x	x	x
14100	*C. obesa*	20 Km west Wiluna, WA	26°34′	150°26′	0	x	x	x
13144	*C. glauca*	Burrill Inlet, NSW	35°24′	150°26′	0	x	x	x
14192	*C. equisetifolia* var. *incana*	NNW Emu Park, QLD	23°13′	150°48′	3	x	x	x
14196	*C. equisetifolia* var. *equisetifolia*	Wangetti Beach, QLD	16°41′	145°35′	2	x	x	x
13515	*C. cunninghamiana*	9 Km SE Mareeba, QLD	17°04′	145°28′	400	x	x	x
13204	*Allocasuarina decaisneana*	NW Hermansburg Mission, NT	23°45′	132°41′	580	x	x	x
13201	*Allocasuarina decaisneana*	Curtin Springs, NT	25°18′	131°42′	440	x	x	x
13225	*Allocasuarina campestris* ssp. *eriochlamys*	Comet Vale, WA	29°56′	121°07′	0	x		
14177	*Grevillea glauca*	Mt Garnett, QLD	17°40′	145°07′	765	x	x	x
14182	*Terminalia platyptera*	Mt Carbine, QLD	17°40′	147°07′	765	x	x	
14044	*E. gamophylla*	West Olgas, NT	25°05′	130° 03′	610	x	x	x
14089	*E. intertexta*	Gary Highway, WA	25°04′	124°59′	500	x	x	x
10433	*E. pachyphylla*	South Tennant Creek, NT	19°34′	134°13′	330	x	x	x
10700	*E. normantonensis*	32 Km west Mt Isa, QLD	20°20′	138°50′	300	x	x	x
14023	*E. mannensis*	West Palmer River, NT	24°34′	132°40′	400	x	x	x
12776	*E. socialis*	20 Km west Wilcannia, NSW	31°32′	143°34′	20	x	x	x
7949	*A. pruinocarpa*	21 Km ESE Meekatharra, WA	–	–	490	x	x	x
13719	*A. aneura*	Vaughan Springs, NT	22°12′	130°55′	600	x	x	x
13780	*A. latzii*	Wallera Range Rd, NT	24°35′	133°02′	430	x	x	x
14104	*A. holosericea*	North Alice Springs, NT	20°05′	134°24′	–	x	x	x
						Planting site		
1986 planting						MV05	MV06	MV08
a) Australian seedlots								
711	*E. camaldulensis*	Small dam on Irvinebank/ Petford Rd, QLD	17°24′	145°09′	206	x	x	x
14094	*E. gibsonensis*	Lake Coen, Gary Highway, WA	24°27′	125°03′	440	x	x	x
13751	*Melaleuca lasiandra*	Vaughan Springs, NT	22°18′	130°52′	600	x		
14095	*M. lasiandra*	East Carnegie Station, WA	25°28′	123°23′	470	x		
13567	*M. leucadendra*	Mareeba, QLD	17°0′	145°30′	500	x	x	
14147	*M. leucadendra*	Weipa, QLD	12°31′	141°48′	10	x	x	x
14150	*M. symphocarpa*	Weipa, QLD	12°13′	141°48′	10	x	x	x
13530	*M. viridiflora*	Iron Range, QLD	12°42′	143°20′	60	x		x
14149	*M. stenostachya*	38 Km southeast Weipa, QLD	12°44′	142°06′	10	x	x	
13144	*C. glauca*	South Burril Inlet, NSW	35°24′	150°26′	–	x	x	x
14192	*C. equisetifolia* var. *incana*	NNW Emu Park, QLD	23°13′	150°48′	3	x		
13149	*C. cunninghamiana*	Uriarra Crossing, ACT	35°14′	148°57′	520	x	x	x
13171	*Allocasuarina huegeliana*	40 Km west Marembeen, WA	32°05′	118°50′	380	x	x	x
14177	*Grevillea glauca*	Mt Garnett, QLD	17°40′	145°07′	765	x	x	
14164	*Grevillea glauca*	Weipa, QLD	12°43′	142°06′	18	x	x	x
14188	*Cassia brewsteri*	Blackwater, QLD	12°33′	141°52′	10	x	x	
14047	*E. oxymitra*	Docker River, NT	24°53′	129°07′	650	x	x	x
14025	*E. oxymitra*	West Palmer River, NT	24°34′	132°40′	400	x	x	x
11731	*E. ochrophloia*	35 Km south Quilpie, QLD	26°53′	144°20′	180	x	x	x
14044	*E. gamophylla*	West Olgas, NT	25°05′	130°03′	610	x	x	x
12541	*E. gamophylla*	Dales Gorge, WA	22°57′	118°38′	500	x	x	x
13713	*E. argophloia*	S.F. 302, Ballon, QLD	26°20′	150°40′	300	x	x	
10433	*E. pachyphylla*	South Tennant Creek, NT	19°34′	134°13′	330	x	x	x

CSIRO Seedlot	Species	Provenance	Lat	Long	Alt (ml)	Planting site		
						MV01	MV02	MV03
13621	*A. mangium*	Piru, Ceram, Indonesia	03°04′	128°12′	150	x	x	x
14183	*A. torulosa*	Northwest Chillagoe, QLD	16°36′	144°07′	275	x	x	x
13864	*A. cincinnata*	Shoteel Landing, QLD	16°57′	145°38′	440	x	x	
7947	*A. pruinocarpa*	21 Km ESE of Meekatharra, WA	–	–	490	x	x	x
7859	*A. pruinocarpa*	Wiluna, WA	26°37′	120°15′	520	x	x	x
13871	*A. polystachya*	Bridle Landing, QLD	16°58′	145°37′	480	x	x	x
13500	*A. polystachya*	McIlwraith Range, QLD	13°42′	143°37′	360	x	x	x
12055	*A. podalyrifolia*	Bundaberg Area, QLD	24°50′	152°40′	100	x	x	x
14003	*A. plectocarpa*	Middle Springs, WA	15°45′	128°40′	50	x	x	x
13962	*A. pendula*	Collie-Triangie district, NSW	31°40′	148°18′	200	x	x	x
13863	*A. crassicarpa*	Shoteel Landing, QLD	16°57′	145°38′	440	x	x	x
13861	*A. crassicarpa*	Mata, PNG	08°40′	141°45′	30	x	x	x
13775	*A. cowleana*	Tanami Bore, NT	19°58′	129°42′	450	x	x	
13774	*A. cowleana*	Vaughan Springs, NT	22°18′	130°52′	600	x	x	x
13861	*A. auriculiformis*	Springvale Holdings, QLD	15°50′	144°55′	500	x	x	x
13854	*A. auriculiformis*	Oenpelli Area, NT	12°20′	133°04′	50	x	x	x
13865	*A. aulacocarpa*	Buckley Landing, QLD	17°19′	145°37′	720	x	x	x
13687	*A. aulacocarpa*	Iokwa, PNG	08°41′	141°29′	35	x	x	x
13481	*A. aneura*	6 Km east Charleville, QLD	26°25′	146°17′	300	x	x	x
13794	*A. ammobia*	Uluru National Park, NT	25°20′	131°12′	580	x	x	x
14139	*A. leptocarpa*	Mt Molloy, QLD	16°40′	145°18′	400	x	x	x
13691	*A. leptocarpa*	Woroi-Wipim, PNG	08°52′	143°03′	30	x	x	x
14104	*A. holosericea*	West Alice Springs, NT	20°05′	134°24′	–	x	x	x
13879	*A. holosericea*	Mt Molloy - Mareeba, QLD	16°46′	145°15′	380	x	x	x
13782	*A. murrayana*	Olgas, NT	25°12′	130°53′	580	x	x	x
13781	*A. murrayana*	Ayers Rock, NT	25°13′	130°53′	580	x	x	x
14008	*A. monticola*	South Broome, WA	18°50′	121°40′	25	x	x	x
13773	*A. monticola*	Wannaby Hill, NT	22°21′	131°18′	700	x	x	x
14176	*A. melanoxylon*	Atherton, QLD	17°17′	145°26′	1022	x	x	x
13630	*A. melanoxylon*	Jeeralangs, VIC	38°25′	146°30′	550	x	x	x
14055	*A. ligulata*	10 Km NW Giles, WA	24°59′	127°16′	520	x	x	
b) Central American seedlots								
12515	*Senna atomaria*	Valle Comayagua, HOND	14°22′	87°39′	600		x	x
12516	*Prosopis juliflora*	Valle Comayagua, HOND	14°21′	87°37′	600		x	x
12519	*Leucaena leucocephala*	Finca San Felipe close Duyure, HOND	13°38′	86°55′	1050		x	x
12520	*L. diversifolia*	Puerto Del Golpe, Montagua Valle, GUAT	15°02′	89°40′	480		x	x
12521	*L. shannonii*	Valle Comayagua, HOND	14°22′	87°39′	600		x	x
12522	*Crescentia alata*	Valle Comayagua, HOND	14°14′	87°36′	700		x	x
12524	*Albizia guachepele*	Montagua Valle, GUAT	14°59′	98°30′	200		x	x
12531	*Alvaradoa amorphoides*	Steep slopes above town La Venta, HOND	14°00′	87°02′	793			x
12533	*Caesalpinia velutina*	Montague Valle round El Rancho, GUAT	14°55′	90°00′	274		x	x
12534	*Casalpinia eriostachys*	El Pallado, HOND	13°23′	87°07′	100			x
12536	*Parkinsonia aculeata*	Flat ground SE Rio San Antonia, NIC	12°23′	86°09′	55		x	x
12540	*Pinus caribaea* var. *honduras*	Guanaja, HOND	16°26′	85°50′	0-300		x	x

Table 3. Summarised results of analysis of variance for height, basal diameter and survival of species trials in Zimbabwe. *, ** and *** indicate significance at the 5, 1 and 0.1% levels respectively. ns indicates not significant at the 5% level.

Trial	Planting date	No. of seedlots analysed	No. of replicates	Age (years)	F. ratio		CV(%)
					Seedlot	Block	
Mean height							
Domboshawa MV 03	3/85	11	2	2	50.51***	0.17ns	9.8
Grasslands MV 02	1/85	10	2	2	18.62***	0.24ns	17.1
Kadoma MV 05	12/85	39	3	1.5	18.81***	22.82***	18.1
Makoholi MV 01	2/85	19	2	2	13.02***	0.31ns	25.6
Makoholi MV 06	1/86	53	3	1.5	9.94***	18.39***	26.5
Middle Sabi MV 08	12/85	29	2	1.5	3.99***	1.15ns	28.2
Mean basal diameter							
Domboshawa MV 03	3/85	11	2	2	1.14ns	6.76***	4.6
Grasslands MV 02	1/85	10	2	2	19.90***	4.77ns	16.1
Kadoma MV 05	12/85	39	3	1.5	17.94***	9.85***	20.6
Makoholi MV 01	2/85	19	2	2	25.41***	2.39ns	20.4
Makoholi MV 06	1/86	51	3	1.5	5.93***	28.06***	28.5
Transformed survival							
Domboshawa MV 03	3/85	11	2	2	9.74***	10.81**	14.3
Grasslands MV 02	1/85	10	2	2	2.45ns	0.94ns	21.6
Kadoma MV 05	12/85	39	3	1.5	6.19***	3.10ns	20.3
Makoholi MV 01	2/85	19	2	2	18.98***	3.38ns	12.6
Makoholi MV 06	1/86	53	3	1.5	4.22***	10.14***	19.6
Middle Sabi MV 08	12/85	29	2	1.5	2.17*	0.035ns	34.1

Table 4. Ranking for mean height (m) 2 years after planting in the species trials in Zimbabwe. The results for Duncan's multiple range test ($P<0.05$) are shown for the first 10 seedlots in each trial.

Makoholi (MV01)			Grasslands (MV02)			Domboshawa (MV03)		
Seedlot	Species*	Height	Seedlot	Species*	Height	Seedlot	Species*	Height
13939	EUCCAM	2.90	13939	EUCCAM	4.28	13939	EUCCAM	2.92
14177	GREGLA	2.61	13515	CASCUN	3.87	13515	CASCUN	2.33
13144	CASGLA	2.20	13144	CASGLA	3.51	13144	CASGLA	2.16
13515	CASCUN	2.19	14104	ACAHOL	3.46	14192	CASEQU	1.60
14192	CASEQU	1.59	10700	EUCNOR	1.91	14104	ACAHOL	1.50
14089	EUCINT	1.21	14192	CASGLA	1.79	10700	EUCNOR	1.41
10700	EUCNOR	1.21	14196	CASEQU	1.58	14196	CASEQU	1.34
14196	CASEQU	1.16	10489	EUCINT	1.36	14089	EUCINT	1.07
10433	EUCPAC	1.08	10433	EUCPAC	1.32	12776	EUCSOC	0.85
13719	ACAANE	1.03	12776	EUCSOC	0.85	14023	EUCMAN	0.68
14023	EUCMAN	0.85				14100	CASOBE	0.52
12776	EUCSOC	0.84						
14100	CASOBE	0.83						
14044	EUCGAM	0.80						
13201	ALLDEC	0.44						
13204	ALLDEC	0.43						
13255	ALLGAM	0.43						
7947	ACAPRU	0.43						
13780	ACALAT	0.32						

* The first three letters are the first three letters of the genus name and the last three letters are the first three letters of the species name.

Table 5. Ranking for mean basal diameter (cm) 2 years after planting in species trials in Zimbabwe. The results for Duncan's multiple range test ($P<0.05$) are shown for the first 10 seedlots in each trial.

Makoholi (MV01)			Grasslands (MV02)			Domboshawa (MV03)		
Seedlot	Species	Diameter	Seedlot	Species	Diameter	Seedlot	Species	Diameter
14177	GREGLA	5.51	13939	EUCCAM	7.00	13939	EUCCAM	3.93
13939	EUCCAM	5.41	13515	CASCUN	5.36	14100	CASOBE	3.64
13515	CASCUN	2.66	14104	ACAHOL	4.76	13515	CASCUN	3.18
14089	EUCINT	2.56	13144	CASGLA	4.34	14089	EUCINT	2.54
13144	CASGLA	2.38	14089	EUCINT	3.07	13144	CASGLA	2.34
13179	ACAANE	2.16	10700	EUCNOR	2.86	10700	EUCNOR	2.30
10433	EUCPAC	1.71	14196	CASEQU	2.21	14104	ACAHOL	2.07
10700	EUCNOR	1.65	14192	CASGLA	2.08	14192	CASEQU	2.00
14192	CASEQU	1.54	12776	EUCSOC	1.85	12776	EUCSOC	1.70
14023	EUCMAN	1.50	10433	EUCPAC	1.59	14196	CASEQU	1.59
14196	CASEQU	1.39				14023	EUCMAN	1.35
13201	ALLDEC	1.12						
13204	ALLDEC	1.12						
12776	EUCSOC	1.10						
14100	CASOBE	1.08						
7947	ACAPRU	1.00						
14044	EUCGAM	0.95						
13225	ALLCAM	0.72						
13780	ACALAT	0.65						

Table 6. Ranking for survival (%) 2 years after planting in species trials in Zimbabwe. The results for Duncan's multiple range test ($P<0.05$) are shown for the first 10 seedlots in each trial.

Makoholi (MV01)			Grasslands (MV02)			Domboshawa (MV03)		
Seedlot	Species	Survival	Seedlot	Species	Survival	Seedlot	Species	Survival
13939	EUCCAM	100	13939	EUCCAM	91.7	13939	EUCCAM	98.6
13144	CASGLA	95.8	13515	CASCUN	87.5	10700	EUCNOR	84.7
13515	CASCUN	93.1	13144	CASGLA	80.6	14089	EUCINT	83.3
10433	EUCPAC	86.1	10700	EUCNOR	77.8	13515	CASCUN	76.4
14177	GREGLA	84.7	14089	EUCINT	72.2	14104	ACAHOL	66.7
14089	EUCINT	83.3	14104	ACAHOL	69.4	14196	CASEQU	59.7
13719	ACAANE	79.2	14192	CASGLA	65.3	12776	EUCSOC	54.2
10700	EUCNOR	73.6	14196	CASEQU	45.8	14192	CASEQU	50.0
14100	CASOBE	65.3	10433	EUCPAC	43.1	13144	CASGLA	44.4
14196	CASEQU	63.9	12776	EUCSOC	29.2	14100	CASOBE	31.9
12776	EUCSOC	62.5				14023	EUCMAN	19.4
14023	EUCMAN	47.2						
14192	CASEQU	45.8						
14044	EUCGAM	36.1						
13201	ALLDEC	30.6						
13225	ALLCAM	30.6						
7947	ACAPRU	22.2						
13780	ACALAT	12.5						
13204	ALLDEC	9.7						

Table 7. Ranking for mean height (m) 1.5 years after planting in species trials in Zimbabwe. The results for Duncan's multiple range test ($P < 0.05$) are shown for the first 10 leading seedlots in each trial.

Kadoma (MV05)			Makoholi (MV06)			Middle Sabi (MV08)		
Seedlot	Species	Height	Seedlot	Species	Height	Seedlot	Species	Height
13861	ACAAUR	2.63	13861	ACAAUR	2.16	12515	SENATO	2.84
13854	ACAAUR	2.63	711	EUCCAM	2.13	12521	LEUSHA	1.75
711	EUCCAM	2.55	14183	ACATOR	2.10	12536	PARACU	1.74
13775	ACACOW	2.35	14139	ACALEP	2.04	12520	LEUDIV	1.69
12055	ACAPOD	2.16	13681	ACACRA	2.03	14055	ACALIG	1.56
14139	ACALEP	2.06	12055	ACAPOD	2.02	13681	ACAAUR	1.55
13567	MELLEU	1.99	14003	ACAPLE	1.91	13774	ACACOW	1.38
14176	ACAMEL	1.91	13854	ACAAUR	1.82	13854	ACAAUR	1.37
13681	ACACRA	1.88	13774	ACACOW	1.65	14139	ACALEP	1.34
13774	ACACOW	1.80	13863	ACACRA	1.61	14104	ACAHOL	1.24
13630	ACAMEL	1.76	13687	ACAAUL	1.56	13794	ACAAMM	1.12
14104	ACAHOL	1.73	14150	MELSYM	1.49	14183	ACATOR	1.16
14147	MELLEU	1.62	13567	MELLEU	1.45	13481	ACAANE	1.12
13864	ACACIN	1.62	13171	ALLHUE	1.43	12533	CAEVEL	1.09
14150	MELSYM	1.56	14147	MELLEU	1.43	13879	ACAHOL	1.05
14003	ACAPLE	1.55	13865	ACAAUL	1.36	7947	ACAPRU	0.95
14149	MELSTE	1.53	14177	GREGLA	1.35	13691	ACALEP	0.94
13863	ACACRA	1.51	13691	ACALEP	1.28	12516	PROJUL	0.90
13691	ACALEP	1.50	13773	ACAMON	1.28	13962	ACAPEN	0.89
13621	ACAMAN	1.46	14176	ACAMEL	1.25	13681	ACACRA	0.87
13773	ACAMON	1.39	13621	ACAMAN	1.24	12524	ALBGUA	0.86
14055	ACALIG	1.21	14094	EUCGIB	1.22	13781	ACAMUR	0.85
14094	EUCGIB	1.16	13775	ACACOW	1.21	13864	ACACIN	0.85
13500	ACAPOL	1.15	13500	ACAPOL	1.19	10433	EUCPAC	0.82
13865	ACAAUL	1.06	14008	ACAMON	1.19	12534	CAEERI	0.80
14008	ACAMON	1.04	13149	CASCUN	1.19	12055	ACAPOD	0.75
14192	CASEQU	0.96	13481	ACRANE	1.19	7859	ACAPRU	0.72
12541	EUCGAM	0.95	13144	CASGLA	1.18	13871	ACAPOL	0.70
13871	ACAPOL	0.92	14149	MELSTE	1.06	12531	ALVAMO	0.62
10433	EUCPAC	0.83	13864	ACACIN	1.02			
13481	ACAANE	0.81	13630	ACAMEL	1.01			
14047	EUCOXY	0.76	13871	ACAPOL	1.01			
13751	MELLAS	0.72	13794	ACAAMM	1.00			
13530	MELVIR	0.68	13879	ACAHOL	0.96			
14044	EUCGAM	0.66	10433	EUCPAC	0.93			
14095	MELLAS	0.58	14047	EUCOXY	0.87			
13962	ACAPEN	0.53	14025	EUCOXY	0.86			
14164	GREGLA	0.38	14055	ACALIG	0.85			
14188	CSABRE	0.35	14104	ACAHOL	0.79			
			12541	EUCGAM	0.76			
			14044	EUCGAM	0.62			
			13713	EUCARG	0.60			
			12540	PINCAR	0.57			
			13782	ACAMON	0.54			
			7859	ACAPRU	0.53			
			14164	GREGLA	0.52			
			11731	EULOCH	0.49			
			12515	SENATO	0.48			
			7947	ACAPRU	0.47			
			13962	ACAPEN	0.27			
			12521	LEUSHA	0.21			
			12522	CREALA	0.31			
			12519	LEULEU	0.11			

Table 8. Ranking for mean diameter (cm) 1.5 years after planting in species trials in Zimbabwe. The results for Duncan's multiple range test ($P<0.05$) are shown for the first 10 seedlots in each trial.

Kadoma (MV05)

Seedlot	Species	Diameter
13854	ACAAUR	4.59
711	EUCCAM	3.98
13861	ACAAUR	3.95
13775	ACACOW	3.65
13681	ACACRA	3.22
13630	ACAMEL	2.84
13567	MELLEU	2.74
14176	ACAMEL	2.74
13864	ACACIN	2.72
14139	ACALEP	2.60
14147	MELLEU	2.52
13863	ACACRA	2.44
14104	ACAHOL	2.39
13774	ACACOW	2.28
12055	ACAPOD	2.18
13621	ACAMAN	1.92
14003	ACAPLE	1.75
14149	MELSTE	1.72
13691	ACALEP	1.71
13865	ACAAUL	1.64
14150	MELSYM	1.54
13500	ACAPOL	1.51
14094	EUCGIB	1.44
13871	ACAPOL	1.43
13773	ACAMON	1.38
14055	ACALIG	1.33
13751	MELLAS	1.26
14008	ACAMON	1.24
13481	ACAANE	1.22
12541	EUCGAM	1.21
14047	EUCOXY	1.14
10433	EUCPAC	1.12
13530	MELVIR	1.06
14044	EUCGAM	1.00
13962	ACAPEN	0.99
14192	CASEQU	0.93

Makoholi (MV06)

Seedlot	Species	Diameter
13861	ACAAUR	3.37
711	EUCCAM	3.36
13681	ACACRA	3.22
13854	ACAAUR	3.07
14177	GREGLA	2.81
14139	ACALEP	2.80
14183	ACATOR	2.72
13687	ACAAUL	2.72
13863	ACACRA	2.62
12055	ACAPOD	2.48
13774	ACACOW	2.40
13865	ACAAUL	2.40
14003	ACAPLE	2.33
13775	ACACOW	2.27
12540	PINCAR	2.25
13871	ACAPOL	2.22
13171	ALLHUE	2.13
14176	ACAMEL	2.12
13621	ACAMAN	2.02
13864	ACACIN	1.93
13500	ACAPOL	1.93
14147	MELLEU	1.89
13567	MELLEU	1.87
13879	ACAHOL	1.85
14104	ACAHOL	1.81
13481	ACAANE	1.81
14094	EUCGIB	1.79
13691	ACALEP	1.69
13773	ACAMON	1.67
14008	ACAMON	1.65
13713	EUCARG	1.64
13630	ACAMEL	1.47
13794	ACAAMM	1.44
14164	GREGLA	1.43
14150	MELSYM	1.42
13149	CASCUN	1.35
14047	EUCOXY	1.34
10433	UECPAR	1.32
7854	ACAPRU	1.26
13782	ACAMUR	1.25
14025	EUCOXY	1.23
12541	EUCGAM	1.23
7947	ACAPRU	1.16
14149	MELSTE	1.14
14055	ACALIG	1.06
13144	CASGLA	1.04
14044	EUCGAM	0.95
11731	EUCOCH	0.75
12522	CREALA	0.66
13962	ACAPEN	0.46
12521	LEUSHA	0.36

Table 9. Ranking for mean height (m) 1.5 years after planting in species trials in Zimbabwe. The results for Duncan's multiple range test ($P<0.05$ are shown) for the first 10 leading seedlots in each trial.

Kadoma (MV05)			Makoholi (MV06)			Middle Sabi (MV08)		
Seedlot	Species	Height	Seedlot	Species	Height	Seedlot	Species	Height
14139	ACALEP	94.7	13854	ACAAUR	98.7	12515	SENATO	92.0
13775	ACACOW	92.0	711	EUCCAM	97.3	14055	ACALIG	72.0
13861	ACAAUR	89.3	13687	ACAAUL	94.7	12524	ALBGUA	72.0
13621	ACAMAN	88.0	13861	ACAAUR	94.7	7947	ACAPRU	66.0
14094	EUCGIB	88.0	13863	ACACRA	93.3	12533	CAEVEL	64.0
14104	ACAHOL	86.7	12540	PINCAR	93.3	12536	PARACU	58.0
14003	ACAPLE	86.7	13871	ACAPOL	93.3	13794	ACAAMM	56.0
13500	ACAPOL	85.3	14150	MELSYM	92.0	13861	ACAAUR	50.0
13871	ACAPOL	82.7	14094	EUCGIB	89.3	7859	ACAPRU	50.0
13863	ACACRA	81.3	13481	ACAANE	88.0	14139	ACALEP	50.0
13691	ACALEP	78.7	14177	GREGLA	88.0	13879	ACAHOL	48.0
13962	ACAPEN	78.7	14176	ACAMEL	88.0	10433	EUCPAC	48.0
13854	ACAAUR	77.3	14055	ACALIG	86.7	14104	ACAHOL	46.0
14055	ACALIG	77.3	14164	GREGLA	86.7	13481	ACAANE	46.0
711	EUCCAM	76.0	14025	EUCOXY	85.3	13774	ACACOW	44.0
13865	ACAAUL	73.3	13681	ACACRA	84.0	12520	LEUDIV	40.0
13774	ACACOW	72.0	13621	ACAMAN	84.0	12516	PROJUL	40.0
13630	ACAMEL	70.7	13691	ACALEP	82.7	13781	ACAMUR	38.0
13864	ACACIN	65.3	13794	ACAAMM	82.7	13871	ACAPOL	34.0
14164	GREGLA	65.3	14183	ACATOR	80.0	12531	ALVAMO	32.0
14008	ACAMON	64.0	14139	ACALEP	80.0	13962	ACAPEN	30.0
14176	ACAMEL	61.3	12055	ACAPOD	77.3	1251	LEUSHA	26.0
14150	MELSYM	57.3	13500	ACAPOL	77.3	13691	ACALEP	18.0
12055	ACAPOD	54.7	14047	EUCOXY	76.0	12534	CAEERI	18.0
14149	MELSTE	52.0	13865	ACAAUL	74.7	13854	ACAAUR	14.0
10433	EUCPAC	52.0	13149	CASCUN	74.7	14183	ACATOR	14.0
14188	CSABRE	50.7	13774	ACACOW	73.3	13681	ACACRA	14.0
14095	MELLAS	49.3	13773	ACAMON	72.0	13864	ACACIN	10.0
13681	ACACRA	48.0	13144	ACAGLA	72.0	12055	ACAPOD	8.0
13481	ACAANE	46.7	10433	EUCPAC	72.0			
14192	CASEQU	42.7	13775	ACACOW	68.0			
13751	MELLAS	40.0	11731	EUCOCH	68.0			
13773	ACAMON	38.7	7859	ACAPRU	66.7			
14147	MELLEU	36.0	13567	MELLEU	66.7			
12541	EUCGAM	36.0	14149	MELSTE	65.3			
13530	MELVIR	30.7	12541	EUCGAM	62.7			
14047	EUCOXY	26.7	13713	EUCARG	62.7			
13567	MELLEU	13.8	7947	ACAPRU	60.0			
14044	EUCGAM	6.7	13171	ALLHUE	58.7			
			13864	ACACIN	58.7			
			14147	MELLEU	57.3			
			14003	ACAPLE	48.0			
			13630	ACAMEL	46.7			
			13879	ACAMOL	46.7			
			14008	ACAMON	45.3			
			13782	ACAMON	45.3			
			13962	ACAPEN	44.0			
			12522	CREALA	44.0			
			14104	ACAHOL	38.7			
			12515	SENATO	34.7			
			12521	LEUSHA	30.7			
			12519	LEULEU	20.0			

new species, but the performance of *Grevillea glauca* at Makoholi was similar to that of *E. camaldulensis*.

The second set of trials had more or less the same seedlots in each trial and the results at Kadoma (MV05) and Makoholi (MV06) were similar. After 1.5 years the most promising species were *E. camaldulensis, A. auriculiformis, A. cowleana, A. podalyrifolia, A. leptocarpa* and *A. crassicarpa*, though the seedlots performances were better at Kadoma. In addition, *A. torulosa* performed well at Makoholi. *Acacia auriculiformis* was the only new species that performed better than the check species (*E. camaldulensis*). At Middle Sabi the best performers were the Central American species *Senna atomaria, Leucaena shannonii, L. diversifolia, Parkinsonia aculeata* and Australian *Acacia ligulata*. The most rapid height growth in these trials was demonstrated by *Senna atomaria* (2.8 m in height at 1.5 years) at Middle Sabi.

Survival of the most promising species was good except for *L. shannonii* (26%), *Parkinsonia aculeata* (58%) and *L. diversifolia* (40%) at Middle Sabi, *C. glauca* (44%) at Domboshawa and *A. podalyrifolia* (54%) and *A. crassicarpa* (48%) at Kadoma. Most of these deaths occurred within the first year after planting.

There was evidence of provenance variation in growth of some species. For example, *Acacia crassicarpa* at Makoholi (MV06) had a mean height of 2.03 m (Papua New Guinea provenance) and 1.61 m (Queensland provenance), whereas *A. leptocarpa* at Kadoma had a mean height of 2.06 m (Queensland) and 1.50 m (Papua New Guinea).

Discussion and Conclusion

The results have shown considerable differences in performance between seedlots within sites and between the same seedlots on different sites. In the first set of trials seedlot performance at Grasslands was superior to those at Domboshawa and Makoholi, probably due to earlier planting, better soil and lower rainfall (in the case of Makoholi). Growth was least at Domboshawa mainly because the trial was planted very late in the rainy season and failed to establish well. In the second set of trials, overall seedlot performance was slightly better at Kadoma than at Makoholi, mainly due to better soils at the former site. Results at Middle Sabi were quite different from the other two, probably because the alkaline soil tended to be ideal for some Central American seedlots. Survival at Middle Sabi was low because of a higher termite population on the site. At this site all the trees of the check species were killed by termites.

In similar trials in Thailand *A. auriculiformis, A. leptocarpa* and *A. crassicarpa* were fast-growing (Pinyopusarerk and Boland 1987), and at Gympie in Australia *A. auriculiformis, A. crassicarpa, A. holosericea, A. leptocarpa* and *A. torulosa* were among the fast growers (Ryan et al. 1987). Although the above are among the most promising species in terms of height and diameter, it would have been more appropriate to select them in terms of biomass since the trees are going to be used for fuelwood and fodder. At the conclusion of the trials, biomass studies should be undertaken to validate the rankings reported in this paper.

Some species had only one seedlot in the trial and this is unlikely to be representative of the species. In general, at least three seedlots per species should have been included per trial. Some of the plantings included species which have a wide distribution and the correct choice of the provenance is important.

Other studies which are important include fuelwood quality demands in the communal areas, fodder quality and coppicing ability. These factors, together with growth rate, need to be examined and assessed before selecting and recommending species for communal planting.

Although the results reported here are early growth, it is encouraging to note that at this stage there are species performing better or nearly as good as *E. camaldulensis*, the most widely planted introduced species in communal areas. If this growth is maintained for at least 5 years, Zimbabwe will have at least four new alternative species suitable for planting in communal areas for fuelwood, poles and agroforestry. Care needs to be taken when extrapolating from early growth to later growth, as relative rankings of species in terms of growth are likely to change with time. Despite this, the study is important in that several previously untried and little-known species are promising for community forestry in Zimbabwe.

Based on the results and knowledge of the trials, the species that warrant further studies are *Acacia auriculiformis, A. holosericea, A. crassicarpa, A. cowleana*, and *A. torulosa*. Less intensive work should be initiated on *C. cunninghamiana, Grevillea glauca, A. mangium, A. aulacocarpa, Senna atomaria, A. farnesiana, A. leptocarpa*, and *Parkinsonia aculeata*. Such trials should investigate provenance variation, fuelwood quality, fodder production and quality and coppicing ability.

Acknowledgments

I wish to thank the Australian Centre for International Agricultural Research for financing the establishment of the trials.

Chapter 13

Response of Australian Tree Species to Nitrogen and Phosphorus in Thailand

R.N. Cromer

Abstract

Acacia auriculiformis and *Casuarina equisetifolia* responded to applications of phosphorus fertiliser at Huai Bong in northern Thailand, but not at Ratchaburi, near Bangkok. A smaller response by *Eucalyptus camaldulensis* at Huai Bong was not significant. Growth was generally poor at Huai Bong compared with Ratchaburi (height of *E. camaldulensis* less than 4 m and approximately 8 m, respectively, in 24 months). No response to nitrogen fertiliser was evident at either site.

Growth responses to phosphorus occurred at Huai Bong, despite high levels of available soil phosphorus and lack of a general increase in foliar concentrations of phosphorus following fertiliser application. However, foliar nitrogen concentrations in *Acacia* and *Casuarina* increased in phosphate-treated plots, suggesting that nitrogen fixation was enhanced by fertilisation with phosphorus. Site characteristics other than phosphorus nutrition probably limited growth rates at Huai Bong.

Application of phosphorus fertiliser at Ratchaburi resulted in substantial and significant increases in foliar phosphorus concentrations in all species despite lack of a (significant) response in growth. Better growth rates at Ratchaburi were probably due to more favourable physical and chemical (other than phosphorus) soil properties and potential access to groundwater.

Results from these experiments demonstrate that soil characteristics of potential plantation sites in Thailand vary considerably, resulting in variable growth rates and response to applied nutrients. Nutrition should receive greater attention when planning research projects with Australian tree species in Thailand.

Introduction

Inherent fertility of many tropical soils is low. In natural forests of tropical regions, most nutrients available for plant growth accumulate in living and dead biomass, and clearing frequently leads to rapid deterioration in carbon, nitrogen and exchangeable cations. Thus, productivity of plantations managed on short rotations may not be sustainable over several rotations. Potential nutritional problems were identified in fast-growing plantations in lowland humid tropical regions by Chijioke (1980), and theoretical nutrient balance sheets calculated for short rotation plantations indicated some soils will require supplemental nutrients (Jorgensen and Wells 1986).

Experience with rubber plantations in Malaysia has shown that use of phosphate fertiliser at planting, followed by application of a complete fertiliser, is required for good establishment (Watson 1973). Sowing of cover crops in conjunction with establishment of rubber plantations helped arrest the process of soil deterioration following clearing (Pushparajah 1983). While evidence of responses to added nutrients in rubber plantations has been available for many years (Pushparajah 1966), research into fertiliser use in tropical plantations intended for wood production in developing countries has been limited due to their high cost.

In Chiang Mai, Thailand, a Thai-Danish Forestry Project started in 1969 and concentrated primarily on introduction and testing of coniferous species. However, in a preliminary trial in northern Thailand with four conifers, responses to both cultivation and fertiliser were demonstrated, almost doubling height growth after 14 months when both treatments were combined (Granhof 1974). Unfortunately, it was concluded that intensive establishment of only one of the four species was profitable, and there seems to have been no further work on nutrition and establishment methods.

Substantial increases in growth of planted tropical and subtropical eucalypts have frequently been reported in field experiments following application of fertilisers, particularly in combination with other intensive establishment techniques (Schoñau 1983; Ward et al. 1985; Yost et al. 1987). Growth responses have most frequently been reported for nitrogen and phosphorus so any new investigation should concentrate first on these elements.

In view of the lack of information on response of trees to nutrient applications in Thailand, preliminary fertiliser experiments were commenced in 1985. These trials were exploratory because seedling stock and space available were limited at the time. The principal aim of the experiments was to determine potential for growth response to phosphatic and nitrogenous fertilisers, singly and in combination, in the three species under test.

Methods

Fertiliser experiments were set out in conjunction with larger trials of Australian species and provenances, established at Ratchaburi and Huai Bong. Three species (*Eucalyptus camaldulensis, Acacia auriculiformis* and *Casuarina equisetifolia*) were chosen because they were expected to perform well in adjacent species trials. Choice of these species enabled both nitrogen-fixing and a nonfixing species to be compared.

The design included six fertiliser treatments (Table 1), with three replicates per site and three species per plot over the two sites. Species were set out in a standard pattern as subplots within fertiliser treatment plots, which were allocated at random within replicates. Subplots consisted of two rows of 10 trees each, providing 20 trees of each species per plot. Spacing was 1 m between trees within rows and 2 m between rows, providing an area of 2 m^2/tree (5000 stems/ha) and 40 m^2 in each subplot.

Triple superphosphate was used as the source of phosphorus (17.5% P), and was applied at three levels in factorial combination with nitrogen as urea (46% N) at two levels (Table 1). A high level of P was included to ensure that adsorption of phosphate did not mask a potential response in P-fixing soils. Fertiliser applications were made in the following way:

Table 1. Fertiliser treatments applied in experiments at Ratchaburi and Huai Bong.

Treatment no.:	1	2	3	4	5	6
N level	N_0	N_0	N_0	N_1	N_1	N_1
P level	P_0	P_1	P_2	P_0	P_1	P_2

Where: $N_0 = 0$ $N_1 = 100$ kg N/ha;
$P_0 = 0$ $P_1 = 100$ kg P/ha; $P_2 = 400$ kg P/ha.

(i) An application of 100 kg/ha of nitrogen required 217 kg/ha of urea, which was equal to 44 g/tree. Since urea is very soluble, it can burn and kill young seedlings if not applied carefully. The total amount was therefore split into two applications, one of 14 g applied at or soon after planting, and the remaining 30 g applied after 3 or 4 weeks. The fertiliser was placed in a small hole 100–150 mm from each seedling, to avoid losses of fertiliser or toxicity to seedlings.

(ii) An application of 100 kg/ha of phosphorus required 571 kg of triple superphosphate/ha which was equal to 114 g/tree. There is little danger to seedlings from superphosphate fertiliser and the full amount was spread over an area of 1 m^2 around each tree. An application of 400 kg/ha of phosphorus required 2286 kg of triple super/ha which was equal to 457 g/tree. This amount of fertiliser was also spread in one application.

Soil samples had been taken from each site prior to installation of experiments and chemical properties were determined by the Royal Thai Forest Department. Measurements of height and diameter were made 6 and 24 months after planting, and samples of foliage were taken from trees for chemical analysis at the same time. An equal mass of young, fully formed leaves from the upper crowns of at least six trees was removed to form a composite for each plot. Samples were oven-dried at 70°C and ground in Thailand, then sent to Canberra for analysis of nitrogen and phosphorus using a sulfuric acid, hydrogen peroxide digest and automated methods of analysis (Heffernan 1985). Results are presented for growth and nutrient concentrations of trees at 24 months of age.

Results and Discussion

Soil texture at Ratchaburi was sandy clay loam to 30 cm, in comparison with Huai Bong which was sandy to 30 cm and graded into loamy sand to 50 cm. Soil chemical data (Table 2) show that while available phosphorus was low at Ratchaburi, cation exchange capacity was satisfactory. By comparison, available phosphorus levels at Huai Bong were high but cation exchange capacity was low. Organic

Table 2. Chemical properties of soils at Ratchaburi (R) to 30 cm and Huai Bong (H) to 50 cm (means of three and four sites respectively). Available phosphorus by Bray No. 2 method, pH by 1:1 in water, organic matter by Walkley and Black and cation exchange capacity using ammonium acetate.

Depth (cm)	Available P (ppm)		pH		Organic matter (%)		Cation exchange capacity (meq/100 g)	
	R	H	R	H	R	H	R	H
0–15	0.78	23.15	5.8	5.6	1.99	0.77	8.6	2.8
15–30	1.14	9.49	5.9	5.6	1.64	0.23	7.8	2.9
30–50	ND	4.84	ND	5.7	ND	0.35	ND	4.0

ND = no data

Table 3. Mean height (h, m) and mean diameter at breast height over bark (d, cm) of *Eucalyptus camaldulensis* (EC), *Acacia auriculiformis* (AA), and *Casuarina equisetifolia* (CE), 24 months after planting at Ratchaburi. Fertiliser treatments as given in Table 1; data represent mean values over the two nitrogen treatments.

	P_0		P_1		P_2		Mean		Sig*	
Species	h	d	h	d	h	d	h	d	h	d
EC	7.64	4.83	8.05	5.24	8.25	5.24	7.98	5.11	ns	ns
AA	4.40	2.99	4.51	3.04	4.79	3.28	4.57	3.10	ns	ns
CE	4.32	2.31	4.73	2.57	4.29	2.51	4.44	2.46	ns	ns

*Significance level of fertiliser effect, within species.

Table 4. Nitrogen (N) and phosphorus (P) concentration (mg/g) in foliage of three species, 24 months after planting at Ratchaburi. Fertiliser treatments as given in Table 1; data represent mean values over the two nitrogen treatments (species as for Table 3).

	P_0		P_1		P_2		Mean		Sig*	
	N	P	N	P	N	P	N	P	N	P
EC	16.0	1.08	15.5	1.10	16.6	1.27	16.0	1.15	ns	0.001
AA	25.7	1.18	24.4	1.23	26.8	1.42	25.6	1.27	ns	0.01
CE	19.4	0.99	19.7	1.25	22.1	1.88	20.4	1.37	0.05	0.001

*Significance level of fertiliser effect, within species.

Table 5. Mean height (h, m) and mean diameter at breast height over bark (d, cm) of three species at Huai Bong, 2 years after planting. Fertiliser treatments as given in Table 1; data represent mean values over the two nitrogen treatments (species as for Table 3).

	P_0		P_1		P_2		Mean		Sig*	
	h	d	h	d	h	d	h	d	h	d
EC	3.23	2.46	4.04	3.38	3.70	2.97	3.66	2.94	ns	ns
AA	0.97	0.54	2.12	1.21	2.14	1.38	1.74	1.04	0.05	ns
CE	0.77	0.48	1.46	0.67	1.42	0.98	1.21	0.71	0.05	0.05

*Significance level of fertiliser effect, within species.

Table 6. Nitrogen (N) and phosphorus (P) concentration (mg/g) in foliage of three species 24 months after planting at Huai Bong. Fertiliser treatments as given in Table 1; data represent mean values over the two nitrogen treatments (species as for Table 3).

	P_0		P_1		P_2		Mean		Sig*	
	N	P	N	P	N	P	N	P	N	P
EC	11.7	1.31	11.4	1.32	11.1	1.47	11.4	1.36	ns	0.05
AA	18.5	2.07	19.3	2.18	21.7	1.71	19.8	1.98	0.05	ns
CE	15.0	2.45	15.2	2.57	16.9	2.57	15.7	2.53	ns	ns

*Significance level of fertiliser effect, within species.

matter content was quite low at both sites, reflecting the fact that native forest had been cleared for some time.

Application of nitrogen had no significant effects on growth of seedlings of any species at Ratchaburi or Huai Bong. Nutrient concentration of foliage was influenced to a small extent by nitrogen application at Ratchaburi but not at Huai Bong (data not presented).

At Ratchaburi, all species showed a positive but nonsignificant response in height and diameter growth with increasing amount of phosphorus fertiliser applied (Table 3). Growth rate of *E. camaldulensis* was substantially faster than the other two species, but as species were analysed separately due to unequal variances (growth data only), the significance of this could not be tested. Application of phosphorus fertiliser produced a substantial and highly significant increase in foliar phosphorus concentrations in all species, and significantly increased nitrogen content in *C. equisetifolia* (Table 4). Phosphorus and nitrogen content of foliage was significantly different between species, with *E. camaldulensis* having the lowest levels.

Growth rate of all species at Huai Bong was about half that recorded at Ratchaburi. Phosphorus fertiliser significantly increased height growth of *A. auriculiformis* and *C. equisetifolia*, and although a similar trend was evident in *E. camaldulensis*, it was not significant (Table 5). There was also a trend towards greater diameter growth with increasing amount of phosphorus fertiliser, but the effect was only significant in *C. equisetifolia* (Table 5). Application of phosphorus fertiliser significantly increased phosphorus concentration in foliage of *E. camaldulensis* but not the other species (Table 6). Phosphorus application significantly increased nitrogen concentration in *A. auriculiformis* and, whilst there was a similar trend in *C. equisetifolia*, it was not significant (Table 6).

Phosphorus concentration in foliage of unfertilised *E. camaldulensis* was 1.08 mg/g at Ratchaburi and 1.31 mg/g at Huai Bong. These compare with levels of 0.6 mg/g in unhealthy trees and 1.2 mg/g in healthy 2-year-old *E. camaldulensis* trees in India (Bhimaya and Kaul 1966). Irrigated trees of *E. camaldulensis* in northern Australia responded to both nitrogen and phosphorus fertiliser and had phosphorus concentrations in foliage of 0.78 and 0.89 mg/g in unfertilised and phosphate-fertilised plots respectively, 28 months after planting (Cameron et al. 1986)

Available phosphorus levels in soil and phosphorus concentrations in foliage indicate that this element should not have inhibited growth at Huai Bong. Indeed, lack of a significant growth response in *E. camaldulensis* at Huai Bong, despite an increase in leaf phosphorus, supports this evidence. However, phosphate fertiliser increased growth of *A. auriculiformis* and *C. equisetifolia* at Huai Bong, even though leaf phosphorus concentrations did not change. Rates of growth and nitrogen fixation were enhanced following phosphorus application to several legumes including *A. pulchella* in Western Australia (Hingston et al. 1982). A significant increase in leaf nitrogen in *A. auriculiformis*, and a similar trend in *C. equisetifolia* following phosphate application, suggest increased nitrogen fixation resulting from increased availability of phosphorus was responsible for improved growth of the two nitrogen-fixing tree species at Huai Bong.

Leaf phosphorus concentrations in unfertilised *E. camaldulensis* at Ratchaburi were lower than those at Huai Bong but substantially lower values have been reported elsewhere. Lack of any significant growth response at Ratchaburi, despite substantial increases in leaf phosphorus following fertiliser application, suggests that phosphate availability was not a major limitation to growth. General conditions for growth appear to have been better at Ratchaburi than Huai Bong. It is likely that trees have access to groundwater at Ratchaburi, whereas steep slopes and shallow soils at Huai Bong would provide high runoff and low infiltration rates, leading to longer periods of water stress and reduced growth.

Trees in this experiment were planted at close spacing, plots were quite small and the different species were allocated to adjacent plots. Absolute growth rates vary considerably between the species (eucalypt > acacia > casuarina) so that competition between adjacent plots will bias results from now on. Although the experiment has provided useful preliminary information, further measurement or sampling is not recommended.

Data presented here provide some interesting implications for research into nutrition of tree plantations in Thailand. Growth responses at a hilly site in northern Thailand appear to be due to enhanced nitrogen uptake by two nitrogen-fixing species, as a result of phosphate application to a soil already high in available phosphorus. In contrast, trees growing on a more favourable lowland site in central Thailand did not respond to added phosphorus, despite increased concentrations of the element in their foliage. It is suggested that nutrition should play a more prominent role in research projects dealing with Australian tree species in Thailand.

Acknowledgments

Design of the trials reported here was undertaken by the author in collaboration with Doug Boland, while field work was handled by the Thai Royal Forest Department under supervision of Khongsak Pinyopusarerk. The project involved collaboration with many scientists and foresters in Australia and Thailand and, in addition to Doug Boland and Khongsak Pinyopusarerk, I would like to thank Emlyn Williams and Wanda Pienkowska of CSIRO, and Bunyarit Puriyakorn, Somyos Kitkha and Somchit Viseskaew of the Royal Forest Department, Thailand.

Chapter 14

Statistical Analysis of Tree Species Trials and Seedlot:Site Interaction in Thailand

E.R. Williams and V. Luangviriyasaeng

Abstract

This paper describes the statistical analysis of a series of species trials. Height data from the 24-month measure of 1985 trials in Thailand are used. Various aspects of the analysis are discussed including the preprocessing of individual tree data, analysis of variance for separate trials and a model for the combination of information over several trials. Genotype × environment interaction was investigated and results on the behaviour of different species are discussed.

Introduction

The field testing and evaluation of species usually involves two distinct phases. Firstly, there are individual field trials where plants from a number of seedlots are laid out using an appropriate experimental design. These trials can then be analysed separately in order to determine the relative performance of the seedlots in each trial. Secondly, there is the problem of combining results of individual trials, normally from a number of different locations. Typically, the relative performance of seedlots varies from trial to trial and this leads to investigation of what is known as genotype × environment (or in our context seedlot × site) interaction. It is extremely important to be able to interpret this interaction. Sometimes geographic factors can be identified as contributing to the differential performances of seedlots. Usually, however, the successful interpretation of genotype × environment interaction is not easy and this has led to the development of a number of approaches to analysis. There are various pattern analysis techniques (Williams 1976) and the singular value decomposition technique of Mandel (1971); these seem to offer little over joint regression analysis introduced by Yates and Cochran (1938) and again by Finlay and Wilkinson (1963).

The aim of this paper is to describe the steps involved in the statistical analysis of individual trials, the combination of results over a number of trials and an intrepretation of seedlot × site interaction. The data used are the 24-month measurements of tree height from the 1985 species trials in Thailand. Details of these trials are given elsewhere in this book; here we will simply describe the analysis of results.

Individual Trials

In 1985 species/provenance trials were laid out at six sites in Thailand (see Chapter 11 for details of these trials). The number of entries ranged from 30 to 42, including a number of local species. The experimental design in each case was a randomised complete block design with three replicates; plots consisted of a 5 × 5 arrangement of trees. Data from the 24-month measurement of these trials were brought to Canberra in October 1987 by Vitoon Luangviriyasaeng of the Royal Forest Department, Bangkok. During a 1-month stay in Canberra (under the supervision of A.C. Matheson and E.R. Williams), the second author carried out individual trial analyses using the MICRO version of the statistical package GENSTAT. Three variates were analysed: tree height, diameter at breast height and diameter at ground level. Here we present results for height only.

The processing of raw data on individual trees through to estimated seedlot means for each trial is fraught with danger. It is so easy for strange data values to go undetected; then when it comes to the combination of results over trials, bogus estimated means will either be identified, resulting in time-consuming reanalysis of raw data, or more likely lead to fanciful theories on the interpretation of seedlot × site interaction. Extra time and care spent in the early stages of data analysis will always be rewarded later on. To assist in effective data handling and screening, the following four points should be noted:

(i) The way that the data are entered into the computer can minimise the amount of file manipulation needed. Advice should be sought before data files are created. Such advice will vary depending on the statistical package that is to be used. In any case there are advantages in ordering the data file according to the field layout: that is, plots which are next to each other in the field should also appear that way in the data file. This, of course, would be the normal order that data are collected but experimenters often then attempt to reorder the data into a specified seedlot order, the same for each replicate. This is not to be recommended for not only is there an increased chance of indexing errors in the preprocessing, but it also becomes much more difficult to produce field plans of residuals (i.e. the remainder after fitting a particular statistical model to the data). Studying the field pattern of residuals is extremely useful in checking the effectiveness of field blocking.

(ii) Once the individual tree data have been entered into the computer, they can be summarised to plot means. We thus obtain the mean of the surviving trees (out of 25), the between-tree, within-plot variance and the survival percentage. These three quantities can then be tabulated into a two-way table of seedlots by replicates. A lot of information can be obtained by studying these tables. Firstly, the table of within-plot variances is very useful in alerting us to incorrectly coded data (leading to a large within-plot variance). The table of survival percentages can point to means that should be excluded from the subsequent analysis of variance. For example, if a species has died in two out of the three replicates and only has 10% survival in the third replicate, it would probably be better to exclude that species from the analysis of variance to avoid inflating the residual mean square.

(iii) The plot variances and means can then be analysed according to the appropriate experimental design. The analysis of plot variances (usually after taking logarithms) helps to identify the need for transformation of the plot means to satisfy the assumption of variance homogeneity made in the analysis of variance. For example, seedlots can differ in the magnitude of their tree-to-tree variation. If these differences are too great, it may not be appropriate to estimate a pooled variance component. Graphs of residuals versus fitted values also help in this regard, as well as giving a further diagnostic check for suspect data values.

(iv) For the analysis of later measures it will be necessary to accommodate the possibility of competition between the seedlots on adjacent plots. Internal plots of (up to) nine trees should then be used for analysis, although it is still advisable to measure the full plot, so that extra analysis of border trees using a neighbour-type model can be carried out to assess the extent of competition.

Numerous other points can arise during the processing of the raw data, but with appropriate attention to detail a set of estimated seedlot means is obtained from each trial. In addition, other important information should be collated such as the mean squares in the analysis of variance table and the average within-plot variance. These quantities allow us to estimate variance components and provide standard errors for the comparison of estimated means.

When there is not 100% survival, the above approach of analysing plot means is strictly speaking only approximate. A 'theoretically exact' analysis would be best carried out on the individual tree data so that the plots could be weighted according to the number of trees in each plot. However, the extra data manipulation and computational problems do not warrant such attention to precision. Provided survival is reasonable, the analysis of plot means is quite adequate; poor survival, especially greatly differing survival of seedlots from replicate to replicate, will cause problems regardless of the approach taken to analysis. A further point of procedure concerns the choice between carrying the plot means forward to a combined analysis over sites, or simply using the estimated means from each individual analysis. Again, any extra accuracy and information obtained by analysing plot means over sites is of questionable value, provided information such as between-replicate mean squares is retained en route. This is particularly the case when incomplete block designs are used, as computer packages and programs for the recovery of seedlot information from block totals are often difficult to generalise.

Results from the analyses of individual sites are summarised in Tables 1–3. Estimated means for the height of seedlots are presented in Table 1. A number of local species which are site-specific are excluded; this has left 37 seedlots in which to investigate the presence and nature of seedlot × site interaction. Seedlot 14176 at Ratchaburi died in all

Table 1. Estimated means for height (cm) at individual sites.

CSIRO Seedlot no.	Species	Site					
		Ratchaburi	Sai Thong	Si Sa Ket	Sakaerat	Chanthaburi	Huai Bong
13877	ACAAUL	334	503	318	304	163	124
13866	ACAAUL	348	413	399	251	205	196
13689	ACAAUL	424	742	606	463	288	254
13688	ACAAUL	439	802	570	375	363	200
13861	ACAAUR	465	858	661	523	379	345
13854	ACAAUR	428	899	658	527	391	316
13686	ACAAUR	567	884	-	-	-	-
13684	ACAAUR	520	941	641	538	363	293
13864	ACACIN	326	571	353	440	166	-
13863	ACACRA	416	799	677	453	-	-
13683	ACACRA	600	1083	738	658	285	-
13681	ACACRA	571	920	679	610	-	272
13680	ACACRA	584	1073	674	-	248	-
14623	ACADIF	608	695	-	-	-	-
14175	ACAFLA	436	624	421	271	137	-
14660	ACAHOL	552	685	415	433	336	251
13691	ACALEP	553	853	735	-	248	167
13653	ACALEP	520	796	658	392	306	189
13846	ACAMAN	473	483	521	283	143	-
13621	ACAMAN	419	497	453	166	187	102
14176	ACAMEL	-	259	158	161	174	134
13871	ACAPOL	256	392	261	177	141	96
14622	ACASHI	467	-	392	205	-	-
13876	ALLLIT	-	444	311	226	84	136
13519	CASCUN	500	434	420	334	186	176
13514	CASCUN	523	349	330	273	143	123
13148	CASCUN	383	267	220	223	132	158
13990	CASEQU	406	263	252	206	171	-
14537	EUCCAM	764	884	839	632	569	337
14106	EUCCAM	764	939	809	575	555	397
12013	EUCPEL	673	894	566	332	382	327
14130	EUCTOR	490	475	414	310	415	237
14485	MELBRA	216	116	97	63	80	75
14166	MELDEA	209	-	338	174	-	119
11935	MELDEA	218	319	363	191	76	100
14170	MELSYM	248	-	353	235	-	142
14152	MELVIR	-	-	288	90	111	113

Table 2. Transformed survival percentages at individual sites.

CSIRO Seedlot no.	Species	Site					
		Ratchaburi	Sai Thong	Si Sa Ket	Sakaerat	Chanthaburi	Huai Bong
13877	ACAAUL	49	47	82	78	33	53
13866	ACAAUL	86	78	86	59	53	74
13689	ACAAUL	72	58	86	73	52	69
13688	ACAAUL	67	81	82	82	53	69
13861	ACAAUR	78	85	90	90	90	90
13854	ACAAUR	85	91	90	90	86	86
13686	ACAAUR	72	57	–	–	–	–
13684	ACAAUR	76	86	90	90	50	78
13864	ACACIN	37	51	76	68	44	–
13863	ACACRA	55	73	82	68	–	–
13683	ACACRA	75	64	85	72	52	–
13681	ACACRA	82	67	77	70	–	45
13680	ACACRA	86	86	85	–	41	–
14623	ACADIF	74	73	–	–	–	–
14175	ACAFLA	65	45	90	50	44	–
14660	ACAHOL	76	63	44	67	69	23
13691	ACALEP	50	63	86	–	67	59
13653	ACALEP	86	69	78	73	72	47
13846	ACAMAN	62	76	82	75	44	–
13621	ACAMAN	55	54	86	58	51	30
14176	ACAMEL	–	44	19	50	17	45
13871	ACAPOL	86	68	83	69	68	60
14622	ACASHI	41	–	81	38	–	–
13876	ALLLIT	–	56	73	77	34	60
13519	CASCUN	79	76	90	81	66	71
13514	CASCUN	81	52	90	82	56	63
13148	CASCUN	62	55	60	84	57	57
13990	CASEQU	29	64	59	60	37	–
14537	EUCCAM	90	88	90	78	77	78
14106	EUCCAM	90	79	90	78	86	79
12013	EUCPEL	69	82	86	55	68	90
14130	EUCTOR	86	76	86	79	73	75
14485	MELBRA	67	58	63	69	47	69
14166	MELDEA	72	–	85	75	–	71
11935	MELDEA	75	63	90	70	45	63
14170	MELSYM	90	–	90	65	–	70
14152	MELVIR	–	–	85	55	24	64

Table 3. Summary of analysis of variance mean squares for height from individual sites.

Site	Mean squares			
	Replicates	Seedlots	Plot residuals	Within-plot residuals
Ratchaburi	34376	63700	3068	364
Sai Thong	85252	216149	5852	846
Si Sa Ket	13521	112267	1866	430
Sakaerat	61971	82155	1804	381
Chanthaburi	128691	58430	3865	526
Huai Bong	256	23893	891	274

three replicates and seedlot 14622 at Sai Thong started with only two replicates, both of which died; these have been excluded from the Tables. Details on survival are given in Table 2; the tabulated quantities are in fact the estimated means from the analysis of variance of angular transformed plot survival percentages. Information on mean squares obtained from the analysis of variance table, as well as the pooled within-plot variances, is summarised in Table 3.

Combination of Trials

The data in Table 1 are a two-way array of estimated mean heights which can now be analysed using the simple model

$$E[y_{ij}] = \mu + \theta_i + w_j, \qquad (1)$$

where y_{ij} is the height for the seedlot i at site j; E is the symbol for the expected value of y_{ij}; μ is a parameter for the grand mean; and the θ_i and w_j are effects for seedlots and sites respectively. The analysis is complicated by the fact that not all seedlots are present at all the sites, but a statistical package such as GENSTAT readily performs the appropriate nonorthogonal analysis. The analysis of variance table is given in Table 4 and is on a plot mean basis. To this table we have appended the pooled within-plot error and plot residual mean square (both obtained from Table 3) and based on a very large number of degrees of freedom. There is a large difference (Table 4) between the seedlot means. The seedlot × site interaction is also highly significant, and we should therefore investigate the nature of this interaction and hopefully interpret the differential behaviour of seedlots over sites, in terms of site characteristics, and also determine which seedlots are contributing most to the interaction.

The most common extension to (1) is the model for joint regression analysis:

$$E[y_{ij}] = \mu + \theta_i + \gamma_i w_j, \qquad (2)$$

where the γ_i are regression parameters for seedlots to try to cater for seedlots behaving differently over sites. For example, in model (1) the parameters are estimated from the margins of the seedlot × site table, but in model (2) the body of the table is used to estimate the γ_i, and so a component of the interaction is being modelled. Model (2) was first introduced by Yates and Cochran (1938) and again by Finlay and Wilkinson (1963). There have been many approaches to the difficult problem of interpreting genotype × environment interaction, but joint regression analysis remains the simplest and most successful.

The application of model (2) to the data in Table 1 is complicated by the fact that the table is incomplete. Therefore, the sequential analysis mapped out by Finlay and Wilkinson where the θ_i and w_j are estimated first and then the γ_i, is only approximate. Instead, the simultaneous analysis presented by Digby (1979) is appropriate. Estimated means for seedlots and sites as well as estimates for the regression parameters γ_i are given in Table 5; the calculations have been carried out using GENSTAT. A problem with joint regression analysis on incomplete tables is that the very instructive analysis of variance table given by Finlay and Wilkinson (1963) is no longer available. This is because the nonorthogonality does not allow the main effect and interaction component sums of squares to be separated out. Instead, we can merely report the success of model (2) over model (1) by the fact that the seedlot × site interaction mean square has been reduced from 27021 to a remainder of 14919. A convenient summary of results is provided in Fig. 1 where the estimated regression coefficients are plotted against the estimated seedlot means; this corresponds to fig. 3 of Finlay and Wilkinson (1963) where the interpretation is discussed in detail. The numbers correspond to the seedlots as in Table 5.

A strong linear relationship is evident between the estimated slopes and means in Fig. 1. Seedlots 29 and 30 are the best in terms of height, but seedlots 11 and 13, whilst also being good performers in terms of height, exhibit a more unstable character as measured by the higher estimated regression coefficients. This means that seedlots 11 and 13 have performed very well at good sites, but relatively speaking not so well at poorer sites. Seedlots with estimated regression coefficients less than one would be termed stable varieties.

Table 4. Analysis of variance table for height.

Source	Degrees of freedom	Mean square	Variance ratio
Site	5	2495163	–
Seedlot	36	367529	–
Seedlot site	150	27021	9.35
Residual	–	2891	–
Within-plot	–	470	–

Table 5. Estimated means and regression coefficients for combined analysis.

No.	Seedlot no.	Code	Species	Estimated mean (cm)	Estimated regression coefficient
(a) Seedlots					
1	13877	a	*Acacia aulacocarpa*	291	0.87
2	13866	a	,, *aulacocarpa*	302	0.62
3	13689	a	,, *aulacocarpa*	463	1.18
4	13688	a	,, *aulacocarpa*	458	1.30
5	13861	b	,, *auriculiformis*	538	1.20
6	13854	b	,, *auriculiformis*	536	1.30
7	13686	b	,, *auriculiformis*	510	1.62
8	13684	b	,, *auriculiformis*	549	1.46
9	13864	c	,, *cincinnata*	336	0.93
10	13863	d	,, *crassicarpa*	465	1.49
11	13683	d	,, *crassicarpa*	594	2.04
12	13681	d	,, *crassicarpa*	572	1.46
13	13680	d	,, *crassicarpa*	519	2.25
14	14623	e	,, *difficilis*	592	0.45
15	14175	f	,, *flavescens*	326	1.33
16	14660	g	,, *holosericea*	445	0.92
17	13691	h	,, *leptocarpa*	500	1.73
18	13653	h	,, *leptocarpa*	477	1.46
19	13846	i	,, *mangium*	342	1.01
20	13621	i	,, *mangium*	304	1.02
21	14176	j	,, *melanoxylon*	179	0.23
22	13871	k	,, *polystachya*	221	0.68
23	14622	l	,, *shirleyi*	304	1.62
24	13876	m	*Allocasuarina littoralis*	246	0.81
25	13519	n	*Casuarina cunninghamiana*	342	0.74
26	13514	n	,, *cunninghamiana*	290	0.66
27	13148	n	,, *cunninghamiana*	231	0.34
28	13990	o	,, *equisetifolia*	250	0.25
29	14537	p	*Eucalyptus camaldulensis*	671	1.24
30	14106	p	,, *camaldulensis*	673	1.26
31	12013	q	,, *pellita*	529	1.33
32	14130	r	,, *torelliana*	390	0.46
33	14485	s	*Melaleuca bracteata*	108	0.14
34	14166	t	,, *dealbata*	226	0.67
35	11935	t	,, *dealbata*	211	0.67
36	14170	u	,, *symphyocarpa*	261	0.67
37	14152	v	,, *viridiflora*	189	0.58

(b) Sites

No.	Site	Estimated mean
1	Ratchaburi	426
2	Sai Thong	621
3	Si Sa Ket	480
4	Sakaerat	359
5	Chanthaburi	258
6	Huai Bong	198

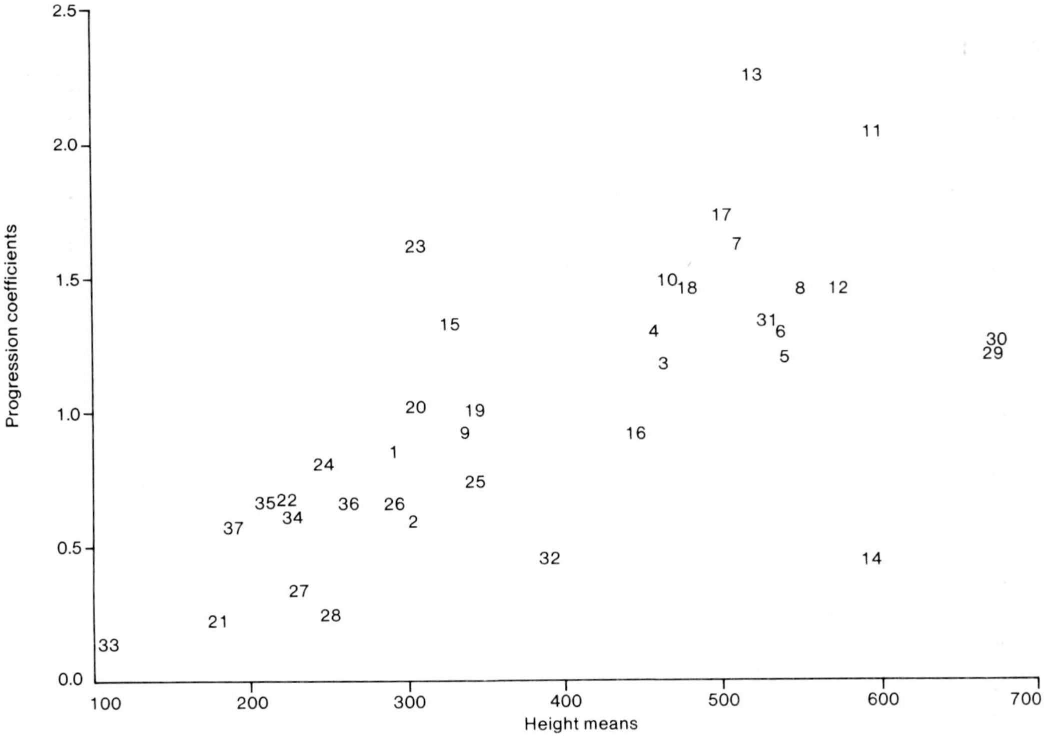

Fig. 1. Plot of slopes versus means for individual seedlots.

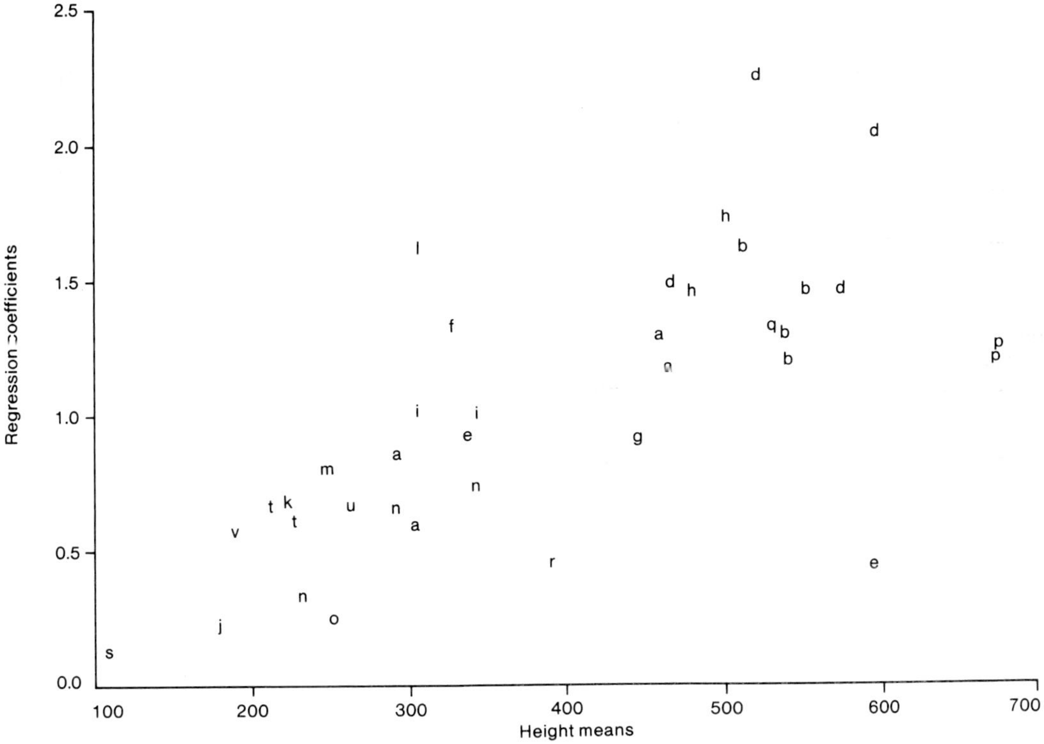

Fig. 2. Plot of slopes versus means for species.

Discussion

The best seedlots in terms of height are shown in Fig. 1, which also gives an indication of the stability of the different seedlots. It is more instructive and illuminating to look at the same data recoded according to the different species comprising the 37 seedlots. This is done in Fig. 2 where the letters represent species as in Table 5. Now it is clear that there is considerable grouping of the species. In particular, *Eucalyptus camaldulensis* (p) has the best growth and is fairly stable across the sites. On the other hand, *Acacia crassicarpa* (d), whilst exhibiting good growth across sites, tends to do particularly well at the good sites, i.e. it is an unstable species. Of interest are the four seedlots of *Acacia aulacocarpa* (a) which fall into two separate groups of two. It is noted that one group (3,4), whose seed origin is Papua New Guinea, grows taller than the other group (1,2) whose seed origin is Queensland (see Chapter 11 for detailed seed sources). These results reflect a clear variation between provenances of the species. There is one seedlot each of *Acacia holosericea* (g) and *Eucalyptus torelliana* (r) in this series of field trials. Their height growth is considered to be intermediate (means 445 and 390 cm, respectively) but is stable across the six sites at which they were planted.

In general, the strength of the groupings of species in Fig. 2 gives us reason to place some confidence in the results. Nevertheless, there is no doubt that more accurate results would have been obtained if extra attention had been paid to the experimental design of the individual trials. This can easily be demonstrated by looking at residual plots from the randomised block analyses carried out. In essence, the problem stems from the size of the replicates in the randomised block layout; for example, with 42 seedlots of 25-tree plots, the land area would be about 0.4 ha, which is normally too large an area over which to make the statistical assumption of equal between-plot variances. Incomplete block designs would be better so that, if there is any field trend, adjustments can be made to increase the accuracy of estimation of seedlot means.

Suitable experimental designs would be those of the generalised lattice type, that is square and rectangular lattices (Cochran and Cox 1957) or the more flexible α-designs (Patterson et al. 1978). Patterson and Silvey (1980) discuss the use of α-designs for statutory field trials in Britain. Neighbour-type models may also be appropriate for such trials but the analysis is more involved.

This paper has concentrated on the height measurements of trials planted in 1985 to highlight some procedural matters for the statistical analysis of data. There are also results for diameter from the 24-month measure of these trials, as well as data from the 12-month measure of the 1986 series of trials planted at seven sites. These results will be reported elsewhere.

Resource Evaluation

Top — Coppice developing on 3-year-old stumps of *Acacia auriculiformis* near Ratchaburi, Thailand. *Bottom* — Airlayers attached to side branches of 2-year-old trees of *Casuarina cunninghamiana* in Thailand. This simple technique of vegetative propagation involves removing the bark to expose the cambium, applying moist clay to the wound then wrapping the clay in moistened coconut fibre. This is then enclosed within a layer of plastic to contain the fibre and moisture.

Chapter 15

Vegetative Propagation of *Casuarina* and *Acacia*: Potential for Success

L.D. Pryor

Abstract

Australian species of *Casuarina* and *Acacia* have been little explored for their capacity to respond to vegetative propagation. Available evidence suggests that this will be possible by a variety of methods, some of which will be suitable for clonal silviculture. Variations must be expected in the response from various species, but this cannot be assessed adequately until more studies of the two genera are undertaken. A vigorous research program is merited to further evaluate the full potential of these two genera for vegetative propagation.

Introduction

Compared with herbaceous plants the vegetative propagation of woody plants is often rather difficult, with fewer methods being available than for herbaceous species.

In silvicultural work the most notable group in which vegetative propagation by the simplest means, stem cuttings, is regularly employed is the Salicaceae, particularly *Populus* (poplar) and *Salix* (willow).

Amongst the evergreen tree genera, both conifers and broadleaved species, few have been successfully propagated vegetatively for field-scale application, the exceptions being *Cryptomeria* in forest plantations in Japan and a few tropical types such as *Erythrina* and *Hibiscus* for amenity use.

The importance of the eucalypts in world plantation silviculture in recent decades has led to intensive research into their vegetative propagation, and since they are considered to be in the 'very hard to vegetatively propagate' class, the solution to the problem has involved some basic aspects which have application much more widely than to the genus itself.

The principal factor in stem propagation of *Eucalyptus* is the recognition that physiological juvenility in the mother shoots is one of the most important features which leads to success. The use of stool beds regularly cut back, so that juvenile shoots coming from near ground level are the source of cuttings, has been an important development.

This aspect applies widely in woody plant genera, and *Acacia* and *Casuarina* are no exception.

Benefits of Vegetative Propagation

The technical advantages of clonal propagation in plantation silviculture have been widely explored in recent decades, and the risks have been recognised. However, the advantages to small-scale operations or village activities have been little emphasised.

If simple methods of vegetative propagation can be devised, it may be possible to dispense with centralised nurseries and an associated financial investment as well as the consequent transport needs, the latter in places where communications often remain difficult.

It would also allow the introduction of improved or new material into cultivation very rapidly, and methods may require no more than making the knowledge available or the provision of materials such as polyethylene sheets.

Both *Acacia* and *Casuarina* are generally less difficult than eucalypts to propagate vegetatively, but within each genus there is a lack of uniformity and some species are more responsive than others.

Methods of Vegetative Propagation

Vegetative propagation of plants has been practiced since antiquity, both in simple and more complex forms.

The main traditional methods are: (1) stem cuttings; (2) root cuttings; (3) root suckers; (4) air-layering (Marcottage, Gootee, Chinese airlayering); (5) layering; (6) grafting — covering an array of types such as tip, cleft, crown, bud, and root graftings. Recent developments include micropropagation and tissue culture. Most plants can be propagated by one or other of the traditional methods, especially if note is taken of the benefits of using physiologically juvenile material.

The recent developments based on rather intricate or even complex technology such as tissue culture (Duhoux et al. 1986; Abo el-Nil 1987) do not offer an especially desirable pathway for clonal propagation in the developing regions of the world. In such areas traditional methods with minor modifications are likely to achieve the same ends in a more appropriate way. Local climate has a marked influence on the outcome, and those places with mild temperatures and humid atmosphere are usually the best for success, so methods must be adjusted to those conditions.

The benefits from a program of reassessment and modification of traditional vegetative propagation methods for use in clonal silviculture would stimulate an area of research development that has been much neglected in recent times. The introduction of such a program is strongly urged.

Experience with Casuarina

Casuarina has been of much more silviculture interest than *Allocasuarina* so that most experience is with the former genus. Whether that information is transferable to the latter genus remains to be explored.

There is an outstanding current use of vegetative propagation developed in Thailand. It is believed that spontaneously occurring hybrids between *Casuarina junghuhniana* and *C. equisetifolia* appeared in the Singapore Botanic Garden. They grow well but do not produce seed. A successful method of raising planting stock is to take stem cuttings from mother plant shoots within reach from the ground and place them in 20 cm diameter pots, enclosing both pot and plant completely with a polyethylene bag supported by a small stake or wire.

The rooting process takes a few weeks and some hardening off in partial shade is necessary to harden plants for field planting. Substantial areas have been planted with such material in Thailand (Chittachumnonk 1981).

A similar method has been successful for *C. equisetifolia* by setting stem cuttings in sandy soil and covering them with a polyethylene 'tent' until rooting and renewed growth is achieved. There are several records of successful propagation of this species by cuttings with minor variations in method (Somasundaram and Jagedees 1977; Halos 1981; Kondas 1981; Lunquist and Torrey 1984).

In trials with a limited number of species, *C. cunninghamiana, C glauca* and the hybrid *C. cunninghamiana* × *C. glauca*, Willing (pers. comm.) found that cuttings taken from near the base of seedlings or the base of a root sucker (in the case of the hybrid) rooted readily under standard greenhouse conditions in 4–7 weeks. The success rate was lower as cuttings were taken further and further from the base of the mother plant, but it was improved somewhat with the proprietary rooting hormone Seradix.

These limited reports indicate that *C. equisetifolia* roots more readily from stem cuttings than *C. cunninghamiana* and that in the latter species more care must be given to taking the cuttings from nearer the base of the mother plant. Variation in ability to strike from stem cuttings given equal environmental conditions must be expected but will be revealed only by systematic screening.

Airlayers

There are several reports of successful airlayering of *Casuarina equisetifolia* and *C. cunninghamiana*. In the case of the latter species it is successful in zones on the stem beyond those that give side shoots that strike successfully as stem cuttings. It is likely that most species will airlayer, as this method of propagation is widely successful with a great array of plants. The method has limitations, however, in that it is time-consuming, and the number of shoots suitable for treatment tends to be limited on the mother plants.

Root Suckers

In those species such as *C. glauca* or the hybrid with *C. cunninghamiana* root suckers taken with a piece of the root attached continue to renew root and shoot growth under greenhouse conditions.

Grafting

There are few reports of grafting trials but *C. cunninghamiana* has succeeded quite well with crown tip grafts. Bottle grafts are easy and reliable in both *C. cunninghamiana* and *C. glauca* which would be a useful adjunct to breeding work (Willing, pers. comm.).

Experience with Acacia

The species of *Acacia* that occasion most interest in the Australian flora are in a distinct taxonomic group of around 900 species which some researchers would prefer to consider a separate genus, *Racosperma*. Almost all of these are in Australia but some are in Papua New Guinea. A few are in Indonesia and odd ones as far afield as Madagascar, Taiwan and Hawaii.

It is the tropical members of this group that have attracted most interest recently, a matter stimulated by the spectacular early success of *Acacia mangium*. It is to be hoped this success will continue.

Some reports have been made of successful vegetative propagation of Australian species of the '*Racosperma* type' of acacias. One of the earliest is that of *Acacia melanoxylon* by root cuttings. Another is of *Acacia mearnsii*, the tanbark wattle that is so widely planted in Africa for tannin production, and which has been subject to much silvicultural research. There has been limited success with stem cuttings of seedlings and airlayering.

Acacia obliquenervia has also been propagated readily from root cuttings, a useful feature in view of the very poor seed setting and equally poor seed germination.

In addition, there has been a long-standing record of the propagation of *Acacia podalyriaefolia* × *A. baileyana* in France by grafting by inarching to stock of the lime-resistant *A. retinodes*. This is done to circumvent the troubles caused by calcareous soil to the hybrid, and has been much used for the production of mimosa flowers for winter decoration in Europe.

The tropical species of interest, however, are in the early stages of survey and assessment in regard to vegetative propagation.

Preliminary trials in Thailand in the RFD/ACIAR species trials at Sakaerat, Nakhon Ratchasima, have shown that three tropical species, *A. cincinnata, A. auriculiformis* and *A. aulacocarpa* respond very well to airlayering with an 80% success rate. There have been successful results from second and third order branches, and in the case of *A. cincinnata* rather better success from the upper part of the crown than the lower (S. Sirilak, pers. comm.). Good success with airlayering was also achieved at Ratchaburi, Thailand, with *A. holosericea, A. polystachya, A. aulacocarpa, A. cincinnata, A. shirleyi, A. crassicarpa* and *A. mangium*, although one trial each with *A. crassicarpa, A. flavescens* and *A. aulacocarpa* failed (B. Puriyakorn, pers. comm.). No hormones were used and all plants were about 2 years old.

In more detailed trials with *A. auriculiformis* using stem cuttings, Simsiri (1988) found that those from seedlings gave distinctly better results than those from more mature parts of the crown, which struck in a limited way only if IBA (indole butyric acid) was also applied. Cuttings from plants 1.5 years old, or from hedged 2.5-year-old plants give around a 30% success rate which was more than doubled with the addition of IBA.

Field observations show that many acacias form root suckers, although others do not. It is very probable that those which do sucker would propagate readily by root cuttings.

There is also evidence from a few species that shoots from near the base of plants will strike as stem cuttings given suitable conditions. Because of the general nature of this phenomenon in woody plants, there is reason to expect that this will apply generally.

Conclusion

Preliminary results and observations to date suggest that many *Casuarina* and *Acacia* species will prove relatively easy to propagate vegetatively. Further research is necessary to assess variation amongst species in ease of propagation, and to develop cheap methods to reproduce large numbers of plants. In addition, there is also a need for growth trials to determine if the form of mother trees can be reproduced in vegetatively propagated progeny.

Chapter 16

Fuelwood Evaluation of Four Australian-Grown Tree Species

K.W. Groves and A.M. Chivuya

Abstract

This chapter consists of a review of standard fuelwood tests and attempts to define what constitutes a good domestic fuelwood in a manner relevant to Third World countries. Four Australian-grown species, *Eucalyptus melliodora, E. blakelyi, Acacia melanoxylon* and *Pinus radiata*, were examined. For each of these, calorific value, density, moisture content and chemical composition were investigated.

Burning tests were also carried out by boiling a fixed mass of water using a known mass of fuelwood under standardised conditions. While calorific value of oven-dry wood is important in defining wood as a fuel, our results show little differences between species. The most important factors were density (either basic or air-dry) and moisture content. In the burning tests, only air-dry samples gave satisfactory results, emphasising that wood should be dried before being used as a fuel. The air-dry samples of the lower-density species ignited more readily, burnt more rapidly without producing embers, and boiled the water more quickly. The higher-density species took longer to ignite, burned more slowly, but produced hot embers, which continued to give off a steady heat long after the flames had died down. Overall, the tests indicated that no one species had all the desirable characteristics of a fuelwood. For quick cooking or heating the less dense species may be preferred. Where cooking must be done slowly over a longish period, dense species, which maintain a steady heat by producing quantities of hot embers, may be better.

Introduction

Selection of species for fuelwood plantations has been largely based on the growth characteristics of those species that are perceived as good for domestic fuelwood, i.e. the faster the growth rate the better. However, what constitutes a good domestic fuelwood has never been clearly defined and the purpose of this Chapter is an attempt to redress this omission in a way that is relevant to Third World countries. This is not to say that people using fuelwood regularly cannot give valid reasons for their preferences (e.g. suitability for cooking favourite dishes, low smoke production, etc.).

The most important properties of wood which may help to determine its quality as a fuel may be divided broadly into two categories: those which can be stated *quantitatively* and those which are more *qualitatively* defined, although perhaps susceptible to some degree of measurement.

Quantitative properties should include calorific value, density, moisture content and drying rate, and finally chemical composition including extractive content. For qualitative properties we may include the ability to: (a) burn slowly and consistently without emitting sparks or excessive toxic smoke; (b) produce persistent residual embers; (c) impart a 'good' flavour to any cooked food; (d) 'burn well' under a variety of conditions without excessive sootiness; and (e) provide a good social atmosphere for family and other groups.

There will be others in this second category depending on local preferences and specific requirements. A more detailed review follows.

Table 1. Calorific value of some heating fuels in MJ/kg (source of data Shepherd 1979).

Kerosene	43.6
Charcoal	29.7
Black coal (New South Wales)	27.9
Brown coal (Victoria)	21.0
Air-dry cow dung	16.7
Air-dry peat	16.7
Oven-dry wood	19.7
Air-dry wood	16.0
Green hardwood at 80% moisture content	10.0

Quantitative Tests

Calorific Value

The gross calorific value or heat of combustion is the amount of heat energy released per unit mass when combustion is complete and the products have cooled to the initial temperature. Common units used are kilojoules per gram (kJ/g) or megajoules per kilogram (MJ/kg). Representative calorific values for a range of common heating fuels are given in Table 1.

While calorific value is useful when comparing different fuel types, it has limited usefulness when comparing different wood species, since the range of variation is rather small except in the case of green wood. Calorific values for some New South Wales (NSW) species are given in Table 2. A favoured species as a fuelwood in those parts of NSW where it grows is red box (*Eucalyptus polyanthemos*) despite apparently having one of the lowest calorific values (see Table 2). Other species, having a similar calorific value (e.g. turpentine, *Syncarpia glomulifera*), make very poor fuelwood — they don't burn well in practice.

Density

Density is mass per unit volume usually expressed either as g/cm^3 or kg/m^3. However, this apparently simple relationship is rather more complicated in wood in that it can be stated in five ways:

(1) *Green density.* The mass of green wood (including water) per unit of green (swollen) volume.

(2) *Air-dry density.* The mass of air-dry wood per unit of air-dry volume.

(3) *Basic density.* The mass of oven-dry wood per unit of green volume.

(4) *Oven-dry density.* The mass of oven-dry wood per unit of oven-dry volume.

(5) *Density of wood substance.* The mass of oven-dry wood substance per unit of volume excluding all the gross capillaries of the wood.

Green density is highly variable largely because of moisture content variations, although basic density and extractives content may also contribute to a variation both between species and within a single tree. Green volume assumes the wood is above fibre saturation point and that no shrinkage has occurred. Green density is important since, in many Third World countries, fuelwood is frequently harvested in the green condition and carried a long way by hand.

Air-dry density is important in that, in practice, wood will burn most efficiently in the air-dry condition. Since the air-dry mass and volume will vary from place to place depending on atmospheric relative humidity and temperature, so will the air-dry density. For making accurate comparisons, therefore, air-dry wood should be at a specific moisture content (e.g. in southern Australia 12% is the standard).

Table 2. Estimated calorific value of some New South Wales (Australia) tree species in MJ/kg (source of data, Bootle 1971).

	Moisture content		
	Oven-dry gross	Air-dry*	Green
Radiata pine (*Pinus radiata*)	20.5	17.9	7.0
Rose she-oak (*Casuarina torulosa*)	20.5	17.9	13.3
Red bloodwood (*Eucalyptus gummifera*)	20.2	17.7	13.7
River red gum (*E. camaldulensis*)	20.2	17.7	11.6
White stringybark (*E. eugenioides*)	20.0	17.5	10.5
River she-oak (*C. cunninghamii*)	19.8	17.2	10.5
Grey box (*E. hemiphloia*)	19.5	17.2	13.3
Tallowwood (*E. microcorys*)	19.5	17.2	12.1
Spotted gum (*E. maculata*)	19.3	17.0	11.2
Red box (*E. polyanthemos*)	19.3	17.0	11.2
Blackbutt (*E. pilularis*)	19.1	16.8	10.7

* 12% moisture content.

Oven-dry density is of no importance in the context of fuelwood.

Basic density is a means of expressing wood density which is reproducible since oven-dry mass and green volume do not vary for any given piece of wood. It is a measure of the actual amount of wood substance present in a given volume.

Air-dry and basic densities are useful criteria for evaluating fuelwoods (i.e. a good species will be one which provides most heat for a given volume). For example, in Australia, eucalypts are generally preferred to pines, despite the higher calorific value of the latter since they are generally denser and give more wood substance, hence more fuel, per unit volume.

The density of wood substance is relatively constant for all species although it will vary according to the method by which it is determined. However, it is generally taken as about 1.5 g/cm^3. It has no practical significance in fuelwood evaluation.

Moisture Content

The moisture content of wood has a marked effect on the amount of effective heat released when it is burnt (Fig. 1). However, the relationship between effective heating value and moisture content can be expressed in several ways. This depends on whether the moisture content is on an oven-dry basis (i.e. weight of moisture as a percentage of wood dry weight) or green basis (i.e. weight of moisture as a percentage of wood green weight) and whether the weight of fuelwood is on an oven-dry or green basis. Thus relationships for effective heating value can be determined for: (1) moisture content on an oven-dry basis and oven-dry fuel weight (Fig. 1a (i)); (2) moisture content on an oven-dry basis and green fuel weight (Fig. 1a (ii)); (3) moisture content on a green wood basis and oven-dry fuel weight (Fig. 1b (i)); and (4) moisture content on a green wood basis and green fuel weight (Fig. 1b (ii)).

In practice, the second of these relationships is the most important since fuelwood is handled on a green weight basis.

It is important to note also that even when oven-dry wood is burnt, the gross energy contained in the wood is not converted completely to available heating energy. The average gross calorific value of 19.5 kJ/g for oven-dry hardwoods has been derived from bomb calorimetry tests in the laboratory. Because the calorimeter is a closed system, all heat generated by the combustion of the wood components including hydrogen (about 6% of wood mass) is captured. However, in practice, fires are open to the atmosphere and the latent heat of vaporisation from the water formed by the combustion of the hydrogen is lost from the system. This loss is equivalent to about 1.4 kJ/g so that the net calorific value of oven-dry hardwood is about 18 kJ/g (Harker et al. 1982; Fung 1984).

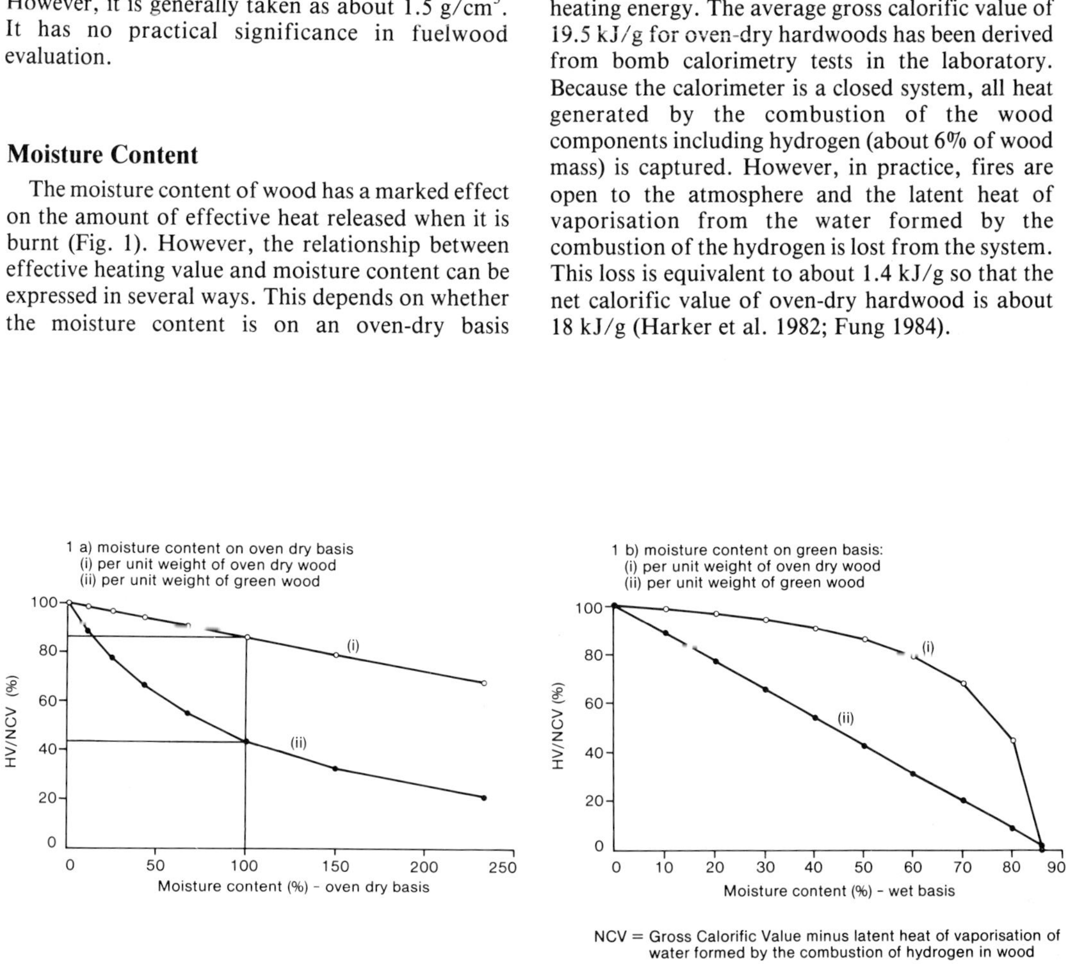

Fig. 1. Effect of moisture content on the heating value (HV) of fuelwood relative to the net calorific value (NCV).

Chemical Composition of Wood

Wood is mainly composed of three elements: carbon, hydrogen and oxygen. These are chemically combined and usually highly polymerised into two main groups of compounds: carbohydrates and phenolics. The former are cellulose and hemicelluloses, the latter are lignins.

Wood also contains substances known as extractives which are not part of the wood structure and which consist of a very large number of compounds of diverse chemical composition such as polyphenols, oils, fats, gums, resins, waxes and starch. These can be extracted from wood by various solvents such as water, methanol, ethanol, benzene, ether, acetone, sodium hydroxide and others without significantly affecting the wood structure.

The extractive content of some *Eucalyptus* species using four different solvents (Hillis 1962) is given in Table 3. Various solvents remove different extractive fractions and, in the case of NaOH, can remove part of the less resistant lignin and some of the carbohydrate (Smelstorius 1971). Although Smelstorius was investigating *Pinus radiata*, it would be prudent to anticipate a similar effect in other species.

The amount and types of extractives vary widely between species, within species and within a single tree. They are invariably more abundant in heartwood than in sapwood and increase with age of the wood. They may affect the calorific value of the wood and its flammability, but the effect is unpredictable. They also contribute to its density.

The calorific value of fuelwood is directly related to its elemental composition. Hence, ultimate analysis, i.e. quantitative estimation of each element, is a possible approach to fuelwood evaluation. Ultimate analysis of a number of species by Arola (1976) suggests that a 'typical hardwood' has less carbon and more oxygen than a typical softwood, and there will be variations within and between species and within the same tree (Table 4). The higher calorific values of softwoods are related to their oxygen content which is lower than in hardwoods. Oxygen is not a fuel; carbon and hydrogen are.

Qualitative Tests

Wood-Burning Tests

Qualitative wood properties are also important in

Table 3. Extractives content of the heartwood of some *Eucalyptus* species (source Hillis 1962).

Species	Number of samples	Mean extractives content as % of initial oven-dry mass after extracting with:			
		Hot water[b]	Ethanol[b]	Ethanol-benzene, then hot water[c]	0.5% NaOH[d]
E. crebra	10	13.4[a] 6.8–20.2	13.5 5.0–18.6		31.3 25.4–34.3
E. diversicolor	1			7.6	20.6
E. delegatensis	6	4.0 2.4–6.5	4.3 1.6–6.5		16.9 14.8–21.5
E. marginata	1			6.4	32.4
E. microcorys	1			17.5	24.8
E. obliqua	20	14.3 7.9–26.6	15.1 10.1–29.1		26.6 20.1–40.8
E. paniculata	12	10.4 7.6–17.5	9.0 5.6–17.9		22.7 18.7–26.2
E. polyanthemos	1			25.2	40.3
E. punctata	10	12.4 9.3–16.1	14.6 10.3–17.7		29.6 24.5–33.3
E. regnans	13	10.2 4.8–15.3	6.7 1.3–16.5		20.1 12.9–29.8
E. robusta	1			18.0	43.3
E. sideroxylon	8	15.0 10.2–19.0	19.1 12.8–23.6		34.0 30.1–38.4
E. sieberi	10	9.2 2.6–15.5	10.1 4.3–17.7		23.7 17.2–29.6

[a] Mean values and range.
[b] Continuous extraction for 24 hours.
[c] Continuous extraction for 24 hours in ethanol-benzene (1:2) followed by hot water for 24 hours.
[d] 2.5 g sample heated in 300 ml of NaOH for 1 hour, filtered, and washed with hot water.

Table 4. Ultimate analysis of some hardwoods and softwoods (source Arola 1976).

	Composition %				
	C	H	O	N	Ash[a]
Hardwood	50.8	6.4	41.8	0.4	0.9
Softwood	52.9	6.3	39.7	0.1	1.0

[a] A small amount of ash remains after combustion made up of inorganic constituents such as calcium and magnesium.

evaluating wood as a fuel. In this context 'wood-burning tests' are a practical method of evaluating species. One major aim of such tests is to measure the 'thermal efficiency' of fuelwood species under comparable cooking conditions. Thermal efficiency is inversely correlated with the mass of wood consumed during a standard test; less wood is required for a species with high thermal efficiency than for one with low.

Wood-burning tests may also evaluate a species in terms of the time required to complete a specific cooking task.

The test criteria outlined before may then be combined with observations on such characteristics as ease of ignition and smoke, spark and soot production in order to obtain some kind of ranking according to what are perceived as desirable qualities.

VITA (1982) describe three tests which can be used to assess fuelwood species:

(1) *Water boiling test:* A fixed mass of water is boiled using a known mass of fuelwood under standardised conditions. Species can be assessed by comparing the amount of fuel consumed during the test and the time taken to boil the water.

(2) *Controlled cooking test:* This test compares the fuel used and time spent in cooking an actual standardised meal (e.g. of rice). The test can be extended to determine whether or not a species can adequately cook the range of typical meals consumed by a defined community.

(3) *Kitchen performance test:* This test compares the wood consumed under normal household conditions within a community. It takes at least 5 days recording in detail each family's consumption of wood.

In addition to the VITA tests outlined above, a *crib test* was designed by Krilov et al. (1986) to evaluate the combustion characteristics of species. These are defined in three phases as follows (see also Fig. 2):

(1) *Ignition*: The ease with which wood ignites is determined. In general, the shorter the ignition phase the better.

(2) *Flaming*: The wood is actually flaming and being consumed rapidly. The relative importance of this phase depends on requirements. If a slow cooking is required, then the shorter the flaming stage the better. However, if an open fire is the only source of light then species that produce a longer flaming phase may be more desirable.

(3) *Embers*: This is the final combustion stage and generally produces the greatest proportion of usable heat energy under household conditions. For some heating purposes, species that produce the greatest quantity of persistent glowing embers may be most desirable.

In the rest of this chapter, tests carried out on four Australian-grown species using traditional criteria of density, moisture content and drying rate, extractives content and calorific value are discussed. Then, using the same four species, evaluations are made of burning tests which incorporate some of the features of the VITA tests and some of the Krilov crib test.

Materials and Methods

Sampling

Four species growing in the Australian Capital Territory (ACT) were selected to cover a wide range of density, initial moisture content and extractive content, and because they were readily available. These were *Eucalyptus melliodora* (yellow box), *E. blakelyi* (Blakely's red gum), *Acacia melanoxylon* (Tasmanian blackwood) and *Pinus radiata* (radiata pine). Only the first two would be regarded as good quality fuelwood in the ACT.

Each species, except *A. melanoxylon*, was collected from four ACT forests: Kowen, Stromlo, Uriarra and Pierce's Creek. *Acacia melanoxylon* was only available from Uriarra and Pierce's Creek. From each site, test samples were collected from a single tree. The diameter at breast height over bark (dbhob) of all trees sampled is given in Table 5. The samples were as follows: (a) 5 cm thick disc at breast height; (b) 40 cm long billet taken from immediately above the disc.

The samples were debarked, sealed in plastic bags and stored in a cold room at 4°C within 2 hours of felling to avoid moisture loss. The billets were subsequently radially sawn into quarters, each quarter sealed in a plastic bag and stored as above.

Determination of Density and Moisture Content

The procedure is as follows: (a) weigh each green disc to the nearest 0.1 g; (b) determine the green volume of each disc by displacement in distilled water using the method of Brown et al. (1952); and

Table 5. Dbhob in centimetres of trees sampled for fuelwood evaluation.

Site	Species			
	E. melliodora	*E. blakelyi*	*A. melanoxylon*	*P. radiata*
Pierce's Creek	22.8	24.6	17.0	23.7
Uriarra	22.6	21.4	19.0	21.7
Stromlo	23.6	20.7	NIL	23.4
Kowen	24.5	22.0	NIL	22.6

(c) oven-dry each disc at 105°C until it attains constant mass (about 48 hours).

Green and basic densities and moisture contents were calculated as previously discussed.

Assessment of Drying Rate

From each billet one of the quarters was resawn into 20 specimens measuring 2 × 2 × 18 cm (280 specimens in all). The specimens were immediately weighed and stored in a conditioning room at 20°C and 54% relative humidity. The specimens were reweighed at 2-day intervals for 20 days and then oven-dried and the oven-dry mass determined. The data were used to calculate the mean moisture loss as a percentage of oven-dry mass and to illustrate drying profiles.

Extractive Content

The discs used for density and moisture content determinations were sawn radially into quarters. Single slivers not more than 2 mm thick were then cut from alternate radial edges of each quadrant. Slivers were then ground, each species and each set of slivers separately, to pass a 20 mesh screen in a Wiley mill (Browning 1967). The number of replicates for each species was 16, except for *A. melanoxylon* for which it was 8.

Because of difficulties in removing some extractives, particularly polyphenols, from many hardwoods including eucalypts, the three hardwood species were treated differently to the *P. radiata*. The procedures were as follows:

(a) *P. radiata*: 2 g of wood flour from each set of slivers (oven-dried at 105°C) were placed in oven-dried cellulose extraction thimbles of known mass and extracted for 8 hours in a Soxhlet apparatus with a 7:3 ethanol:toluene mixture. After extraction the thimbles and wood flour were oven-dried for 12 hours and reweighed. The mass of extractives removed was determined as the difference between the initial mass of the wood flour plus thimble and the final mass. The extractive content was expressed as a percentage of the initial oven-dry mass of the wood flour.

(b) *Hardwoods*: 2 g of oven-dried wood flour of each species and each set of slivers were transferred into 3 × 300 cm^3 tall form beakers and 100 cm^3 of 0.5 *M* NaOH solution added to each. Each beaker was covered with a watch glass and heated in a bath of boiling water. Each mixture was stirred every 15 min and, after 1 hour, filtered by suction into a sintered glass crucible of known mass. Each residue was washed with hot water and 50 cm^3 of 10% acetic acid. Each crucible and contents were oven-dried for 12 hours and reweighed. The extractive content for each species was calculated in the same way as for *P. radiata*.

Calorific Value

The extracted wood flour obtained from the extractive content determinations and further samples of unextracted wood flour from each set of slivers were used to determine calorific values of the four species. Six samples from each of the four sites (two sites in the case of *A. melanoxylon*) were compacted into small cylindrical pellets and oven-dried at 105°C for 24 hours.

A Gallenkamp CB-370 bomb calorimeter was used for determining calorific values. It was calibrated with melted benzoic acid having a known calorific value of 26.48 MJ/kg. Samples (0.1, 0.2, 0.3, 0.4, 0.5 and 0.6 g) of benzoic acid were fired with oxygen at 25 atm in the calorimeter. Galvanometer readings were observed and the calibration constant (Y) calculated as follows:

Galvanometer deflection due to benzoic acid $= Q\ (Q_1, Q_2$ etc.)

Heat release from M kg of benzoic acid $= 26.48 \times M$ MJ

$$\therefore Y = \frac{26.48 \times M}{Q}.$$

After calibration, each wood flour pellet was weighed and transferred in a crucible to the calorimeter. Each pellet was tested separately. After each test the calorimeter was washed out thoroughly, dried and returned to the same conditions as when it was calibrated.

The calorific value of each pellet was calculated as follows:

Galvanometer deflection due to test pellet $= q$.

Heat release from m kg of test sample $= q \times Y$ MJ

$$\therefore \text{Calorific value of test sample} = \frac{q \times Y}{m}.$$

Water Boiling Test

The burning characteristics of a fuelwood are important in assessing both its performance and likely acceptance in domestic fireplaces. Water boiling tests were used to compare *E. melliodora, E. blakelyi, A. melanoxylon* and *P. radiata*, using a technique adopted from VITA (1982).

Quarters (3 × 40 cm long) of each of the sample billets (i.e. 42 quarters in all) were used for the tests. These had been kept green in the cold room. Each quarter was sawn into specimens 2 × 2 cm square by 18 cm long giving 20 specimens per quarter or 840 in all. One set, 280 specimens, derived from 14 quarters was sealed in a plastic bag and stored in a freezer to prevent loss of moisture; the second set was dried to 30% moisture content; the third set was dried to 12%. All sets were then further resawn to give a final accurate specimen size of 1 × 1 cm square by 18 cm long and were then stored by sets in plastic bags prior to testing.

The apparatus used was developed by the Queensland Department of Forestry and consists of the following: (1) a 20-l steel drum mounted on a balance with a full-scale capacity of 16 kg and accurate to 1 g. The drum was used as a combustion chamber with circular vent holes around its circumference and near the base. The balance was protected from heat by three layers of fibre/cement board. (2) A 4-l metal can with a lid and filled with 2 kg of distilled water. A thermometer and a thermocouple were taken through the lid to about 1 cm above the base of the can and were thus immersed in the water. The can was suspended at an exact height above the fuel bed. The fuelwood was supported on a grate of steel rods inserted into the combustion chamber (see fig. 1 in Chapter 18 of this Monograph).

The test apparatus was surrounded by a windbreak and all tests were carried out on clear days between 12 noon and 4 pm when air temperatures were between 20 and 25°C.

The wood samples were arranged in a standard criss-cross crib, total mass of each crib for each species being 351 g. Some adjustment in the size of one of the samples in each crib was necessary to obtain equal masses per crib.

The wood was ignited with 25 ml of kerosene which was poured carefully over the wood. The test started with ignition of the kerosene.

The reduction in fuel mass was recorded against time. The water temperature was recorded against time on a chart connected to the thermocouple. After reaching 60°C water temperature was recorded at 5C° intervals using the thermometer and checked against time to ensure the chart recorder was correctly calibrated.

The time taken to reach boiling point and the mass of fuel consumed during that time was also recorded.

Exactly 10 min after the boiling point was reached the metal can and the combustion chamber plus contents were weighed in order to calculate the mass of evaporated water and the total mass of wood consumed.

Results and Discussion

Densities and Moisture Content

Green, air-dry and basic densities and initial (green) moisture content are given in Table 6 for wood of the four species.

The green, air-dry and basic densities in descending order of magnitude are: *E. melliodora, E. blakelyi, A. melanoxylon* and *P. radiata* except for green density, where *A. melanoxylon* and *P. radiata* are reversed because of the very high moisture content of *P. radiata*.

Since high densities and low initial moisture contents are traditionally preferred for fuelwood, and since basic and air-dry densities are initially useful criteria for evaluating fuelwoods in a more scientific way, the four species must be rated in

Table 6. Mean density values and green moisture contents of the four species of wood.

Species	No. of samples	Green density (g/cm^3)	Basic density (g/cm)	Air-dry density (g/cm^3)[a]	Green moisture content (%)	CSIRO data[b]	
						No. of samples	Basic density (g/cm^3)
E. melliodora	4	1.261	0.785 (0.006)	0.977	60.7 (5.4)	12	0.899 (0.009)
E. blakelyi	4	1.188	0.698 (0.102)	0.854	70.2 (15.5)		
A. melanoxylon	2	0.965	0.519*(0.099)	0.624	85.9 (1.5)	45	0.546 (0.010)
P. radiata	4	1.001	0.395*(0.026)	0.477	153.5*(25.8)	10[c]	0.404 (0.010)

* Significant difference $P < 0.05$

Standard deviations are given in parentheses.

[a] Estimates based on unit shrinkages given in Kingston and Risdon (1961) and an air-dry moisture content of 12%.

[b] Kingston and Risdon (1961)

[c] 10–20 years old from South Australia.

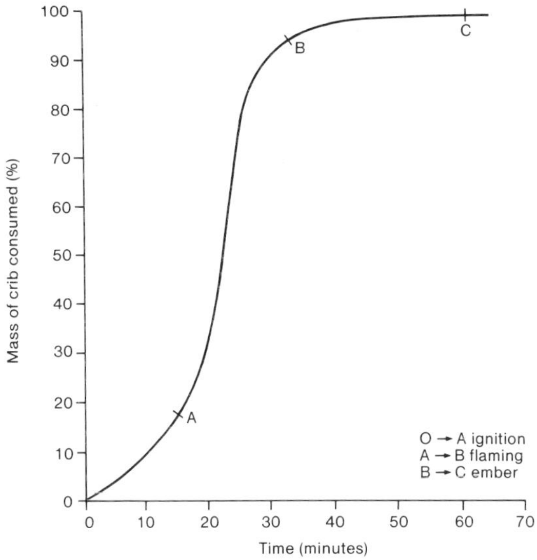

Fig. 2. Phases of combustion of wood (source Krilov et al. 1986).

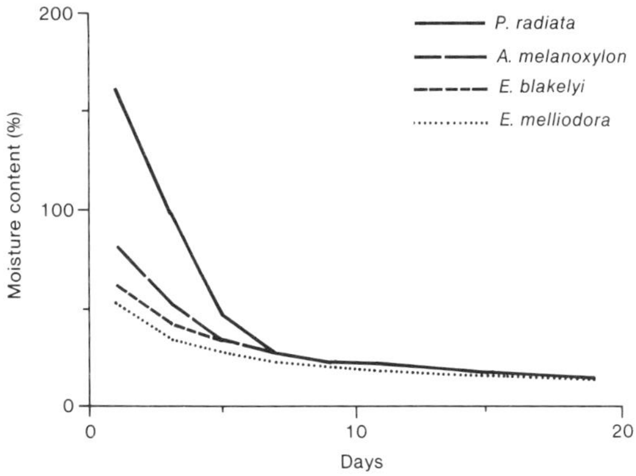

Fig. 3. Drying profile.

descending order of desirability as follows: *E. melliodora, E. blakelyi, A. melanoxylon, P. radiata.*

The high-density species have a major advantage in that, although the heating value per unit mass may not be so very different between the species, the heating value per unit volume will be substantially greater (e.g. *E. melliodora* will have roughly twice the heating value of *P. radiata* per unit volume).

Drying Rate Assessment

Although high initial moisture content is undesirable for fuelwood, this disadvantage may be offset if the drying rates of low-density species are faster than those of high-density species. Figure 3 summarises the data for the four species.

The initial drying rate is, in descending order of magnitude: *P. radiata, A. melanoxylon, E. blakelyi* and *E. melliodora*. However, the fibre saturation point (about 28–30% moisture content) is reached about day eight by all four species. Thereafter, the drying rate is virtually identical.

The faster initial drying rate of the less dense species compensates for the higher initial moisture content. Thereafter, however, the differences between the species are negligible.

In practical terms, species characteristics with respect to drying rate would seem to be not very important if the fuelwood pieces are small enough to allow rapid drying as in the work reported here. Larger sizes would show somewhat different drying characteristics and actual drying times could be markedly different between the species. In other words, when using wood as fuel, pieces should be as small as practicable to encourage rapid drying.

Extractive Content

The mean values for the extractive contents of the samples of the four species tested are shown in Table 7.

The data for the two eucalypts are of the same order of magnitude as those given in Table 3 for a range of eucalypts when using 0.5% sodium hydroxide as solvent.

Apart from *E. melliodora*, there is a wide variation in the extractives content from the mean value as shown in Table 7. *Pinus radiata* is apparently significantly lower than the rest. However, the extraction procedure for *P. radiata* was very different to that for the other species and valid comparisons cannot be made. On the other hand, the very low extractives content of radiata pine may stem from the age of the material (from 0 to about 20 years old) implying a high proportion of sapwood. Where *P. radiata* has been extracted with 0.1 *M* sodium hydroxide then loss of mass during treatment was 16.7% of the initial oven-dry mass of the wood (Smelstorius 1971). We may therefore conclude that in order of decreasing extractive content the four species tested may be ranked as in Table 7.

Table 7. Extractives content as a percentage of the unextracted oven-dry mass of wood.

Species	No. of samples	Extractives content (%)
E. blakelyi	4	33.8(10.19)
E. melliodora	4	27.4(3.88)
A. melanoxylon	2	23.2(12.19)
P. radiata	4	3.0*(3.05)

* Significant difference $P < 0.05$.
Standard deviations are given in parentheses.

Table 8. Mean gross calorific values of unextracted and extracted oven-dry wood in MJ/kg.

Species	No. of samples	Unextracted oven-dry wood	No. of samples	Extracted oven-dry wood
E. melliodora	24	21.12 (1.10)	24	21.22 (0.47)
E. blakelyi	24	20.89 (1.19)	24	20.95 (1.02)
A. melanoxylon	12	21.00 (0.85)	12	20.76 (0.86)
P. radiata	24	21.58 (0.85)	24	22.04 (1.26)

Standard deviations are given in parentheses.
Differences between values are not significant.

We cannot, at this stage, draw any conclusions as to the effect of extractives on the fuelwood properties of the four species.

Calorific Value

Mean gross calorific values for unextracted and extracted oven-dry wood, determined in the Gallenkamp bomb calorimeter, are given in Table 8. There were no significant differences between species with respect to these values. The data also suggest that extractives have no importance in determining calorific value. This is contrary to Howard (1973) and Wang and Huffman (1982) who found calorific value and extractive content were positively correlated. Except for *A. melanoxylon*, the presence of extractives would appear to diminish the calorific value. However, whilst having apparently little effect on calorific value, extractives may have some effect on the combustion characteristics of the wood (e.g. the duration of the ignition, flaming and ember phases respectively).

Since wood used as a domestic fuel would not be burnt in the oven-dry condition in practice, calorific values derived from the bomb calorimeter would have to be adjusted for moisture content. Estimates for unextracted wood, derived from first principles and from the original data in Table 8, are given in Table 9.

Moisture content is obviously important in determining effective heating value. In the extreme case of *P. radiata* the calorific value of green wood is only 36% that of air-dry wood. Moisture content will also have an effect on other combustion characteristics of wood.

Table 9. Estimated mean calorific values in MJ/kg adjusted for moisture content (unextracted wood).

Species	Moisture content (%)		Green values based on % moisture content
	Air-dry (12%)	Green	
E. melliodora	18.57	12.15	60.7
E. blakelyi	18.37	11.20	70.2
A. melanoxylon	18.47	10.08	85.9
P. radiata	18.98	6.88	153.5

The heat energy available from wood will also be determined by the efficiency with which heat is transferred from burning wood to the food to be cooked or to the people to be warmed. In many Third World households the transfer process is very inefficient so that further substantial losses occur. How to increase the efficiency of the transfer process is outside the scope of this paper but its importance should not be overlooked. It is much more important than considerations of calorific value.

Water Boiling Tests

None of the species ignited in the green condition and only *P. radiata* ignited at 30% moisture content. These tests were abandoned; all the data were derived from wood at 12% moisture content. The results for the boiling water tests are summarised in Table 10.

The tests were continued until all the wood was deemed to have been consumed, the loss in mass being determined at frequent intervals. Combustion profiles for each species are shown in Fig. 4. These profiles are similar in general outline to Fig. 2.

Subjective assessments of combustion were also made with respect to ease of ignition, smoke emission, flaming and production of embers.

Although the test results relate solely to wood at 12% moisture content, it may be inferred that the moisture content of wood is important not only in terms of available energy but also through its effect on ease of ignition. Furthermore, it would seem that a low basic density species such as *P. radiata* will ignite more readily than high basic density species at about 30% moisture content (i.e. approximately at the fibre saturation point). What the effect would be at higher moisture contents needs further investigation. Basic density and moisture content tend to be inversely correlated so that high moisture content may offset the effect of low basic density. In practice, of course, the size and shape of the fuelwood is also important. Hence the use of 'kindling' to start a fire.

From subjective assessments *P. radiata* and *A. melanoxylon* at 12% moisture content ignited easily; *E. melliodora* and *E. blakelyi* did not.

Table 10. Water boiling tests.

	E. melliodora	*E. blakelyi*	*A. melanoxylon*	*P. radiata*
SFC[a]	0.11 (0.016)[d]	0.11 (0.008)	0.13 (0.007)*	0.14 (0.017)*
SSC[b]	1.96 (0.429)	2.15 (0.469)	1.92 (0.198)	2.50 (0.652)
Time to BP[c]	19.40 (9.281)	17.20 (5.430)	8.90 (2.298)*	7.40 (0.657)*

[a] Specific fuel consumption in kilograms of wood required to bring 1 kg of water to boiling point.
[b] Standard specific consumption in kilograms of wood required to evaporate 1 kg of water (covering the entire test period) i.e. from starting the fire to 10 min after the boiling point was reached.
[c] The time to reach boiling point in minutes.
[d] Figures given are the mean of four replicates from each test site with standard deviations in parentheses.
* Significant differences at $P < 0.05$.

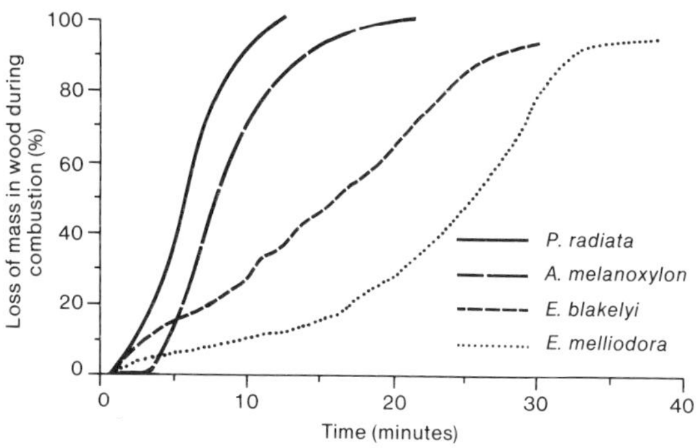

Fig. 4. Combustion profile (loss in mass of wood during combustion vs time).

According to the data in Table 10, *E. melliodora* and *E. blakelyi* are the most efficient fuelwoods in terms of specific fuel consumption (SFC) in an open fire (SFC is the mass of wood required to bring 1 kg of water to boiling point). This would make them more desirable if fuelwood was in short supply. However, both species took much longer to bring water to the boil than *A. melanoxylon* and *P. radiata*. The wood substance of the higher density species is consumed more slowly than that of the lower density species so that energy release is slower.

Where quick cooking is required, low-density species in the air-dry condition may be most effective. Where slow cooking (e.g. simmering or baking in an oven) is required, high-density species may be most effective. After reaching an appropriate cooking temperature glowing embers will maintain that temperature, whereas fires of low-density, fast-burning species will require more frequent refuelling. The rate of combustion can, of course, be controlled in slow or 'controlled' combustion stoves.

Standard specific consumption (SSC) is the mass of wood required to evaporate 1 kg of water. Equal masses of wood contain roughly equal calorific values, so that SSC data should be about the same for each species. This is not markedly confirmed by the data in Table 10. Indeed one would not expect this since combustion was uncontrolled; there are energy transfer processes involved and substantial energy losses. However, there are no significant differences between the species suggesting that, although they burnt differently, they would rank equally using SSC as a criterion.

From subjective assessments, *P. radiata* and *A. melanoxylon* not only ignited more easily than the other two species but their smoke emissions were low. They burnt quietly with large flames but did not form hot embers. *Eucalyptus melliodora* and *E. blakelyi* were very smoky during the early stages of combustion and only provided good flames just before the water reached boiling point. However, they both produced good hot embers.

In general, the experimental procedures for determining SFC and SCC need to be improved. The net calorific value of the 0.22 kg of air-dry *E. melliodora* or *E. blakelyi* consumed in bringing 2 kg of water to the boil is 3.7 MJ; the net calorific value of 25 ml of kerosene is about 20% of the total energy available to bring the water to the boil. The SFCs are, therefore, probably understated by about 20%, the SCCs by a smaller percentage.

Conclusions

The ideal fuelwood should have high calorific value, high density, low initial moisture content and a rapid rate of seasoning from green. However, gross calorific value of oven-dry wood is not important in comparative evaluations. In the four species tested there were no significant differences in this value. Even when extractives were removed there were no significant differences in the gross calorific value of extracted and unextracted wood of the four species tested. However, effective calorific value depends also on the moisture content. The higher the moisture content the less efficient the wood as a fuel since the net calorific value available for heating is reduced. All wood used as a domestic fuel, especially in countries suffering a fuelwood shortage, should be air-dried.

Tests conducted on drying rate showed that the fibre saturation point for the four species (about 28–30%) was reached after 8 days irrespective of the initial moisture content and that, thereafter, the drying rates were virtually identical. The sample pieces were, however, small and results would become very different as size increased. Therefore, when using wood as a fuel, the pieces should be as small as practicable to encourage rapid drying. They should also be protected from rain since smaller pieces will also increase in moisture content more rapidly if wetted. In the burning tests, moisture content was also shown to have an important effect on ignition. The tests of green wood and wood at 30% moisture content were abandoned because, under the standard ignition conditions adopted, none of the test species could be ignited at 30% moisture content except *P. radiata*. In practice, of course, virtually all species, no matter what their moisture content, can be ignited provided sufficient kindling is available. However, it would take much more time to get the fire started and a great deal of smoke would be generated.

High basic (or air-dry) density is a useful criterion in predicting fuelwood properties. High-density wood yields more heat per unit volume and is, therefore, better in a practical sense for storage, handling and transport purposes. It may also reduce the time and cost of obtaining the wood. However, high-density wood is more difficult to ignite and, in the water boiling tests reported in this Chapter, took much longer to bring the water to the boiling point. On the other hand the same air-dry mass burned for much longer in high-density than in low-density wood and produced hot embers, so that for certain cooking requirements and for room heating these may be very desirable characteristics.

Overall, the tests indicate that no one species has all the desirable fuelwood characteristics. If cooking must be done quickly, less dense species that burn quickly and consequently generate heat more rapidly may be preferred. On the other hand, cooking rates can also be varied by altering the piece size of a single fuelwood species. In this case, however, low-density wood will not give the range of variation which high-density wood can provide. Where cooking must be done slowly dense species, which maintain a steady heat over a long period by producing quantities of hot embers, may be preferred.

Finally, it should be emphasised that all the tests were carried out under conditions of uncontrolled combustion. Testing needs to be done under conditions of both uncontrolled and controlled combustion in order to elucidate the advantages to be derived from the latter. Where fuelwood plantations are planned for a particular population a sample of households could be supplied with several candidate species. The people could then be asked to rank them according to their preferences and to provide their answers in a standard questionnaire.

Acknowledgments

The authors thank Mr Andy McNaught, Queensland Department of Forestry, for compiling data for Fig. 1.

Chapter 17

Fuelwood Evaluation Using a Simple Crib Test

W.D. Gardner

Abstract

A simple crib test was developed to evaluate several fire performance properties of fuelwood. Twelve Australian-grown species were evaluated during three burning phases (ignition, flaming and ember) and for residual matter. Species have been ranked for each phase and assessments made for most desirable fuelwoods using up to two phases in combination.

Introduction

The scientific evaluation of tree species for fuelwood has been based traditionally upon a knowledge of their basic properties including basic density, ash content, carbon content and volatile matter content (Krilov et al. 1986). However, these basic properties should not be used alone or combined to predict the likely performance of tree species when they are used for fuelwoods, as no relationships have been established between basic properties and fuelwood performance. Indeed, the value of a tree species as a fuelwood may be dependent upon the manner in which it is used (e.g. in an open fire, conventional stove, slow combustion stove, etc.). It is therefore important that a test be developed for wood that will determine the fire performance properties relevant to a specific fuelwood utilisation.

A simple crib test has been developed to determine several fire performance properties of wood that are relevant to fuelwood being used in an open-fire situation. The aim of the experiment is to examine the fuelwood properties derived from the crib test for a select range of Australian-grown timbers. It is felt that the crib test has special relevance to people in developing countries.

Materials and Methods

Sample Preparation

Twelve hardwood and one softwood timber species were tested (Table 1). Test pieces, 85 × 12 × 12 mm, were sawn by band-saw from air-dried boards of each species. The test pieces were conditioned to equilibrium moisture content in an atmosphere of 20 ± 2°C and 65 ± 5% relative humidity before testing.

Determination of Air-Dry Density

The air-dry densities of the test materials were determined when equilibrium moisture content had been achieved in the conditioning atmosphere. The equilibrium moisture content (determined by oven drying at 102°C) and air-dry density data are given in Table 1.

Crib Construction

Cribs were constructed from test pieces of all species. Cribs are an ordered arrangement of cross-piled pieces of wood. For this study 24 test pieces were used in each crib. Each crib contained 6 rows of 4 test pieces. The cribs were built on two 10 × 10 × 85 mm steel runners in a 65 × 85 × 5 mm metal tray. The steel runners were used to ensure that the ignition fluid did not contact the test pieces in the crib.

The cribs were built on the pan of a Mettler P2210 balance. The balance was protected from the heat generated during the test by two separated layers of 5 mm insulating millboard. A metal beaker containing 2 l of tap water at ambient temperature was placed 105 mm above the top surface of the crib

Table 1. Common and scientific names, nature, air-dry density and moisture content (oven-dry at 102°C) of tested timber species.

Species	Nature (hardwood = H; softwood = S)	Air-dry density (kg/m^3)	Moisture content (% mass/mass)
Blackbutt (*Eucalyptus pilularis*)	H	799	12.1
Broad-leaved tea-tree (*Melaleuca quinquenervia*)	H	753	9.4
Brush box (*Lophostemon confertus*)	H	776	11.4
Forest oak (*Allocasuarina torulosa*)	H	832	11.4
Grey ironbark (*E. paniculata*)	H	1099	11.7
Northern wattle (*Acacia crassicarpa*)	H	604	9.3
Parinari (*Parinari nonda*)	H	816	9.9
Radiata pine (*Pinus radiata*)	S	565	10.4
Red ironbark (*E. sideroxyolon*)	H	1063	11.9
Spotted gum (*E. maculata*)	H	1040	10.0
Swamp oak (*C. glauca*)	H	895	12.0
Sydney blue gum (*E. saligna*)	H	911	12.4
Turpentine (*Syncarpia glomulifera*)	H	985	12.4

Fig 1. A carefully constructed crib for controlled fuelwood tests is built over a tray containing ethanol and a weighing balance. The balance is protected by two separated layers of insulated millboard and the test performed in a draught-free environment. Photograph courtesy of the Forestry Commission of New South Wales.

to simulate the influence of a suspended cooking utensil that would normally be used with an open fire, on the burning performance of the crib. The assembled test is shown in Fig. 1.

The ignition fluid was introduced into the metal tray by pipette. The ignition fluid was ethanol and initially 20 ml was used. Subsequently the tests were repeated using 10 ml. The cribs were tested in duplicate with 20 ml of ethanol and singly with 10 ml of ethanol.

Test Observations

The balance reading was recorded at 1-min intervals after the fluid was ignited. The test was terminated at 60 min or after constant weight was recorded for three consecutive 1-min intervals, whichever occurred first. The cribs always collapsed between the time they stopped flaming and the time the test was terminated. When the cribs collapsed the residual material was piled in the metal tray to cover approximately the same areas as the original crib. Supplementary observations were made during the test, including the time at which the crib ignited, time at which flaming of the crib ceased and the time at which the crib collapsed.

Assessment of Fire Performance Properties

The following parameters were used to determine fire performance properties:

(a) Ease of Ignition

The ease of ignition was determined by whether the crib ignited and sustained combustion with either 10 or 20 ml of ethanol (Table 2). It is classified into three classes:

Table 2. Ease of ignition of test cribs.

Species	Ignition fluid (ethanol)	
	10 ml	20 ml
Blackbutt	Ignition	Ignition
Broad-leaved tea-tree	Not tested	Ignition
Brush box	Ignition	Ignition
Forest oak	Not tested	Ignition
Grey ironbark	No ignition	Ignition
Northern wattle	Ignition	Ignition
Parinari	Ignition	Ignition
Radiata pine	Ignition	Ignition
Red ironbark	No ignition	Ignition
Spotted gum	No ignition	Ignition
Swamp oak	Ignition	Ignition
Sydney blue gum	No ignition	Ignition
Turpentine	No ignition	No ignition

Class 1 — easiest to ignite (ignited with 10 ml of ethanol); *Class 2* — moderate ease of ignition (ignited with 20 ml of ethanol but not 10 ml); and *Class 3* — most difficult to ignite (could not be ignited with 20 ml of ethanol).

(b) Percentages of Cribs Consumed in the Flaming Mode

The percentages of the cribs consumed in the flaming mode were calculated on an oven-dried (OD) crib mass basis to remove the variation that would result from testing cribs at different moisture contents. The following formula was used:

OD weight of crib consumed
= OD weight of original crib - weight remaining when flaming ceased

% weight of crib consumed

$$= \frac{\text{OD weight of crib consumed }\%}{\text{OD weight of original crib}}.$$

(c) Percentages of Cribs Consumed in the Ember Mode

The percentage (mass/mass) of the cribs consumed in the ember mode is the difference between 100% and the sum of the percentage of the cribs consumed in the flaming mode and the percentage remaining as residual matter.

(d) Percentages of Residual Matter After Combustion of the Crib

At the termination of each test the percentage (mass/mass) of residual matter was calculated. The calculations were based on the oven-dried weight of the crib. The residual matter was a combination of inorganic and unburnt organic materials.

(e) Maximum Consumption Rates of Cribs

The maximum weight loss over any 3-min interval was determined for each crib. The 1 min average of that 3 min weight loss was the maximum consumption rate for the crib.

(f) Data Analyses

The data obtained for percentages of cribs consumed in the flaming mode, percentages of cribs consumed in the ember mode, percentage of residual matter after combustion of cribs and maximum consumption rates of cribs were analysed using Duncan's multiple range test.

Results

Ease of Ignition

Results are given in Table 2. There were insufficient samples of the broad-leaved tea-tree and forest oak to determine whether ignition could be achieved with 10 ml of ignition fluid. Of the 13 species tested, turpentine was the only one that was not ignited with 20 ml of ethanol. Of the 11 species that were tested with 10 ml of ignition fluid only six achieved ignition. The species can be assigned to three classes according to their ease of ignition:
Class I — easiest to ignite: blackbutt, brush box, northern wattle, parinari, radiata pine, swamp oak; *Class II* — moderate ease of ignition: broad-leaved tea-tree, forest oak, grey ironbark, red ironbark, spotted gum and Sydney blue gum; *Class III* — most difficult to ignite (could not be ignited with 20 ml of ethanol): turpentine.

The species listed in Class I are in the air-dry density range of 565–895 kg/m^3, while those in Class II, with the exception of broad-leaved tea-tree and forest oak, are in the range of 911–1099 kg/m^3. It is probable that broad-leaved tea-tree and forest oak would have been assigned to Class I had they been tested as their densities are 753 and 832 kg/m^3 respectively (i.e. they are in the density range of the species assigned to Class I for ease of ignition).

Percentages of Cribs Consumed in the Flaming Mode

The percentages (mass/mass) of the cribs consumed in the flaming mode are given in Table 3. The data were generated for duplicate tests with 20 ml of ignition fluid for all species. When crib combustion was achieved with 10 ml of ignition fluid the data for the percentage consumed in the flaming mode were analysed with that for the 20 ml ignition fluid tests.

The data were analysed using Duncan's Multiple Range Test to determine whether the means were significantly different at the 5% level of significance. The means were found to be in three significantly different groups:

Group I	*Group II*	*Group III*
Broad-leaved tea-tree	Parinari	Red ironbark
Radiata pine	Northern wattle	Blackbutt
	Forest oak	
	Swamp oak	
	Brush box	
	Sydney blue gum	
	Spotted gum	
	Grey ironbark	

The test species are arranged above in deceasing order of percentages of the cribs consumed in the flaming mode. Thus significantly greater percentages of the species in Group I were consumed in the flaming mode than those listed in Group II or Group III.

Percentages of Cribs Consumed in the Ember Mode

The data are given in Table 3. When Duncan's Multiple Range Test is applied, the means can be placed into three groups that are different at the 5% level of significance.

Group I	*Group II*	*Group III*
Brush box	Parinari	Forest oak
	Swamp oak	Radiata pine
	Spotted gum	Sydney blue gum
	Grey ironbark	
	Red ironbark	
	Northern wattle	
	Broad-leaved tea-tree	
	Blackbutt	

The species are arranged above in decreasing order of percentage consumed in the ember mode. Thus a greater percentage of the brush box cribs were consumed in the ember mode than those in Group II or Group III.

Percentage of Residual Matter After Combustion of Cribs

The data for the test cribs are given in Table 3. When the data were analysed using Duncan's Multiple Range Test it was found the means could be placed in three groups that were significantly different at the 5% level of significance.

Group I	*Group II*	*Group III*
Forest oak	Red ironbark	Parinari
Sydney blue gum	Grey ironbark	Brush box
Blackbutt	Spotted gum	
	Radiata pine	
	Northern wattle	
	Broad-leaved tea-tree	
	Swamp oak	

The species are arranged in decreasing order of residual matter from Group I to Group III. Thus significantly greater quantities of residual matter remained after combustion of cribs made from the species in Group I than those in Group II and Group III.

Maximum Consumption Rates of Cribs

The data for the maximum consumption rates of the cribs are given in Table 4. Analysis of the data using Duncan's Multiple Range Test indicates that

Table 3. Percentages of cribs (mass/mass) consumed in the flaming and ember modes and the percentages of residual matter after combustion of the cribs.

	Crib consumed					
	Flaming		Ember		Residual matter	
Species	Range	Mean	Range	Mean	Range	Mean
Blackbutt	75.3–77.3	76.1	11.1–15.1	13.7	9.1–11.6	10.1
Broad-leaved tea-tree	81.4–84.7	83.1	13.9–14.3	14.1	1.0–4.7	2.9
Brush box	77.6–79.1	78.5	20.2–21.9	21.0	0.5–0.7	0.6
Forest oak	78.3–82.8	80.6	6.0–10.4	8.2	11.2–11.3	11.3
Grey ironbark	76.5–78.3	77.4	14.0–15.0	14.5	6.7–9.5	8.1
Northern wattle	79.0–84.3	81.0	8.5–17.4	14.2	3.6–7.2	4.8
Parinari	78.9–82.3	81.0	16.6–20.9	18.3	0.2–1.1	0.7
Radiata pine	80.8–84.5	82.6	10.9–11.7	11.4	3.9–8.3	6.0
Red ironbark	75.3–77.0	76.2	14.3–14.5	14.4	8.5–10.4	9.5
Spotted gum	77.7–77.9	77.8	14.0–17.7	15.9	4.6–8.1	6.4
Swamp oak	78.2–80.2	79.5	17.9–18.4	18.1	1.0–4.7	2.4
Sydney blue gum	74.8–80.9	77.9	5.7–16.9	11.3	8.3–13.4	10.9

Table 4. Maximum rate of consumption of cribs.

Species	Maximum consumption rate (g/min) Range	Mean
Blackbutt	26.7–35.0	30.6
Broad-leaved tea-tree	38.0–40.3	39.2
Brush box	24.7–41.7	33.0
Forest oak	31.3–35.3	33.3
Grey ironbark	26.3–30.3	28.3
Northern wattle	34.3–34.7	34.4
Parinari	36.3–40.7	38.4
Radiata pine	24.0–28.9	26.3
Red ironbark	22.3–26.3	24.3
Spotted gum	33.0–41.0	37.0
Swamp oak	28.3–42.7	36.6
Sydney blue gum	25.7–29.3	27.5

the means can be placed in three groups that are different at the 5% level of significance.

Group I	*Group II*	*Group III*
Broad-leaved tea-tree	Spotted gum	Radiata pine
Parinari	Swamp oak	Red ironbark
	Northern wattle	
	Forest oak	
	Brush box	
	Blackbutt	
	Grey ironbark	
	Sydney blue gum	

The species are arranged in decreasing order of maximum consumption rates in the above groups.

Discussion

The crib test data can be used to rank the relative values of timber species as fuelwoods. The ranking may be based upon one fire performance property or a combination of properties and may only be relevant to fuelwoods that are used in an open-fire situation with a suspended cooking utensil.

The ease of ignition and the amount of residual matter remaining after combustion are the most important fire performance properties to be considered when efficient fuelwood utilisation is important. The energy required to ignite the fuelwood and maintain the early stages of combustion will not be available for heating or cooking and the unburnt organic materials in the residual matter remaining after combustion will not have been converted to heat energy. Therefore, ideal fuelwoods should ignite easily and produce very little residual matter. Cribs made from six of the test species — blackbutt, brush box, northern wattle, parinari, radiata pine and swamp oak — achieved ignition with 10 ml of ethanol and are preferable to those species that required 20 ml of ethanol to achieve ignition. The turpentine crib could not be ignited with 20 ml of ethanol and was the least suitable species for use as a fuelwood. Of the species that ignited with 10 ml of ethanol parinari and brush box produced significantly less residual matter than the other species. Blackbutt produced significantly greater amounts of residual matter than the other species that achieved ignition with 10 ml of ethanol. Therefore, if ease of ignition and least residual matter remaining after combustion are to be the two critical fire performance properties, brush box and parinari were the best performers and turpentine the worst.

If it is considered desirable to have fuelwoods that are largely consumed in the flaming mode, broad-leaved tea-tree and radiata pine are the best performing species. If the desired fire performance is to have most of the crib consumed in the ember mode, brush box is the preferred species.

If high energy production rates in the flaming mode are required to reduce the time available for heat losses from the cooking utensils, broad-leaved tea tree and parinari are the preferred species.

Thus the combinations of fire performance properties required will allow the most suitable timber species to be determined.

An equally important aspect is the ability to identify timber species that would not be suitable for use as fuelwoods. These would include those that are difficult to ignite and/or produce large quantities of residual matter. With these criteria, turpentine, forest oak, Sydney blue gum and blackbutt would not be suitable as fuelwood species.

As stated earlier, the prediction of the fuelwood suitability of the timber species may vary if a different fire scenario was tested, or if the parameters of the reported crib test (e.g. test piece size, number in crib, arrangement of test pieces in crib, etc.) were changed.

The crib test should be repeated using the same tree species and crib design but with varying conditions of insulation and air supply. This would allow the fuelwood value of the species to be determined when they were used in stoves, ovens and other cooking and heating appliances.

Fuelwood testing unit developed by the Queensland Department of Forestry. Robyn Bell is shown with the test rig, consisting of a weighing balance, protective asbestos sheeting, a small drum to contain the fire and another small drum containing water inside the fire drum. The temperature of the water is monitored. Photograph courtesy of the Queensland Department of Forestry.

Chapter 18

Drying and Burning Properties of the Wood of Some Australian Tree Species

D.K. Gough, R.E. Bell, P.A. Ryan and C.T. Bragg

Abstract

Material 2.5 years old, from 15 species established in trial plots in southeast Queensland, was sampled for drying and burning studies. For the drying studies, 0.6 m lengths were dried under cover and weighed periodically until their weight approached stability. Each of the species was then tested as a fuel in the burning studies, using standardised simulated cooking fires.

Drying models were derived in which initial moisture content, basic density, piece diameter and a developed drying factor were included as variables. The drying factor was found to have the greatest influence on drying rate. Data are presented on the initial moisture content, green density, basic density and computed drying times to 24% moisture content for each species.

In each burning study, 800 g of air-dried fuel was burnt in a 20-l fire bucket to heat 4 l of water. The rate of fuel consumption, rate of temperature rise of the water and the heat energy used by the water were obtained for each species. The burning trials revealed that all species tested should be acceptable as fuelwood.

It was concluded that the emphasis in future studies should be on the drying behaviour of species rather than on the development of detailed quantitative information on burning properties. These properties are more appropriately described in terms of qualitative attributes such as the capacity of the wood to burn evenly, without smoke, crackling or sparking.

Introduction

An ideal domestic fuelwood burns slowly, producing long-lasting embers, without smouldering or emitting sparks or smoke. The best wood for domestic fuel is generally accepted as being dry, dead wood from mature trees of high-density species. However this type of wood is seldom available where fuelwood is scarce and has to be harvested from planted trees. In such situations the wood is likely to be from living trees, no more than a few years old, mostly sapwood, of high moisture content and small dimensions. Differences in ash content, volatile matter, carbon content and calorific value have been reported between species but these differences are usually relatively minor and of no real significance in fuelwood selection. While one important consideration in selecting species for fuelwood plantings is productivity, expressed as dry weight yield of fuelwood per hectare per year, others include the drying rate of the wood and its burning properties when used in domestic cooking fires.

These two properties, drying rate and burning properties, were evaluated for 15 Australian species included in the ACIAR species trials described by Ryan and Bell (Chapter 5).

Materials and Methods

Sample material for the study was obtained from 2.5-year-old trees. A minimum of three trees per species was sampled and where more than one provenance was available, all provenances contributed to a species sample. Stems and branches were cut into pieces about 1.2 m long and replicates obtained over the range of piece diameter classes for each species. Pieces >25 mm diameter were sampled at mid-length for moisture content (expressed as a percentage of oven-dry weight) and green and basic density determination (bark included). Each half piece was numbered, weighed and the mid-point diameter measured. Twigs <25 mm diameter were bulked for each species and weighed. All material was stacked on bearers for drying under cover and protected from rain. Each piece was weighed at weekly intervals until no significant weight losses were recorded over two consecutive weeks. Weather was warm to hot initially and cool at completion; daytime humidity levels were generally 50–70% (Table 1).

Initial moisture content determinations, combined with the weekly weight data, were used to compute moisture contents over time for 325 pieces, 25 mm or greater (the bulked twig material was excluded from the analysis). A simple, negative exponential, drying model was fitted to the moisture content data for each piece and the effect of species, piece thickness, initial moisture content and density on the drying model parameters was examined.

Burning tests of a minimum of three replicates per species commenced once weight loss had stabilised. These were performed on clear days in a temporary laboratory which provided protection from wind, while maintaining good ventilation. Only one replicate per species was tested on any particular day.

The burning test used was an adaptation from Krilov et al. (1986), in which mature wood samples were sawn to precise dimensions to obtain constant volume and were burnt under controlled conditions. This study sought to simulate the type of cooking fire that may be used where the test species are grown for firewood. The test material was constant in weight, covered a range of piece sizes and consisted of juvenile wood and bark. A further modification included measurement of the rate of heating of a standard volume of water (4 l).

The test rig consisted of a fire bucket with a container holding the water suspended at a set height within the bucket, the whole rig being mounted on a 30-kg top-loading scale separated from the fire by several layers of fireproof material (Fig. 1). The container was fitted with a thermocouple to measure water temperature.

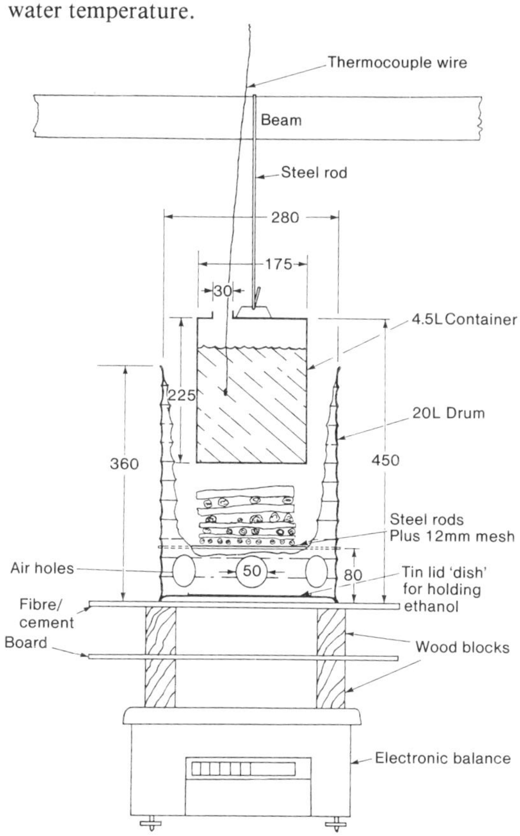

Fig. 1. Test rig for fuelwood evaluation trials (all dimensions in millimetres).

Table 1. Weather data for the period of the drying study.

	Air Temperature (°C)			Relative humidity[a] (%)	
	Mean max	Mean min	Wet days	Range	Average
January	33.7	21.8	11	35 to 90	62
February	30.4	20.2	15	45 to 97	67
March	31.1	17.7	6	31 to 100	60
April	26.1	16.5	15	35 to 100	77
May	25.6	13.4	9	21 to 100	71
June	22.0	11.6	17	31 to 98	72
July	23.0	5.3	7	19 to 96	64
August	23.5	9.1	9	21 to 99	61
September	26.0	9.7	4	15 to 78	50

[a] Derived from 9 am, 3 pm readings.

Test burns were carried out under ambient conditions although trials were not carried out when conditions were excessively dry, windy or humid. Initial water temperature, air temperature and relative humidity were measured, while samples of the test material were obtained across the range of diameter classes for moisture content determination. These uncontrolled variables were recorded prior to each burn to enable their effects to be tested in covariate analysis. Large-diameter material was split and crib-style fires were built by placing small pieces at the bottom, graduating to larger at the top. The weight of fuel used for each burn was 800 g.

Each test was started by igniting 40 ml of ethanol placed in a shallow container at the base of the fire bucket. Weight loss (i.e. weight of fuel burnt) and water temperature were recorded at 30-sec intervals for the first 5 min and then every minute for 30 min. The time taken for the water to boil and the starting time of ignition, flaming and ember phases were recorded as well as qualitative observations such as spark emission, crackling, smokiness and evidence of unpleasant odours. At the end of the test the water container was reweighed to determine water loss. The residue was divided into ash and charcoal by passing through a 5 mm sieve and each component was weighed.

The rate at which energy was used in heating the water in the container per unit of fuel consumed was derived by the equation:

$$E = [(100-t)*w*K + L*H]/W \quad \text{... Eqn 1}$$

where

E = rate of energy use (kJ/g)

t = initial temperature of the water (°C)

w = initial weight of water (g)

K = conversion factor (= 0.0042 kJ/cal)

L = weight loss of water (g)

H = latent heat of vaporisation of water (= 2.257 kJ/g)

W = wood weight (g).

It should be noted that this measure of energy specifies the energy used in heating the water for 30 min and not the total heat energy released by the burning wood.

Analysis of variance was used to test for differences between species. Least significant differences were derived for the time taken to boil the water and for the rate of energy use. Correlations between these variables and basic density were also determined.

Results

Drying Study

It was considered that the drying behaviour of a fuelwood is of most significance in the region from green to 24% moisture content. Drying to about

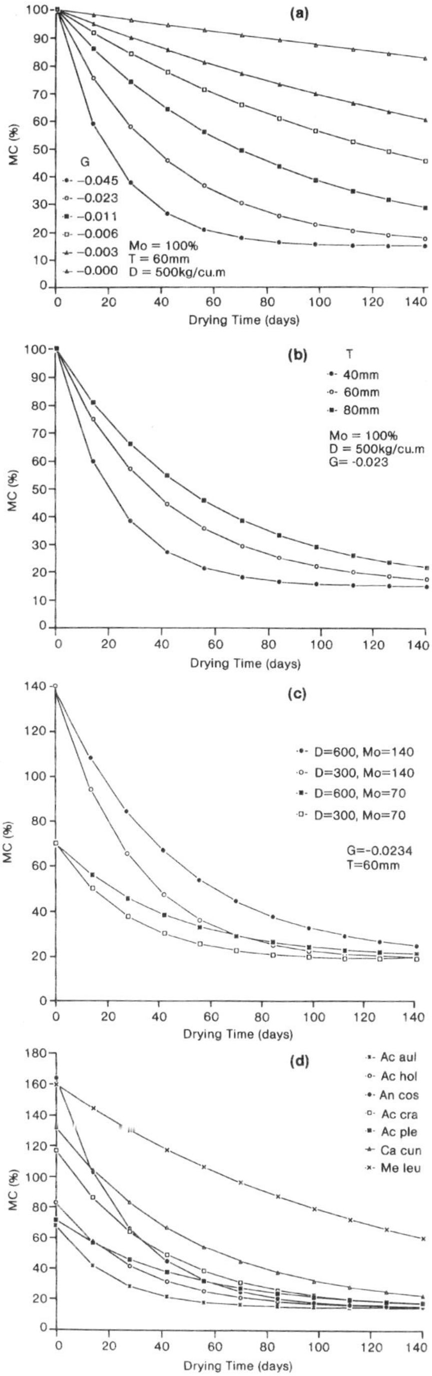

Fig. 2. Using values calculated from Eqn 4, (a) effect of variation of group intercept (G) on drying weight of wood; (b) effect of piece thickness (T) on drying rate of wood; (c) effect of initial moisture (Mo) content and basic density (D) on drying rate of wood; and (d) drying patterns of 60 cm thick pieces of selected species.

24% involves the removal of mostly free water and is relatively fast when compared with the rate of drying below this point (Fig. 2 a–d). Drying below about 24% takes progressively longer as it involves the removal of bound water. In practice there is little point in waiting the considerably extra time for progressively smaller degrees of drying, particularly since the gains in calorific value from additional drying are comparatively small (see Chapter 16). For this reason, only the data down to 24% were used to develop the drying model and in this region of drying, a simple, negative exponential, single parameter, drying model was considered appropriate. The model assumed that the moisture content of all pieces would eventually approach an equilibrium of 15%. In practice the final moisture content of such a large number of diverse species is spread around 15%, and censoring the data at 24% overcamc problems associated with lack of fit, where the observed equilibrium departed from 15%. The drying model took the form:

$$Mt = (Mo-15)*e^{kt} + 15 \qquad \text{... Eqn 2}$$

where
Mt = moisture content at time t
Mo = initial moisture content
k = constant
t = time.

The single parameter k in the above equation was estimated for each of the 325 drying model pieces. In general the drying model approximated the censored drying data well. Table 2 shows the distribution of r^2 values and it also lists the means and range in the number of data units in each set, within r^2 classes.

Initial moisture content, basic density, reciprocal of piece thickness and species, were then examined as possible predictors for k using step-wise linear regression. All variables except initial moisture content were found to be significant. Initial moisture content alone was significant but not in the presence of the other variables. The predictive function derived for k was of the following form:

$$k = S_i-1.243/T + 0.0000381*D \qquad \text{... Eqn 3}$$
$$r^2 = 0.80$$

where
S_i = the intercept for species i
T = piece thickness (mm)
D = basic density (kg/m^3).

The intercept terms for various species were similar and species with similar terms were therefore grouped to simplify the presentation of results. Grouping was such that the S_i terms within a species group were not significantly different, while significant differences existed between the S_i terms for species in different groups.

The final model embodied five species groups as shown in Appendix 1 and was of the following form:

$$k = G_i-1.243/T + 0.0000381*D \qquad \text{... Eqn 4}$$
$$r^2 = 0.78$$

where
G_i = the intercept for species group i.

It is important to note that the r^2 values in equations 3 and 4 do not relate to the full variation observed in the drying rate data but only to the variation in k values for each of the 325 censored drying curves.

There was considerable variation in drying rate between species and the effects of differences in initial moisture content, basic density, piece thickness and the group intercept (G) on drying rates are illustrated in Fig. 2a–d, 3 and Table 3. Differences in the S_i or species grouped G_i terms reflect differences in the drying rate of species not explained by density and piece thickness. Factors such as pit aspiration and the nature and amount of extractives are known to have a major influence on drying rates. Initial moisture content varied slightly within species with a tendency to decrease with increasing piece size, exceptions being *Acacia auriculiformis*, *A. plectocarpa*, *Casuarina cunninghamiana* and *Melaleuca leucadendra*. There was a slight tendency for basic density to increase with increasing piece size but this was variable also (see Appendix 1 for details).

While the general form of the drying model is nonlinear, the log of Mt–15 is linear in G, $-1/T$ and D. Thus increasingly longer drying times could be expected as all factors increase. The magnitude of G appears to have the greatest influence on drying rate. This can be illustrated by comparing drying times to 24% moisture content of *Angophora costata* and *M. leucadendra* (Appendix 1). It can be seen that for pieces of the same thickness and comparable initial moisture content, *M. leucadendra* takes much longer to dry than

Table 2. Distribution of r^2 values for equation 2 including the mean and range in the number of data units in each set, within r^2 classes.

	r^2				
	<0.8	0.8–0.9	0.9–0.95	0.95+	Total
Frequency	5	26	42	252	325
Mean no. of data units per set	5.0	6.3	8.5	8.3	8.1
Range in no. of data units per set	3–8	3–19	2–21	3–21	2–21

Table 3. Species tested in the drying and burning studies and average values for initial moisture content, density and group intercept.

	Average initial moisture content (%)	Average green density (kg/m^3)	Average basic density (kg/m^3)	Value of group intercept G
Group 1				
1 *Acacia aulacocarpa*	68	839	499	-0.04564
2 *A. elata*	84	834	455	
Group 2				
3 *A. holosericea*	83	1003	548	-0.0327
4 *A. podalyriifolia*	74	1015	581	
5 *A. saligna*	90	928	489	
6 *A. melanoxylon*	75	837	479	
7 *Angophora costata*	164	1110	412	
Group 3				
8 *Acacia mangium*	130	756	329	-0.02338
9 *A. crassicarpa*	117	1044	483	
10 *A. auriculiformis*	119	843	404	
11 *A. plectocarpa*	72	1040	604	
12 *A. cincinnata*	99	1012	508	
Group 4				
13 *A. leptocarpa*	127	1032	462	-0.01752
14 *Casuarina cunninghamiana*	132	1160	501	
Group 5				
15 *Melaleuca leucadendra*	160	835	326	0.00007

A. costata, even though the basic density of the *M. leucadendra* is much lower than that of *A. costata*. These differences increase in magnitude with increasing piece thickness. Thus variations in piece size, initial moisture content and basic density have relatively minor effects on the drying times of species having low values of G, but the effects are major for species with high values of G.

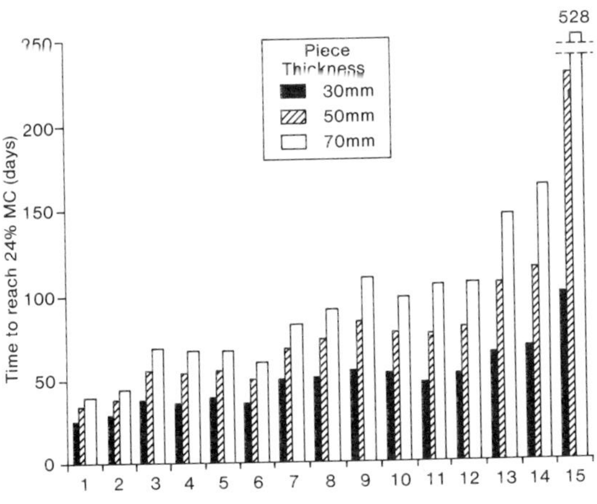

Fig. 3. Computed time to dry (from Eqn 4) to 24% moisture content by species as piece thickness varies.

Burning Study

Casuarinas make excellent fuelwood, being easy to split, burning slowly with a lot of heat but little smoke and leaving only a fine white ash (U.S. National Research Council 1984). Among the casuarinas, *Allocasuarina inophloia*, *A. luehmannii* and *A. torulosa* were singled out by Swain (1928) as being outstanding as fuelwoods. The fuelwood properties of *C. cunninghamiana* were not commented on by Swain but it is also said to have excellent burning properties (Midgley et al. 1983). These observations were based on the burning properties of mature wood; the properties of the juvenile wood (density in particular) are likely to be different to some extent. Nevertheless, because the *C. cunninghamiana* in the study was grown with the other species, it was used as a standard with which to compare and evaluate the performance of the other species.

The acacias were noisy at ignition, crackling and emitting small sparks and fine ash during the flaming phase. *Acacia podalyriifolia*, *A. elata* and *A. saligna* produced the most airborne ash during this stage, whereas *A. saligna* and *A. podalyriifolia* were quite smoky initially. The burning bark of *M. leucadendra* produced an acrid sooty smoke though the wood burnt without smoke. All species burnt to ash with low levels of charcoal; *C. cunninghamiana* left the greatest amount of residue. Details of the characteristics recorded for each species are listed in Appendix 2.

There were significant differences between species in the time for the water to start boiling and in the rate of energy use (Table 4). In the latter case, the value for *C. cunninghamiana* was generally low in comparison with the other species. However, differences were significant only for *A. saligna* and *A. plectocarpa* (higher) and *A. cincinnata* (lower). Similarly, water was slower to boil on the *C. cunninghamiana* fire in comparison with most other species and significantly slower than on the *A. aulacocarpa*, *A. holosericea*, *A. plectocarpa* and *A. saligna* fires. The effects of uncontrolled variables (initial water temperature, fuelwood moisture content, relative humidity and air temperature), were not significant in the covariate analysis.

The fuel consumption curves for all species were similar (Fig. 4). This result is generally consistent with the results of Metz (1963, cited in Shelton 1983) who reported that mass burning rate was independent of density over a wide range (from about 160 kg/m^3 to about 1250 kg/m^3). There was no correlation between basic density and any of the other measure parameters (Table 5).

Table 4. Boiling test data and residue at completion of combustion for each of the species in the burning study.

Species	Rate of energy utilisation (kJ/g)	Boiling time (min)	Residue Ash (%)	Residue Charcoal (%)
Acacia saligna	2.417	11.5	1.05	0.05
A.plectocarpa	2.399	11.3	1.21	1.33
Angophora costata	2.330	13.4	1.87	0
Acacia mangium	2.307	12.3	1.12	0.16
A.podalyriifolia	2.304	13.3	0.83	0.08
A.auriculiformis	2.284	14.2	1.67	0
A.holosericea	2.282	10.2	0.75	0
A.aulacocarpa	2.281	10.7	0.81	0.37
A.leptocarpa	2.210	16.3	1.08	0.12
A.elata	2.200	12.3	0.83	0.12
A.crassicarpa	2.178	12.7	0.81	0.06
C.cunninghamiana	2.170	15.4	2.25	0.06
A.melanoxylon	2.136	13.5	0.96	0.96
M.leucadendra	2.096	13.0	1.7	0
A.cincinnata	1.997	17.5	0.88	1.54
AOV	**	*	n.a	n.a
LSD (P=0.05)	0.170	3.6	—	—

Significance of F in AOV: * <0.05; ** <0.01; n.a. = not analysed.

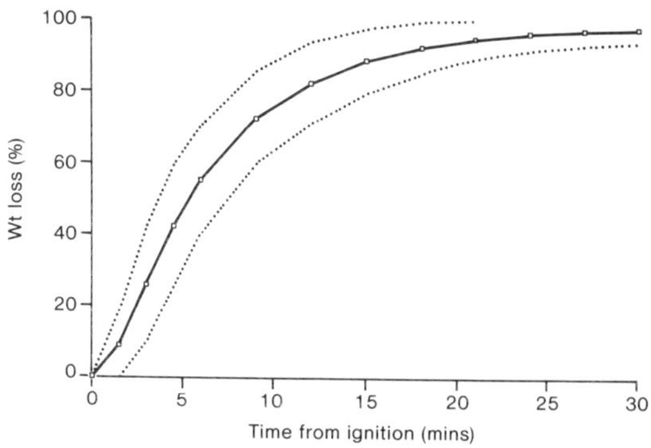

Fig. 4. Rate of fuel consumption (mean weight loss ± 95% confidence intervals) during combustion.

Table 5. Correlation matrix of basic density and burning parameters measured.

	Boil time	Energy capture	Basic density
Boil time	1.0000		
Energy capture	-0.6326*	1.0000	
Basic density	-0.1369	0.2145	1.0000

* Significant at $P<0.05$.

Discussion

Green wood at 80% moisture content has only about 44% of the net heating value of the same weight of oven-dry wood. This rises to about 78% at 25% moisture content and 85% at 15% moisture content (see Chapter 16). The calorific value of wood per se varies only slightly over a wide range of species (Harker et al. 1982; Shelton 1983). Thus moisture content, reflecting as it does the ratio of water to wood substance in a fuel, is the prime determinant of heat yield. Therefore the rate at which a species dries is an important factor in determining its utility for firewood. This is particularly the case where fuelwood resources are scarce and material may need to be burnt soon after harvesting.

All the species tested in the burning study burnt well and should provide acceptable fuelwood where only heat production needs to be considered. This supports the contention of Shelton (1983) that the heat output does not depend critically on the kind of wood used. Providing a species burns reasonably well, fire behaviour itself may not be so important in determining acceptability, although smoke, odour, sparks and explosive combustion may be. While the information on burning characteristics (Appendix 2) shows some differences between species, undue emphasis should not be placed on these differences. However, the smoke and sparks produced by *A. podalyriifolia* and *A. saligna*, although relatively minor, may limit their use in some situations, and in the case of *M. leucadendra* bark would need to be removed before burning. Bark could be retained on the other species since the calorific value of bark is slightly higher than wood, although the levels of ash are also higher (Ince 1977).

Any study attempting to develop reliable quantitative information on the burning properties of wood is likely to be slow and of little practical value in evaluating the fuelwood potential of large numbers of species. Precise measurements of combustion, enabling meaningful comparisons between species, can be obtained only under highly controlled test conditions, and the relevance of such information is likely to be limited because of the necessarily artificial constraints imposed. While we attempted to overcome some of these constraints in our study, we do not consider this to be a satisfactory alternative method since:

(a) testing was slow (five test burns per day maximum);

(b) it was difficult to confine variation within reasonable limits; and

(c) energy utilisation (2.24 kJ/g) was low (only 15% of the net calorific value of the wood of about 14.9 kJ/g at 18% moisture content), although this could be improved by redesigning the fire bucket for improved heating efficiency.

While there is little variation in calorific values between species, there are some species which are nevertheless difficult to burn (e.g. *Syncarpia hillii*, *S. glomulifera*). Thus information on the combustibility of species is essential in evaluating fuelwood potential. Information on attributes such as spark emission and the amount and odour of smoke is also necessary in evaluation, although the level of importance attached to these characteristics may vary considerably depending on location, culture and fuelwood use. While these properties cannot be quantified simply, they can be described qualitatively. Unless efficient quantitative tests can be developed, the most effective way to screen a large array of species for their fuelwood potential may be by observation and comparative qualitative assessment of test fires. More rigorous quantitative testing of a limited number of species could be used to verify the information from qualitative tests.

Provision of information on drying characteristics, however, is essential. Drying rate depends on initial moisture content, piece thickness and basic density and this relationship can be described by generalised models. However, drying rate also depends on less easily measured factors such as pit and lumen geometry and the presence of extractives and deposits. In this study these factors were expressed in terms of the group intercept G. Species-dependent parameters such as G will always need to be determined by experimentation. Accelerated drying tests under controlled conditions using standard piece sizes may be the quickest and simplest method for ranking species drying behaviour. Methodologies need to be developed and tested to ensure that any rankings derived reflect the usual air-drying situation. Other factors such as piece size, length, temperature, humidity, removal of bark and splitting pieces prior to drying may affect the rate of drying and require study also.

The emphasis in future work should be on defining the drying characteristics of potential fuelwood species. At this stage we see little value in developing detailed information on fire behaviour at least at species level, since the process is time consuming and yields little information of practical value. Such studies, however, may be worthwhile in describing the generalised fire behaviour of broad groups particularly where accepted good and poor fuelwood species are included as standards. Unless quick and efficient methods can be developed for providing quantitative data on burning properties of individual species, screening for fuelwood potential should be restricted to qualitative assessment of test fires.

Acknowledgments

Grateful acknowledgment is made of the valued assistance and advice of Mr J.M. Richolson, U.S. Navy Testing Laboratory (retired), and from Mr M. Nester and Mr A. Ward, Forest Research Centre, Queensland Forestry Department, Gympie.

(Appendixes follow)

Appendix 1

Initial moisture content, basic density and computed time to reach 24% moisture content of sample material in the drying study.

	Size class mid diam (mm)	No. of pieces	Initial moisture content (%)		Basic density (kg/m^3)		Average time to reach 24% moisture content (days)
			Mean	Range	Mean	Range	
Group 1							
Acacia aulacocarpa	<40	6	69	60–76	497	455–554	26
	40–60	14	68	60–72	498	453–547	34
	60–80	2	65	65–66	517	509–525	39
	>80	0					
Acacia elata	<40	9	88	69–103	459	427–509	30
	40–60	11	84	76–87	445	429–471	38
	60–80	8	80	76–82	462	440–483	43
	>80	0					
Group 2							
Acacia holosericea	<40	7	88	74–96	554	492–695	39
	40–60	10	80	74–86	543	478–695	54
	60–80	2	81	76–87	549	541–557	67
	>80	1	87		541		80
Acacia podalyriifolia	<40	6	77	75–78	556	539–601	36
	40–60	8	74	66–78	589	561–605	54
	60–80	2	68	66–69	603	601–605	65
	>80	1	67		634		78
Acacia saligna	<40	8	92	83–110	481	415–506	38
	40–60	8	92	83–99	466	437–506	54
	60–80	6	87	80–96	503	473–532	66
	>80	9	88	83–96	507	473–539	77
Acacia melanoxylon	<40	4	83	71–101	450	392–495	35
	40–60	5	74	72–78	471	458–485	47
	60–80	5	73	72–78	468	458–480	57
	>80	11	73	67–81	499	465–526	68
Angophora costata	<40	4	170	144–215	426	369–450	49
	40–60	7	168	139–197	393	373–440	67
	60–80	5	171	138–204	411	391–430	82
	>80	7	152	124–186	423	391–440	90
Group 3							
Acacia mangium	<40	5	135	115–152	284	249–347	48
	40–60	5	123	103–144	326	291–367	69
	60–80	8	137	103–155	334	294–356	92
	>80	5	123	109–139	368	318–411	107
Acacia crassicarpa	<40	3	127	118–141	452	440–472	53
	40–60	6	114	102–141	474	329–548	79
	60–80	2	118	111–125	472	435–508	105
	>80	6	114	106–124	510	464–553	135
Acacia auriculiformis	<40	6	118	104–129	411	321–473	50
	40–60	11	119	95–139	402	321–473	74
	60–80	3	122	115–132	388	357–419	94
	>80	1	132		419		121

	Size class mid diam (mm)	No. of pieces	Initial moisture content (%)		Basic density (kg/m^3)		Average time to reach 24% moisture content (days)
			Mean	Range	Mean	Range	
Acacia plectocarpa	<40	3	72	69-76	588	580-605	44
	40-60	10	72	66-77	606	560-632	73
	60-80	7	72	65-79	608	573-630	103
	>80	2	77	76-79	598	577-619	134
Acacia cincinnata	<40	1	102		488		49
	40-60	4	97	85-114	501	468-518	76
	60-80	7	99	95-107	513	500-525	103
	>80	2	102	96-107	513	500-525	129
Group 4							
Acacia leptocarpa	<40	2	142	142-142	366	343-390	59
	40-60	4	126	119-131	443	438-450	99
	60-80	3	123	121-125	449	447-450	137
	>80	2	119	119-119	520	520-520	213
Casuarina cunninghamiana	<40	4	133	132-134	514	493-534	65
	40-60	6	125	117-138	520	474-542	111
	60-80	4	126	118-140	508	467-537	158
	>80	9	139	118-152	480	442-532	201
Group 5							
Melaleuca leucadendra	<40	7	149	135-154	345	160-480	98
	40-60	9	166	139-189	330	268-410	231
	60-80	7	163	149-174	307	268-377	466
	>80	5	163	147-189	321	278-378	1857

Note: Drying time for pieces <40 mm based on 30 mm thickness and for pieces >80 mm on 90 mm thickness.

Appendix 2

General characteristics of the test species as firewood.

Species	General Characteristics
Acacia aulacocarpa	Wood easy to split. Sparks and smoke at ignition. Some ash and smoke in flaming phase.
A. elata	Wood very easy to split. Some crackling and sparks in early flaming phase — considerable fly ash during flaming phase.
A. holosericea	Wood easy to split. Crackles and sparks at ignition. Some fly ash and smoke during flaming phase.
A. saligna	Crackles and sparks in early flaming phase, smoky to very smoky during flaming phase with plenty of fly ash.
A. podalyriifolia	Very knotty wood. Some pieces difficult to split. Lots of crackles and sparks at ignition. Plenty of fly ash, some smoke and occasional sparks during flaming phase.
A. melanoxylon	Crackling, sparks and sooty smoke in early flaming phase.
Angophora costata	Wood easy to split. Some crackling and sparks at ignition. Smoky in early flaming phase.
Acacia mangium	Light wood. Crackling and sparks at ignition. Some fly ash and sooty smoke in flaming phase.
A. crassicarpa	Sparky at ignition. Some fly ash and smoke during flaming phase.
A. auriculiformis	Wood easy to split. Crackles and sparks at ignition. Plenty of fly ash throughout flaming phase with occasional sparks and some smoke.
A. plectocarpa	Wood distinctly heavy compared to other species. Sparky at ignition with occasional sparks and smoke during flaming phase. Abundant ash emitted during flaming phase.
A. cincinnata	Crackles and sparks at ignition. Some fly ash and smoke during flaming phase.
A. leptocarpa	Pops, crackles and sparks at ignition. Some fly ash, smoke during flaming phase with occasional sparks.
C. cunninghamiana	Very quiet at ignition. Small amounts of fine fly ash emitted with occasional sparks in flaming phase.
M. leucadendra	Very light wood, quick to ignite. Sooty acrid smoke initially from the burning bark.

Chapter 19

Fodder Value of Selected Australian Tree and Shrub Species

T.K. Vercoe

Abstract

Foliage from 39 Australian tree and shrub species cultivated in field trials near Gympie in Queensland, Australia, were analysed for digestibility, protein content, phosphorus, potassium, sodium, calcium, magnesium, copper, zinc and manganese concentrations. Twenty-five are recommended for further study.

Introduction

The idea of using trees and shrubs for fodder production is attractive for a number of reasons. Two important factors are: access to water and nutrient resources which are often unavailable to pasture species, and ability to provide protein and nutrients at times of the year when pastures are dormant.

The problems associated with shortages of animal fodder in various regions around the world have been well documented (Mahadevan 1981; Le Houerou 1980) but information on tree species suitable for forage production is scarce. This scarcity has fuelled doubts about the value of trees for fodder. Dann and Low (1988) concluded that growth rates of some of the more common native fodder species made their use as anything other than drought reserve fodder somewhat tenuous. More information is needed to provide a better basis on which to assess the value of trees for fodder.

The limited availability of information is probably due to difficulties in assessing the 'fodder value' of a species. Studies (McLeod 1973; McDonald and Ternouth 1979) have concentrated on the chemical composition and digestibility of foliage because they are relatively easy to assess and provide repeatable results. However, the most important measure of fodder value is the weight gain produced by animals feeding on leaf material.

Most fodder research is concentrated on species that have been traditionally used for browse at the expense of faster-growing multipurpose trees that are not normally available to stock. However, other species may prove to be useful if made available to browsing animals. An example of this is a reported observation of dairy cattle consuming the foliage and ripe seed pods of *Acacia auriculiformis* that had been lopped for seed collection in the Northern Territory (Maurice McDonald 1987 pers. comm.). This species is not known for its fodder value in Australia, but is well known in tropical regions of the world for its fast growth and its promise for pulpwood production (Turnbull 1986).

Many Australian species have proven valuable around the world for growth under difficult conditions, and for providing a range of products like timber, fuelwood and shelter. This study is aimed at adding to the limited information on the fodder value of some species.

Materials and Methods

Species Selection

Species for study in the Gympie trials were selected in consultation with staff at CSIRO

Division of Forestry and Forest Products and the Queensland Forestry Department, on the basis of reported palatability and potential for fast growth rates, or other attributes of use in fuelwood and agroforestry systems. Over half the species selected were acacias, reflecting the variability in the genus and the emphasis placed on it in the ACIAR trials. A list of species selected for the study is given in Table 1.

Sampling

Trees and shrubs were sampled from ACIAR species trials (Ryan et al. 1987) situated near Gympie in southeastern Queensland. Selections were made from the 1984, 1985 and 1986 plantings.

Foliage samples were collected from five trees/shrubs per plot at four (roughly) equidistant points around the crown of each. Leaves/phyllodes and twigs (for some species) were removed between 10 and 30 cm back from the branch tips to include expanded and expanding foliage. Crowns were sampled approximately half way up the green portion which involved severing branchlets with a pole pruner on taller species.

Processing

The samples were stored on dry ice in the field and placed in a drying oven at 70°C on return to the laboratory. Drying continued for 24 hours after which the samples were ground through a 1 mm sieve in a stainless steel hammer mill.

Analyses

Analysis of the concentrations of nitrogen, phosphorus, potassium, sodium, calcium, magnesium, copper, zinc and manganese was carried out using a quantometer for multielement analysis (Johnson et al. 1985).

Dry matter digestibility (DMD) was estimated using the in vitro pepsin-cellulase method of Minson and McLeod (1978). Standard samples of known in vivo digestibility for comparison in the DMD tests were *Leucaena leucocephala* and a mixture of *Acacia aneura* and sorghum stubble (*Sorghum* sp.). The material was run twice and the average of the two runs has been reported in the results.

Table 1. List of species selected for analysis.

Species	Corresponding no. used in graphs	No. of provenances
Acacia ampliceps[a]	1	3
A. aneura[a]	2	3
A. auriculiformis[a]	3	3
A. cowleana	4	1
A. crassicarpa	5	3
A. deanei	6	1
A. elata	7	1
A. flavescens	8	1
A. glaucocarpa	9	1
A. holosericea[a]	10	3
A. hylonoma	11	1
A. leptocarpa	12	3
A. maconochieana[a]	13	1
A. mangium[a]	14	2
A. melanoxylon	15	6
A. monticola	16	1
A. murrayana	17	2
A. neriifolia	18	2
A. parramattensis	19	1
A. plectocarpa	20	1
A. rothii	21	1
A. salicina[a,d]	22	2
A. saligna[a]	23	1
A. shirleyi	24	2
A. simsii	25	2
A. stenophylla[a]	26	1
A. storyi	27	1
A. torulosa	28	1
A. tumida	29	1
Albizia procera	30	1
Allocasuarina littoralis	31	1
Alphitonia excelsa	32	1
Casuarina cristata[a,c]	33	1
C. cunninghamiana	34	2
Cassia brewsteri	35	1
Dodonea viscosa[a,b]	36	1
Grevillea robusta	37	1
Melia azedarach[a,d]	38	1
Terminalia platyphylla	39	1

[a] Species observed browsed by livestock.
[b] Species reported as having low palatability by Wilson and Harrington (1980).
[c] Species reported as having high palatability by Wilson and Harrington (1980).
[d] Species reported as having poisonous fruit by Everist (1969).

Results

The results of the major DMD and nutrient analyses are given in Table 2. The protein and digestibility contents of the species (mean for all provenances) are given in Fig. 1. The species which lie in the top right quadrant of the graph (digestibility >40% and protein content >10%) deserve further attention. Their digestibility and protein contents suggest they may be suitable for

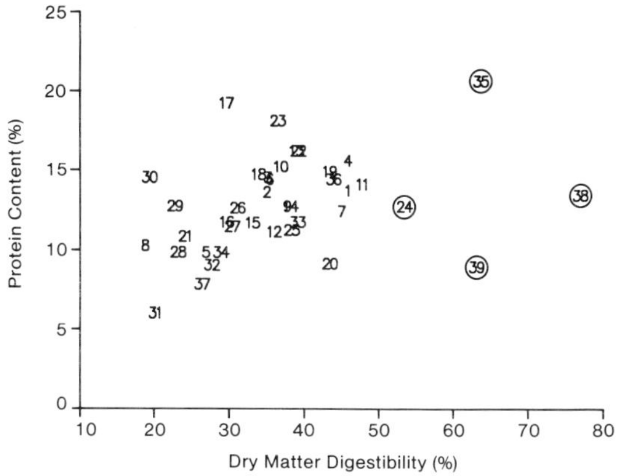

Fig. 1. Protein content vs DMD.

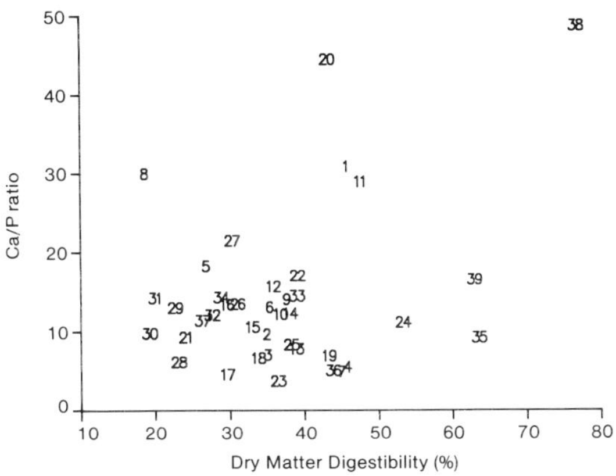

Fig. 2. Calcium/phosphorus ratio vs DMD.

fodder at below maintenance requirements with the addition of some protein supplement. The species marked with circles could provide subsistence forage.

The calcium/phosphorus ratio is plotted (Fig. 2) against the digestibility (mean for all provenances) which is important in determining the response of animals to supplements of phosphorus. Ca:P ratios below about 10:1 (and preferably 5:1 for many animals) allow the greatest use to be made of phosphorus supplements (Underwood 1981). Uptake of micronutrients such as copper, zinc and manganese (important in enzyme pathways) may also be affected at ratios greater than 10:1 for ruminants.

Discussion

The most impressive species from the laboratory work were *Melia azedarach*, *Cassia brewsteri*, *Terminalia platyphylla*, *Acacia shirleyi*, *A. cowleana* and *A. hylonoma*. Their digestibilities and protein contents were high and they contained acceptable levels of other nutrients. Species that showed good digestibilities, protein contents and calcium/phosphorus ratios were *A. cowleana*, *A. elata*, *A. parramattensis*, *A. shirleyi* and *Dodonea viscosa*.

Biomass production will be an important criterion in selecting a species for fodder production. Detailed work in this area by Bell and Ryan (unpublished data) has provided useful data for several species. Observations of other species in the trial suggest their lack of biomass production will negate the positive aspects of high laboratory nutrient content (e.g. *Cassia brewsteri* which is growing very poorly in the trials).

Acacia simsii seed was tested as well as foliage as it produces seed prolifically from an early age and may provide a useful high-protein supplement. The seed is easily collected and cleaned and could be stockpiled against periods of feed shortage. The analyses of the seed show that it is high in protein and phosphorus, and easily digested if some processing to break the hard seed coat is carried out.

A number of provenances of *A. melanoxylon* was available for testing in the trial, and it was decided to use this species to see if major provenance differences in fodder characteristics could be detected. In this case, there was very little difference between provenances for all analyses. However other species (e.g. *A. holosericea*, *A. salicina* and *A. simsii*) showed relatively large variation between seedlots.

The two samples of *Casuarina cunninghamiana* collected for analysis came one each from *Frankia* inoculated (13134/84) and uninoculated (13511/84) plots. The results show very little difference in foliar nitrogen concentrations for the two treatments, although differences in growth rate between the two are thought to be attributable to increased nitrogen availability from more effective nodulation.

The variation in the phosphorus uptake by different species and provenances is an important consideration. This element is one of the critical limiting factors in animal production in Australia and in some areas overseas. Species with high relative phosphorus uptake may be useful in supplementing pastures deficient in this element, or in 'retrieving' leached phosphate fertiliser from below the root zone of pasture plants.

The information contained in Table 2 and Fig. 1 and 2 gives an indication of the chemical suitability of these species for fodder. However, they provide no indication of how a species will be accepted by grazing animals. Important information on the acceptability of various species to grazing animals is infrequently reported and often the information conflicts, with species being browsed in one area but not in another.

Table 2. Results of dry matter digestibility and nutrient analysis for 39 Australian tree and shrub species.

Species	Seedlot/ planting year	Predicted in vivo DMD (%)	Crude protein (%)[a]	P (%)[b]	K (%)[c]	Na (%)[d]	Ca (%)[e]	Mg (%)[f]	Ca/P ratio[g]	Cu (ppm)	Zn (ppm)	Mn (ppm)
Acacia												
ampliceps	14668/86	43.4	10.9	0.08	0.76	0.05	3.40	0.38	42.50	4.4	31.0	113.0
ampliceps	14643/86	45.6	15.6	0.12	0.90	0.07	2.59	0.46	21.58	5.4	35.0	144.0
ampliceps	14486/86	48.6	15.3	0.12	0.71	0.06	3.55	0.45	29.58	5.5	38.0	192.0
aneura	13481/84	36.5	14.9	0.10	0.73	0.03	0.72	0.25	7.20	0.9	44.0	125.0
aneura	13480/84	30.8	13.8	0.10	0.64	0.04	1.19	0.38	11.90	3.7	36.0	92.0
aneura	13719/84	37.8	12.8	0.10	0.58	0.08	1.08	0.29	10.80	0.6	35.0	48.0
aneura *	13481/84	17.3	7.7	0.08	0.32	0.00	0.59	0.08	7.37	1.3	16.0	75.0
auriculiformis	13861/84	33.5	13.8	0.11	0.72	0.30	0.77	0.20	7.00	3.8	32.0	50.0
auriculiformis	13854/84	34.9	14.0	0.06	0.45	0.49	0.52	0.24	8.67	4.4	31.0	53.0
auriculiformis	13686/84	36.9	16.2	0.09	0.62	0.54	0.59	0.18	6.56	1.9	25.0	29.0
cowleana	14621/86	45.8	15.8	0.10	0.72	0.04	0.59	0.31	5.90	5.8	28.0	120.0
crassicarpa	13863/84	28.4	11.8	0.06	0.51	0.29	0.88	0.33	14.67	1.6	32.0	84.0
crassicarpa	13681/86	27.1	9.1	0.04	0.30	0.48	0.99	0.27	24.75	1.5	24.0	213.0
crassicarpa	13682/86	25.2	9.8	0.04	0.65	0.27	0.65	0.19	16.25	2.6	21.0	88.0
deanei	14739/86	35.4	14.6	0.08	0.58	0.03	1.07	0.40	13.38	8.0	36.0	200.0
elata	9972/84	45.0	12.6	0.07	0.43	0.08	0.78	0.37	11.14	1.5	34.0	24.0
flavescens	14175/85	18.8	10.4	0.04	0.22	0.19	1.21	0.84	30.25	2.2	49.0	123.0
glaucocarpa	14763/86	37.6	12.9	0.08	0.47	0.16	1.15	0.36	14.37	1.5	32.0	96.0
holosericea	13879/86	35.6	12.8	0.06	0.73	0.14	0.83	0.30	13.83	3.0	26.0	142.0
holosericea	14660/86	46.0	19.3	0.11	0.48	0.01	1.14	0.45	10.36	10.0	45.0	202.0
holosericea	13583/84	38.1	14.2	0.10	0.59	0.04	1.34	0.47	13.40	5.6	39.0	244.0
hylonoma	14197/85	47.8	14.3	0.04	0.38	0.16	1.17	0.56	29.25	1.9	38.0	477.0
leptocarpa	14139/86	38.0	10.5	0.04	0.39	0.48	0.90	0.37	22.50	2.0	30.0	117.0
leptocarpa	13691/84	33.3	11.2	0.04	0.34	1.09	0.54	0.27	13.50	6.1	30.0	61.0
leptocarpa	13652/84	36.8	12.1	0.05	0.51	0.50	0.60	0.25	12.00	1.5	25.0	61.0
maconochieana	14676/86	39.0	16.4	0.16	0.78	0.27	1.31	0.38	8.19	3.8	36.0	201.0
mangium	13846/84	38.4	11.5	0.07	0.39	0.81	1.13	0.28	16.14	2.7	26.0	165.0
mangium	13460/84	38.0	14.3	0.09	0.62	0.48	0.82	0.23	9.11	1.9	25.0	120.0
melanoxylon	13630/84	30.6	13.5	0.09	0.47	0.11	0.56	0.31	6.22	5.5	32.0	343.0
melanoxylon	13944/84	34.1	11.4	0.06	0.40	0.18	0.86	0.33	14.33	1.8	41.0	403.0
melanoxylon	12986/85	32.5	13.3	0.10	0.47	0.07	0.73	0.46	7.30	2.9	35.0	575.0
melanoxylon	13944/85	34.8	10.8	0.06	0.42	0.14	0.76	0.24	12.67	9.7	23.0	258.0
melanoxylon	14176/85	33.1	11.0	0.05	0.45	0.18	0.68	0.29	13.60	5.8	41.0	317.0
melanoxylon	14766/86	33.7	11.1	0.07	0.40	0.18	0.77	0.29	11.00	3.0	27.0	460.0
melanoxylon +	13944/84	28.1	9.6	0.05	0.35	0.30	1.27	0.36	25.40	2.3	29.0	438.0
monticola	14008/85	29.6	11.9	0.09	0.36	0.03	1.23	0.55	13.67	5.1	46.0	87.0
murrayana	13781/85	30.5	19.3	0.11	0.56	0.19	0.56	0.44	5.09	5.5	38.0	78.0
murrayana	13781/85	28.8	19.4	0.12	0.70	0.09	0.57	0.39	4.75	3.0	36.0	97.0
neriifolia	14759/86	32.9	13.8	0.05	0.27	0.06	0.43	0.34	8.60	1.5	28.0	97.0
neriifolia	14735/86	34.9	16.1	0.08	0.58	0.18	0.43	0.26	5.38	2.1	25.0	86.0
parramattensis	14767/86	43.4	15.1	0.10	0.74	0.16	0.73	0.34	7.30	2.2	26.0	64.0
plectocarpa	14003/85	43.4	9.3	0.04	0.44	0.21	1.79	0.36	44.75	5.2	34.0	66.0
rothii	14140/85	24.2	11.0	0.05	0.37	0.30	0.48	0.48	9.60	2.0	34.0	70.0
salicina	13501/84	34.6	16.6	0.18	1.55	0.08	1.36	0.42	7.56	2.7	40.0	34.0
salicina	14592/86	43.7	16.3	0.10	1.45	0.08	2.73	0.28	27.30	5.4	26.0	161.0

Species	Seedlot/ planting year	Predicted in vivo DMD (%)	Crude protein (%)[a]	P (%)[b]	K (%)[c]	Na (%)[d]	Ca (%)[e]	Mg (%)[f]	Ca/P ratio[g]	Cu (ppm)	Zn (ppm)	Mn (ppm)
salicina +	13501/84	28.4	13.1	0.15	1.25	0.14	1.22	0.27	8.13	3.8	29.0	56.0
saligna	13651/84	36.5	18.3	0.18	1.14	0.13	0.74	0.66	4.11	7.8	54.0	177.0
shirleyi	14622/86	51.6	12.3	0.07	0.34	0.09	0.83	0.39	11.86	1.8	33.0	71.0
shirleyi	14753/86	55.2	13.5	0.08	0.68	0.03	0.89	0.25	11.13	5.3	31.0	50.0
simsii	13960/84	43.0	12.6	0.09	0.38	0.08	0.60	0.49	6.67	2.1	38.0	128.0
simsii	13690/84	33.7	10.1	0.06	0.22	0.08	0.64	0.36	10.67	1.9	29.0	142.0
simsii ×	13960/84	61.0	24.4	0.20	0.68	0.06	0.38	0.46	1.90	4.0	39.0	79.0
stenophylla	14670/86	31.1	12.8	0.10	1.01	0.39	1.38	0.43	13.80	1.0	31.0	206.0
storyi	14760/86	30.4	11.6	0.05	0.30	0.15	1.09	0.50	21.80	2.5	34.0	120.0
torulosa	14141/86	23.2	10.0	0.06	0.48	0.17	0.39	0.16	6.50	3.0	25.0	101.0
tumida	14675/86	22.8	12.9	0.08	0.47	0.01	1.06	0.60	13.25	1.3	47.0	138.0
Albizia												
procera	14213/85	19.4	14.7	0.09	0.55	0.00	0.91	0.40	10.11	2.1	34.0	50.0
Allocasuarina												
littoralis	13133/86	20.2	6.2	0.04	0.41	0.20	0.58	0.12	14.50	4.4	16.0	387.0
Alphitonia												
excelsa	14186/85	27.6	9.2	0.07	0.65	0.04	0.87	0.26	12.43	2.0	26.0	285.0
Casuarina												
cristata	14843/86	39.1	11.9	0.10	0.70	0.22	1.48	0.34	14.80	0.6	37.0	1103.0
cunninghamiana	13134/84	28.8	10.4	0.08	0.47	0.12	0.92	0.24	11.50	3.2	32.0	133.0
cunninghamiana	13511/84	29.1	9.6	0.06	0.37	0.13	1.06	0.28	17.67	1.2	40.0	61.0
Cassia												
brewsteri	14188/85	63.6	20.8	0.14	0.63	0.11	1.35	0.45	9.64	3.9	31.0	207.0
brewsteri *	14188/85	34.5	9.4	0.09	0.57	0.12	1.06	0.27	11.78	2.8	27.0	59.0
Dodonea												
viscosa	13755/86	43.9	14.6	0.16	1.17	0.03	0.87	0.22	5.44	4.1	36.0	112.0
Grevillea												
robusta	11706/84	26.4	8.0	0.09	0.35	0.00	1.05	0.28	11.67	1.5	21.0	249.0
Melia												
azedarach	14500/86	77.0	13.6	0.11	0.49	0.03	5.39	0.60	49.00	53.8	44.0	152.0
Terminalia												
platyphylla	14182/85	63.0	9.1	0.09	0.42	0.09	1.52	0.60	16.89	3.7	46.0	152.0

* — twigs; + — leaves and twigs; × — seed.

[a] Calculated at N × 6.25 Optimal value >7.2 (Milford and Haydock 1965)

[b] Optimal value >0.15

[c] Optimal value >0.7

[d] Optimal value >0.18

[e] Optimal value >0.08

[f] Optimal value >0.07

[g] Optimal value <10

Values from Underwood (1981) except where marked.

Toxic responses have been recorded for a couple of the species in this trial: *A. salicina* is reported to contain high levels of tannin which may have caused poisoning of hungry cattle, and *Melia azedarach* fruits are poisonous, especially when fed to pigs (Everist 1969). No adverse effects have been recorded for other species in this study.

One of the important features of this study is the similarity of the conditions under which the sample material was grown. This similarity allows comparisons between species to be made more readily than if the material had come from natural populations spread over a wide area. This was a limitation in the study outlined by Vercoe (1987).

Recommendations

The following species are recommended for further study for their performance in the laboratory study: *Acacia cowleana*, *A. elata*, *A. parramattensis*, *A. shirleyi*, *Cassia brewsteri*, *Dodonea viscosa*, *Melia azedarach* and *Terminalia platyphylla*.

Other species which come close to the minimum requirements for certain nutrients and warrant further investigation are: *Acacia ampliceps*, *A. auriculiformis*, *A. deanei*, *A. glaucocarpa*, *A. holosericea*, *A. hylonoma*, *A. leptocarpa*, *A. maconochieana*, *A. mangium*, *A. neriifolia*, *A. plectocarpa*, *A. salicina*, *A. saligna*, *A. simsii* and *Casuarina cristata*.

The species recognised in this report as having potential for fodder should be field-tested in animal trials. Some species are reported to be useful forage plants overseas and should be tested in pen trials.

Possible deleterious substances (such as tannins in the acacias) need to be identified and their effect on fodder value gauged.

Methods of managing species for fodder production need to be evaluated so that species characteristics such as coppice and root suckering ability can be used to advantage.

Acknowledgments

Funding for the study was made available through the ACIAR project on Australian Hardwoods for Fuelwood and Agroforestry. Staff at the Queensland Department of Forestry Research Centre in Gympie provided invaluable assistance. Laboratory tests were supervised by Dr D.J. Minson, Dr M.N. McLeod and Mr A.D. Johnson of CSIRO Division of Tropical Crops and Pastures. I would like to thank Lindy Hart and Bryn Gullen for their assistance with sample collection and processing.

Chapter 20

Leaf Essential Oils of *Melaleuca* and *Leptospermum* Species from Tropical Australia

J.J. Brophy, D.J. Boland, and E.V. Lassak

Abstract

The contents of the essential oils from sixteen species of *Melaleuca* and three species of *Leptospermum* growing mainly north of the Tropic of Capricorn have been determined. The oils range from those containing almost exclusively terpenes (either mono-, sesqui- or both) to those that contain exclusively aromatic compounds. The existence of chemotypes has been shown in *Melaleuca citrolens*, *M. cajuputi* and *M. leucadendra*.

Introduction

The genera *Melaleuca* and *Leptospermum* belong, like *Eucalyptus*, to the large Australian plant family Myrtaceae. In Australia there are about 200 *Melaleuca* species (Barlow, pers. comm.) and about 80 *Leptospermum* species (J. Thompson, pers. comm.). Both genera are predominantly temperate to subtropical in character but contain a small number of tropical species occupying wet to dry habitats and ranging from tree to shrub in form.

To date, most work on leaf essential oils in Australia has been directed towards temperate *Eucalyptus* species and no systematic studies have been attempted on the essential oils of either *Melaleuca* or *Leptospermum*. This is somewhat surprising as several species of *Melaleuca* are harvested commercially for their essential oils; e.g. *M. alternifolia* is harvested in northern New South Wales for terpinen-4-ol while overseas *M. quinquenervia* and *M. cajuputi* are harvested in New Caledonia and Indonesia respectively for nerolidol and 1,8-cineole. Some *Leptospermum* species are known to contain useful oils (see Lassak and Southwell 1977), but no commercial harvesting has so far been undertaken in Australia.

Our interest in *Melaleuca* oils arose from an ACIAR project on Australian Hardwoods for Fuelwood and Agroforestry, managed by the Division of Forestry and Forest Products, CSIRO. This project was directed towards exploring the potential of lesser-known tropical and subtropical Australian tree species for use in developing countries. From 1985 to 1987 an opportunity existed to assess the leaf oils of species being grown in field trials near Gympie, Queensland, under a related ACIAR project managed by the Department of Forestry, Queensland. These trials afforded a low-cost opportunity to assess leaf oils of species which would have been otherwise difficult to acquire because of their occurrence in remote parts of northern Australia. Since we started the project we have been able to obtain some additional wild material as well as get access to previously unpublished data from the Museum of Applied Arts and Sciences, Sydney. Our main objective was to seek useful value-added products (leaf oils) from potential fuelwood tree species for the tropics.

The aim of this study was to survey the leaf essential oils of the tree species (individuals >5 m) of *Melaleuca* and *Leptospermum*, with distributions found mainly north of the Tropic of Capricorn in Australia. In our study we gave greater importance to surveying broad-leaved melaleucas as opposed to those with very small leaves (<0.5 cm long) (e.g. *M. foliolosa*, *M. minutifolia*, *M. tamariscina*, and *M.*

A small eucalypt oil distillation unit near Kunming, Yunnan province, People's Republic of China. The extraction of oils from *Eucalyptus globulus* leaves is a popular cottage industry in this region of China. Photographed April 1988.

punicea). In total there are about 35 *Melaleuca* species and seven *Leptospermum* species (three undescribed, J. Thompson, pers. comm.) that occur mainly north of the Tropic of Capricorn (see Table 1.)

For completeness, Table 1 includes all species of *Melaleuca* and *Leptospermum* that occur in our region of interest plus a guide to their distribution by States. We have found this is important, as leaf oils in some species show geographic variation. Table 1 also provides an indication of where further work could be directed. This article is largely based on our published work but does include some previously unpublished work. Where appropriate, reference is made to other published work for completeness of the survey.

Materials and Methods

The range of material collected during our studies is included in Tables 1 and 2. Greater detail on the material collected from the ACIAR Gympie trials in Dinna State Forest is included in Table 2. All oils were examined by gas liquid chromatography (glc) and/or combined gas chromatography-mass spectrometry (gc/ms). In a report of this nature it is not possible to list the detailed results of analyses performed on each sample. Such results can, if needed, be obtained from the authors. Each species tested is reviewed in alphabetical order commencing with *Melaleuca* species. Our usual practice was to collect fresh leaves from two single trees plus an additional bulk lot (5 trees) and extract the oils by steam distillation, usually within 2–3 days.

Isolation of the Oils

Leaves, either air-dried or fresh, were steam-distilled for various lengths of time (depending on the contents and yield of oil) with cohobation in an apparatus modified to give lower phase return. For leaves which were rich in monoterpenes the distillation was carried out for about 7 hours or until it was obvious that no more oil was being distilled. For leaves which seemed to have a poor yield of oil or were rich in sesquiterpenes the distillation was carried out for a longer time, up to 24 hours, or until no more oil appeared to be being distilled. In the case of species (*Melaleuca styphelioides* and *M. dealbata*) that contained little oil (and this usually meant that it was rich in sesquiterpenes) prolonged distillation sometimes resulted in the appearance of a white precipitate in the collecting region of the apparatus. Subsequent analyses of these white solids by mass spectrometry showed that they were composed of the long chain fatty acids, palmitic acid, palmitoleic acid, myristic acid and lauric acid; the former two predominated.

For species that produced oils of density greater than water (*Melaleuca leucadendra* and *M. bracteata*), about 5 ml of pentane was added to the oil collection area to help in the trapping of the oil.

Once the oil had been distilled, it was extracted with pentane (usually about 3 ml). This pentane solution was dried over sodium sulfate and the solution decanted into a storage bottle and a 250 ml beaker upended over it. The pentane was allowed to evaporate overnight at room temperature, and the bottle weighed the next morning. Subsequent gas chromatography showed that there was very little pentane remaining.

Identification of Components

Analytical gas liquid chromatography (glc) was carried out on a Shimadzu GC6 AMP gas chromatograph. A SCOT column of SP 1000 (85 m × 0.5 mm) which was programmed from 65°C to 225°C at 3°C/min was used with helium carrier gas. For combined gc/ms the gas chromatograph was connected to an AEI MS12 mass spectrometer through an all glass straight split interface. The mass spectrometer was operated at 70 eV ionising voltage and 8000V accelerating voltage with the ion source at 200°C. Glc conditions for combined gc/ms were the same as for the analytical glc. Mass spectra were acquired every 6 sec and processed by a VG Display Digispec data system. Glc integrations were performed on a Milton Roy CI-10 electronic integrator.

Compounds were identified by their identical glc retention time to known compounds and by comparison of their mass spectra with either known compounds or published spectra (Stenhagen et al. 1974; Heller and Milne 1978, 1980, 1983).

Results

Melaleuca Species

All species examined or for which published information is available are treated alphabetically, with *Melaleuca* preceding *Leptospermum*.

Melaleuca acacioides The oil from *M. acacioides* ssp. *acacioides*, obtained in 0.3–0.8% yield, was sesquiterpenoid in character. The major components were α and β-selinene in the ratio of 2:1, and these two compounds accounted for almost 80% of the oil. The next most abundant compound was selin-11-ene-4-ol, present in approximately 7% of the oil. There were 26 other unidentified sesquiterpene hydrocarbons and alcohols (mostly alcohols) present which accounted for about 10%. Also present were caryophyllene, δ-cadinene and globulol each approximately 1%. Monoterpenes were almost entirely absent.

Table 1. *Melaleuca* and *Leptospermum* species in tropical Australia distributed mainly north of the Tropic of Capricorn (23°27′S).

Species	Australian State Qld	NT	WA	Typical form
M. acacioides subsp. *acacioides*	+ *(f)	+	—	shrub
subsp. *alsophila*	—	+	+ *(w)	tree
M. angustifolia	+	—	—	tree
M. arcana	+ *(f)	—	—	shrub/tree
M. argenta	+	+	+	tree
M. arnhemica	—	+	—	shrub
M. bracteata	+ *(f)	+ *(f)	+	tree
M. brassii	+	—	—	tree
M. cajuputi subsp. *cajuputi*		+ *(w)	+	tree
subsp. *platyphylla*	+ *(f)	—	—	tree
M. citrolens	+ *(w)	+	—	tree
M. cornucopiae	—	+	—	shrub
M. dealbata	+	+ *(f)	+	tree
M. dissitiflora	+	—	—	shrub
M. foliolosa	+	—	—	shrub
M. kunzeoides	+	—	—	shrub
M. lasiandra	+	+ *(f)	+	shrub
M. leucadendra	+ *(f)	+	+ *(f)	tree
M. linariifolia	+ *(f)	+	—	tree
M. linophylla	—	—	+	shrub
M. magnifica	—	+	—	shrub
M. minutifolia subsp. *minutifolia*	—	+	+	shrub
subsp. *monantha*	+	—	—	shrub
M. nervosa	+ *(f)	+	+	tree
M. punicea = *Regelia punicae*	—	+	—	shrub
M. quinquenervia	+ *(f)	—	—	tree
M. saligna	+	—	—	tree
M. sericea	—	—	+	shrub
M. stenostachya	+ *(f)	+	—	tree
M. stypheloides	+ *(f)	—	—	shrub/tree
M. symphyocarpa	+ *(f)	+ *(f)	—	tree
M. tamariscina subsp. *tamariscina*	+	—	—	shrub/tree
subsp. *pallescens*	+	—	—	shrub
subsp. *irbyana*	+	—	—	shrub
M. viridiflora	+ *(f)	+	+	tree
M. viminalis (syn. *Callistemon viminalis*)	+	—	—	tree
Leptospermum flavescens	+ *(f)	—	—	
L. longifolium	+ *(f)	+	+	
L. petersonii	+ *(f)	—	—	
L. wooroonooran	+	—	—	
L. sp. *z*	+	—	—	
L. sp. *j*	+	—	—	
L. sp. *k*	+	—	—	

Code: + presence in Australian State, — absence from State.
* essential oil tested from particular State.
(f) ACIAR field trial material, (w) wild material.

Table 2. *Melaleuca* and *Leptospermum* species sampled from the ACIAR trials near Gympie for leaf essential oil analyses.

Species	Year planted	Trial plot no. (Dinna)	Seedlot no.[a]	Oil yield[b] %	Seedlot source[c]
Melaleuca acacioides	1985	17	S14146	0.3–0.75	SE Weipa
M. arcana	1986	80	S14866	0.6–1.0	NNE Tozer's Gap
,,	1986	?	S14876	0.01	NW of Cooktown
M. bracteata	1986	59	S14903	0.06–0.1	W Lakeland Downs
,,	1985	50	S14485	1.3–2.2	N of Alice Springs
M. cajuputi	1986	14	S14450	0.1–0.3	SE Daintree
,,	1986	?	S14878	0.5–1.1	N of Mossman
M. dealbata	1984	77	S11935	0.06–0.1	Humpty Doo, NT
M. lasiandra	1984	18	S13751	0.8–1.3	Vaughan Springs
,,	1984	12	S13752	0.3–1.1	Rabbit Flat
M. leucadendra	1984	73	S13532	1.3–1.7	Iron Range
,,	1984	13	S13567	1.0–2.0	Mareeba
,,	1985	49	S14147	0.9–1.7	Weipa
,,	1985	53	S13567	1.3–2.4	Mareeba
,,	1986	6	S13567	0.8–1.3	Mareeba
M. linariifolia	1986	22	S14979	1.4–4.5	The Lynd
M. nervosa	1984	1	S13440	0.13–0.16	Lake Buchanan
,,	1986	68	S13440	0.1–0.3	Lake Buchanan
M. quinquenervia	1986	35	S14902	0.9–1.3	NW of Mt Molloy
M. stenostachya	1985	40	S14149	1.2–1.8	Weipa
M. styphelioides	1984	45	S7177	0.04–0.1	not known
M. symphyocarpa	1985	2	S14150	1.6–2.5	Weipa
,,	1985	23	S14170	2.7–4.3	Weipa[d]
,,	1986	19	S14495	3.6–4.1	Daly River Mission
M. viridiflora	1984	80	S13530	0.4–0.9	Iron Range
,,	1986	11	S14589	1.0–1.9	NNW Rockhampton
,,	1986	85	S14558	0.8–2.1	NW Chillagoe
Leptospermum flavescens	1984	43	S13955	2.9–3.2	Nowra
L. longifolium	1985	21	S14144	0.5–0.6	Weipa
,,	1986	91	S14900	0.7–0.9	30 Km NW of Laura
L. petersonii	1986	39	S14555	0.5–0.8	SW Atherton

[a] Australian Tree Seed Centre, Division of Forestry and Forest Products, CSIRO, Seedlot Number.
[b] Based on dry weight of leaves.
[c] With the exception of Alice Springs, Rabbit Flat, Daly River Mission — all in the Northern Territory — and Nowra in New South Wales; all other sites are in Queensland.
[d] Material not vouched for by either Barlow or Thompson.

The oil from this species has a distinctive pleasant aroma which is associated with the sesquiterpene alcohol fraction. It would depend very much on the advice of perfumers if there is any commercial potential for this oil.

The oil of *M. acacioides* ssp. *alsophila* from northwestern Australia is monoterpenoid in character with an oil yield of 0.2%. The principal components p-cymene, terpinen-4-ol and citral, each approximately 20%, make this oil a possible alternative to the oil of *M. alternifolia*, known as the medicinal Tea Tree Oil. The yield of oil would have to be improved for commercial production (Brophy et al. 1987).

Melaleuca arcana Oil from this species had a quite pleasant aroma and contained mainly α-pinene and 1,8-cineole with the former compound being the more abundant component. These two compounds usually accounted for more than 50% weight of the oil. Accompanying these two compounds were smaller amounts of the usual monoterpene hydrocarbons. There were small amounts (usually <5%) of the monoterpene alcohols terpinen-4-ol and α-terpineol and trace amounts of other cyclic and alicyclic monoterpene alcohols.

Sesquiterpenes accounted for only approximately 10% of the weight of oil. The principal components were germacrene-D, α-amorphene, bicyclogermacrene and δ-cadinene. Some sesquiterpene alcohols were detected, the most abundant being α-cadinol at approximately 1%. Components accounting for approximately 1% of the oil remain unidentified and these were mostly sesquiterpenes. It is hard to envisage much commercial potential for this oil except as a 'bulk' perfume.

The yield of oil from Tozer's Gap, Queensland, material was about 1% and this was considerably greater than that of the oil from leaves obtained from Cooktown. This latter batch of trees contained practically no oil, though its composition (with the exception of α-farnesene present in one sample only) was similar to that of the Tozer's Gap material (Brophy et al. 1988).

Melaleuca bracteata The oil obtained from this species was largely composed of aromatic compounds, with terpenoid compounds accounting for less than 5% in one case (S14485, from north of Alice Springs), while in the other sample (S14903 west of Lakeland Downs) terpenoid compounds accounted for approximately 30% of the oil. The first sample consisted mainly of *trans*-methyl isoeugenol (43–76%), with lesser amounts of methyl eugenol (18–46%), *trans*-methyl cinnamate (2–8%), elemicin (0.1–1%) and isoelemicin (0.2–9%). There were small amounts of monoterpenes with limonene (0.01–1.4%) being the principal member but α-pinene, linalool, *p*-cymene and α-phellandrene all were present in the range of 0.1–0.5%

The second sample of *M. bracteata* (S14903) contained relatively less aromatic compounds. Elemicin (9–66%) was in most cases the major component and *trans*-isoelemicin (0.7–45%) the next most abundant. There were small (<1.5%) amounts of methyl cinnamate and *trans*-methylisoeugenol. This sample was, however, much richer in terpenes. Caryophyllene (7–22%) was in most cases the main member of this group while α-phellandrene (2–13%) was the next most abundant. There were smaller, though significant, amounts of α-pinene, β-*trans*-ocimene, terpinolene, humulene, germacrene-D, α-farnesene, δ-cadinene and caryophyllene oxide. This particular sample produced a very poor yield of oil (0.05–0.1%).

Melaleuca bracteata has been mentioned as a source of the aromatic ethers methyleugenol, methylisoeugenol and elemicin, but the poor oil yield mitigated against its commercial use (Penfold and Morrison 1950). Higher oil-yielding forms of this species have also been reported (Lassak and Southwell 1977). Certainly the oil from seedlot S14903 falls into this category. The other material (S14485), however, is of much higher yield and may be of commercial importance. The existence of chemotypes in this tree, however, means that great caution has to be exercised in the collection of the leaf material.

Melaleuca cajuputi This species is known to occur in three discrete subspecies, viz. *platyphylla*, *cajuputi* and *cu* (Barlow, pers. comm.). The *cajuput* oil produced in the Indonesian area presumably comes from subsp. *cu*. The trees examined in our project belong to subsp. *platyphylla* which occur in northern Queensland and Papua New Guinea and subsp. *cajuputi* from the Northern Territory.

The oil from *M. cajuputi* subsp. *platyphylla* originating from two sources, viz. from southeast of Daintree and north of Mossman, was monoterpenoid in character. The largest component, α-pinene was an order of magnitude larger than the next most abundant monoterpene, 1,8-cineole. The monoterpene alcohols were present but in somewhat larger than trace quantities.

The remainder of the oil of *M. cajuputi* subsp. *platyphylla* consisted of sesquiterpenes with caryophyllene (7–13%) and humulene (4–7%) being the major components. There were small but significant amounts of α- and β-selinene (2%) present in this oil, as well as small quantities of globulol, viridiflorol and spathulenol. The yield of oil varied from 0.1 to 1%. Altogether components accounting for approximately 2.5% of the oil remain unidentified. These were sesquiterpenes (Brophy et al. 1988a).

The high percentage of α-pinene in this oil (which also gives it a characteristic pleasant smell) means it may also have some potential as a source of this compound. The yield of oil would, however, have to be increased before it was of any commercial importance.

The oil from the third subspecies, subsp. *cajuputi*, has also been separately examined and is rich in sesquiterpenes, almost to the exclusion of monoterpenes. The major compounds were α- and β-selinenes and spathulenol (Brophy et al. 1988a).

Melaleuca citrolens Barlow (1986) has recently separated this species from *M. acacioides*. The oil of this species bears a close relationship to that obtained from *M. acacioides* subsp. *alsophila*. Two definite chemotypes have been found for *M. citrolens* and there is a possibility of a third chemotype. Two of these chemotypes are characterised by a strong lemon scent, while the other has a much higher 1,8-cineole percentage and none of the lemon-scented components.

Chemotype 1 (from 0.5 km north of Koolburra Creek on the Peninsular Development Road) contains citronellal (5–12%), the isopulegols (1–7%), citronellyl acetate (7–11%), citronellol (9%), as well as 1,8-cineole (22%), α-terpineol (3%), as the major monoterpenes and caryophyllene (2%) and bicyclogermacrene (4%) as the major sesquiterpene hydrocarbons. The principal sesquiterpene alcohols were globulol, viridiflorol and spathulenol (each in 1–3%). In total, sesquiterpenes accounted for less than 10% of the oil. The yield of oil from this lemon-scented type was 2.5% on a fresh weight basis.

The possible second lemon-scented chemotype (from the same location as chemotype 1), which also produced oil in 2% yield (fresh weight basis), contained all of the above compounds together with geranial (7%) and also possessed more β-farnesene and caryophyllene than the first chemotype.

The third chemotype from 5.5 km south of the Laura River Crossing on the Peninsular Development Road (which gave an oil in 1.5% yield, on a fresh leaf weight basis) was not lemon-scented. It contained 1,8-cineole in approximately 60% amounts as well as α-pinene (5%), β-pinene (5%), limonene (5%), α-terpineol (7%) as the main monoterpenes. Sesquiterpenes accounted for no more than 10% of the oil. The principal members were bicyclogermacrene (2%) and the alcohols globulol (4%), viridiflorol (2%) and spathulenol (1%).

Melaleuca dealbata This species yielded an oil containing approximately 1% monoterpenes, the major components being 1,8-cineole and α-terpineol. The remainder of the oil consisted of sesquiterpenes with caryophyllene, at 34%, being by far the major component. Other sesquiterpene hydrocarbons present included aromadendrene, α-bulnesene, alloaromadendrene, humulene, viridiflorene, α- and β-selinene and calamenene. All of these compounds were present in amounts between 1 and 4%.

The principal oxygenated sesquiterpenes were caryophyllene oxide and globulol, accounting for 11% of the oil. There were also smaller amounts of viridiflorol, spathulenol, T-cadinol and T-muurolol, and a trace amount of farnesol. Seventeen percent of the oil consisted of compounds, mostly oxygenated sesquiterpenes in small amounts, which could not be identified. The yield of oil from this species was 0.1% (Brophy et al. 1988a). No trace was found of the previously reported tetraketone leptospermone (Lassak and Southwell 1977). In view of the poor oil yield and its complex composition, no commercial use can at present be suggested for this oil.

Melaleuca dissitiflora The results of oil analyses on this species have been published (Brophy and Lassak 1983). This species exists in two chemical forms. One form, from Ekedra and Bonney Well in the Davenport Ranges, NT, is high in 1,8-cineole (approximately 65%) and contains lesser amounts of α-pinene (2%), limonene (5%), γ-terpinene (0.5–6%) and terpinolene (3%). The alcohols were mainly terpinen-4-ol (2–6%) and α-terpineol (5–9%). There were virtually no sesquiterpenes present in this oil, which was obtained in approximately 2% (based on air-dried leaves).

The second chemotype, from Charles River, in the vicinity of Alice Springs, NT, is much lower in 1,8-cineole (2–7%) and contains substantial amounts of terpinen-4-ol (23–52%). Other constituents were α-pinene (2–10%), β-pinene (0.5–14%), α-terpinene (4–10%), γ-terpinene (12–18%) and *p*-cymene (2-4%. α-terpineol was also present (2%). This latter chemotype has potential as a source of terpinen-4-ol, particularly in view of its oil yield (2–4%, based on air-dried leaves).

Melaleuca lasiandra The oil from this species was rich in monoterpenes. In this case the principal components were α- and β-pinene and limonene in 30, 12 and 30% respectively. These three components accounted for approximately 70% of the oil. There were only very small quantities of the monoterpene alcohols, with α-terpineol being the principal member. A significant amount of benzaldehyde (1–7%) was detected in the oil due to the decomposition of mandelonitrile or its glycoside during the steam distillation.

Sesquiterpene alcohols were more abundant than the hydrocarbons with α-, β- and γ-eudesmols, together with globulol, being the principal alcohols. The major sesquiterpene hydrocarbons were caryophyllene, aromadendrene and viridiflorene. Total sesquiterpenes were less than 15% of the oil (Brophy et al. 1988a).

The yield of oil and its high pinene content could make this oil of some use in the perfume or disinfectant area, though it may be necessary to remove the sesquiterpenes first.

Melaleuca leucadendra This species occurs in the two northern Australian States and the Northern Territory though the samples used in this study came only from northern Queensland. The oil from this species is at least 93% aromatic (Lassak and Southwell 1977). There appear to be three chemotypes of this species; one that contains mostly methyl eugenol, one that contains mostly methyl isoeugenol and one whose oil is mostly terpenoid in character. Both forms that contain principally aromatic compounds were represented in this study, though unexpectedly only the material from Mareeba (S13567) contains trees in which methyl isoeugenol predominates.

The samples from seedlots S14147 and S13532 contained methyl eugenol as the major component (94–97%). The remainder of the oil was made up of at least 35 compounds, obviously in small amounts, though *trans*-methyl isoeugenol (0.4–2.6%) was the next most abundant compound. Virtually no monoterpenes were detected and the sesquiterpene hydrocarbons had germacrene-D as the major contributor (0.03–1%). Caryophyllene, aromadendrene, bicyclogermacrene, δ-cadinene, cadina-1,4-diene and calamenene (each at <0.3%) were also major compounds. The sesquiterpene alcohols were unidentified, but the sum of the 10 components accounted for <0.5%.

The other chemotype, represented by S13567, contained up to 89% *trans*-methyl isoeugenol, together with methyl eugenol (7–24%) and *cis* methyl isoeugenol (0.1–0.7%). The bulk sample of trees from this seedlot indicated that not all the trees were of the high methyl isoeugenol type but also contained some of the other type (high methyl eugenol). The methyl isoeugenol type does not breed true, though the methyl eugenol type does. The range of terpenes was similar in this latter type to the former type and in no case amounted to more than 3% (Brophy and Lassak 1988).

There is obvious commercial potential for the oil from this species, both the methyleugenol and the methyl isoeugenol form, in the flavour and fragrance industry. If the desired end product was methyl isoeugenol, then no particular care need be taken during the collection of leaf, as all the methyl eugenol could be readily interconverted to methyl isoeugenol. If on the other hand the individual products were wanted, then great care would be needed during leaf collection.

A third chemotype of *M. leucadendra* has recently been found in northwestern Australia. This chemotype (yielding ~ 0.5% oil) contains almost exclusively terpenes. The oil's major components are γ-terpinene (14%), terpinolene (9%), bicyclogermacrene (13%) and globulol (7%). There is the usual range of both mono- and sesquiterpenes present in the oil but only 1.6% of methyl eugenol (Brophy, unpublished results; Brophy and Lassak 1988). It has, however, been pointed out to us (Barlow, pers. comm.) that it is very difficult to distinguish *M. leucadendra* from the narrow-leaved *M. viridiflora* especially in the Kimberley region of Western Australia.

Melaleuca linariifolia There are two known chemotypes of this species. One is rich in 1,8-cineole while the other is rich in terpinen-4-ol (Penfold and Morrison 1950). This latter chemotype is sometimes used as a commercial source of terpinen-4-ol. The sample in this study appeared to be yet another chemotype, though it was closer to the cineole chemotype. This sample had a low terpinen-4-ol content (0.3–2%) but the trees appeared to be of two types, one containing a high terpinolene content (75%) while the other had a high cineole content (54%). The bulk sample, with these two compounds accounting for 37 and 42% respectively, appeared to be a mixture of these two types. The usual monoterpene hydrocarbons were present in small (<3%) amounts while α-terpineol, at 3–7%, was the only other alcohol of any account. There were no sesquiterpenes present though traces of eugenol and methyl eugenol were also detected.

The oil with the high terpinolene content has a pleasant aroma and as such could be (because of its 1.4% yield) of potential use in the perfumery area. The cineole rich oils were often obtained in higher yields and may be of use in the germicides or soaps.

Melaleuca nervosa The oil obtained (in 0.1% yield) from this species contained approximately 2% monoterpenes; the major contributor being limonene (1%) with smaller amounts of camphene and terpinen-4-ol. The remaining 98% of the oil was a complex mixture of sesquiterpenes. The major sesquiterpene hydrocarbon was caryophyllene (18%) with lesser amounts (<3%) of aromadendrene, alloaromadendrene and calamanene and trace amounts of humulene, viridiflorene, α- and β-selinene and α-copaene.

The oxygenated sesquiterpene components were dominated by spathulenol with up to 40% in one tree. There were smaller quantities of caryophyllene oxide, globulol and viridiflorol. Approximately 24% of the complex mixture of sesquiterpenes, the majority of which were present in <0.5% amounts, remains unidentified. The low oil yield of this species and its complex sesquiterpene nature seems to preclude any development unless as a source of spathulenol, but this is more readily available from *Eucalyptus spathulata* (Bowyer and Jefferies 1962, 1963).

Melaleuca quinquenervia There are two known chemotypes of this species, both of which occur in Australia (Lassak and Southwell 1977). One is rich in nerolidol, while the other chemotype is rich in 1,8-cineole and sometimes viridiflorol. This latter type is the chemotype from which Niaouli oil is obtained. The sample in this survey, *M. quinquenervia* affinity *quinquenervia*, belongs to the latter chemotype. The major component by far was 1,8-cineole (52–65%), with smaller amounts of α-pinene (2–8%), myrcene (1%) and limonene (7%). The oxygenated monoterpenes α-terpineol (5–9%) and terpinen-4-ol (1%) were present.

The sesquiterpenes, though numerous, were each present in small (<3%) quantities. Caryophyllene, aromadendrene, viridiflorene and globulol each in the range 1–3% were the major compounds detected, while there were smaller quantities of bicyclogermacrene, ledol and spathulenol. Altogether 24 sesquiterpene hydrocarbons and alcohols, accounting for <5%, remain unidentified. This chemotype may have potential as a source of cineole-rich oil though superior alternative sources exist.

Both chemotypes of *M. quinquenervia* have been planted on the campus of the University of New South Wales. The chemotype that contains nerolidol is very rich (>90%) in that compound. The other chemotype contains approximately 30% 1,8-cineole and approximately 60% viridiflorol.

Melaleuca stenostachya The leaves of this species yielded 1.5% oil and the major compounds were 1,8-cineole (53%) and α-pinene (24%). There were much smaller amounts (<5%) of β-pinene and limonene as the next most abundant monoterpenes. Of the monoterpene alcohols only α-terpineol at 2% was of any consequence. Altogether monoterpenes accounted for over 90% of the oil.

Small amounts of sesquiterpenes were present with caryophyllene, at approximately 6%, by far the largest component. There were small quantities (each <0.7%) of humulene, globulol, spathulenol and a compound which, from its mass spectrum, was assumed to be a caryophyllene alcohol. Approximately 1% of the oil was due to components (most sesquiterpenes) which remain unidentified (Brophy et al. 1988a).

It is possible to conceive of the use of this oil for the same uses as ordinary *Eucalyptus* oil. It has a reasonable yield.

Melaleuca styphelioides This species produces a very low yield of oil (0.04–0.1%). As has been found for other *Melaleuca* species, a poor oil yield indicated that it was mainly composed of sesquiterpenes. In this case at least 95% of the oil was sesquiterpenoid in character. The major compound was suspected of being a caryophyllene alcohol of formula $C_{15}H_{24}O$ at approximately 35%. There were lesser amounts of caryophyllene (2–10%), alloaromadendrene (1–13%), β-farnesene (3%), α-santalene (trace–1%), caryophyllene oxide (1–2%), globulol (5%), viridiflorol (0.3–1%) and spathulenol (1–5%). Approximately 30 of the 64 compounds detected remain unidentified.

The monoterpene components were represented by α-pinene (14%) in one tree, β-pinene, sabinene and limonene linalool, α-terpineol and *p*-cymene-8-ol each in <0.5%. Also detected in trace quantities in one tree were *cis*-hex-3-enyl alcohol, its acetate and *n*-hexanol.

It was found that if the leaves were steam-distilled for >10 hours, a white solid was formed on top of the distillate. Mass spectrometry of this solid indicated that it was composed of palmitic, palmitoleic, myristic and lauric acids. The small quantity of these acids recovered (in the order of 1 mg/100 g leaf) would not indicate this to be a useful source of these fatty acids. It is hard to imagine any use for the leaf oil from this tree.

Melaleuca symphyocarpa The oil from this species was predominantly monoterpenoid in character with 1,8-cineole (46–65%) being by far the largest component. Smaller, though significant, amounts of α-pinene (~9%), β-pinene (1%) and limonene (1%) were the only monoterpene hydrocarbons of note present. α-terpineol (2%) and to a lesser extent terpinen-4-ol (0.5%) were the only oxygenated monoterpenes in any significant quantity.

The principal sesquiterpene present was caryophyllene (15%) in greater quantity than all of the remaining sesquiterpene hydrocarbons and alcohols combined. There were smaller amounts (<0.5%) of aromadendrene, alloaromadendrene, α-gurjunene, humulene, bicyclogermacrene and viridiflorene present but only trace quantities of the various sesquiterpene alcohols (Brophy et al. 1988b in prep.; Lassak and McCarthy 1983). The high oil yield and predominance of cineole in this species makes it an attractive source of an oil of the *Eucalyptus* type.

Melaleuca viridiflora There are two known chemotypes of this species. One is basically terpenoid in character while the other contains methyl cinnamate and β-*trans*-ocimene (Hellyer and Lassak 1968). The three lots of *M. viridiflora* examined in this study all belonged to the former chemotype, though there were quite considerable variations between trees within each seedlot.

The oil from *M. viridiflora* from north west of Chillagoe (S14558) was over 98% monoterpenoid in character. Of the monoterpenes, the two major components were γ-terpinene and terpinolene, each approximately equal and together accounting for over 70% of the oil. Other major compounds were α-pinene (9%), α-phellandrene (3%), α-terpinene (8%), limonene (2%) and *p*-cymene (2%). There was very little 1,8-cineole (0.3%), small amounts of terpinen-4-ol (1%) and only a trace of α-terpineol.

The major sesquiterpene was caryophyllene (0.7%) with humulene being the next most abundant hydrocarbon (0.2%). Six oxygenated sesquiterpenes were identified, accounting for approximately 1% of the oil, but all were unidentified. The perfume of this oil is not remarkable but the oil, which is of reasonable yield, could be used as a solvent.

The oil of *M. viridiflora* from north-northwest of Rockhampton (S14589), while being of similar constituents to the previous sample, was quantitatively different. This oil contained approximately 10% sesquiterpenes. The major component of the oil was 1,8-cineole (31–58%) with other monoterpene hydrocarbons being of minor quantity (α-pinene, 6%; β-pinene, 1.5%; myrcene, 1%, and limonene 9%). Of the monoterpene alcohols α-terpineol (8%) was the major component, followed by terpinen-4-ol (0.5%) and benzaldehyde (0.3%).

Of the sesquiterpenes present in this oil, viridiflorene (3%) was the major component while there were lesser amounts of caryophyllene (1.5%), aromadendrene (1%) and humulene (0.5%). The major sesquiterpene alcohols were ledol (2%) and viridiflorol (4–9%) while there were trace quantities of another 10 alcohols. The previous remarks about the commercial potential of this oil also apply here.

The third sample of *M. viridiflora*, from Iron Range (S13530), contained a variable but lower amount of 1,8-cineole and a higher percentage of sesquiterpenes. The major component in this sample was α-pinene (1–29%) with trace amounts of the other monoterpene hydrocarbons and 11–48% of 1,8-cineole (the tree with the lowest α-pinene content had the highest cineole content). The sample also contained small percentages (<5%) of linalool, terpinen-4-ol and α-terpineol.

The major sesquiterpenes in this sample were the alcohols, with spathulenol (4–15%), globulol (2%) and two unidentified sesquiterpene alcohols each of approximately 6% being the major members. Caryophyllene (2–9%) was the major sesquiterpene hydrocarbon; most of the many hydrocarbons identified in the previous sample (S14589) also being present. Altogether about 36 sesquiterpenes, accounting for approximately 25% of the oil, were unidentified. In view of the yield and composition of this oil, it is difficult to see any possibility of commercial exploitation.

Leptospermum Species

The oil of three tropical *Leptospermum* species was examined: *L. flavescens*, *L. longifolium* and *L. petersonii*. Of these three *L. flavescens* was the only species to produce oil in any reasonable quantity.

Leptospermum flavescens Sesquiterpene alcohols accounted for over 50% of the oil of this species. The principal components were α-, β- and γ-eudesmols in 19, 24 and 20% respectively. There were also smaller (<2%) amounts of spathulenol, viridiflorol and globulol. The major sesquiterpene hydrocarbons were caryophyllene (1%), bicyclogermacrene (2%), viridiflorene (0.5%) and aromadendrene (0.6%). A large number (20) of other sesquiterpene hydrocarbons and alcohols were detected but all were in the range 0.01–0.3%.

The principal monoterpene was α-pinene (10%), with lesser quantities of β-pinene (6%), γ-terpinene (1%) and *p*-cymene (2%). Terpinen-4-ol (5%) and α-terpineol (1%) were the main alcohols. In this study, no trace was found of the β-triketones, flavesone and leptospermone, previously reported from *L. flavescens* (Hellyer 1968).

There is potential for the use of the three eudesmols as fixatives in perfume mixtures, particularly in view of the reasonable yield of oil and their abundance in the oil.

Leptospermum longifolium Two samples of *L. longifolium* were examined (S14144 and S14900), and each produced an oil yield of 0.5–1.0%. There were qualitative similarities between this oil and that of *L. flavescens*. Both samples of *L. longifolium* contained almost equal amounts of both mono and sesquiterpenes. The major members of the monoterpenes were α-pinene (19–34%), β-pinene (5–19%) and cineole (1–10%), with the usual range of hydrocarbons present in <0.5% amounts. Terpinen-4-ol and α-terpineol were both present in small amounts (<2%).

The major sesquiterpenes were caryophyllene and humulene. Both seedlots varied in the amount of humulene present in the oils, the range being 0.6–32%. The two bulk samples contained 20% (S14144) and 10% (S14900) respectively. The amount of caryophyllene varied from 3 to 14%. Other sesquiterpene hydrocarbons present in significant amounts were α-gurjunene (0.7%), aromadendrene (1–3%), viridiflorene (0.1–3%), germacrene-D (0.5–2%), bicyclogermacrene (1.5–7%), δ-cadinene (1.3–2%), cadina-1,4-diene (1–2%) and calamenene (2–6%). The sesquiterpene alcohols were present in lesser quantities, with the major members being globulol (1–4%), viridiflorol (0.7–2%), spathulenol (2–6%), γ-eudesmol (0.2–2%), α-eudesmol (0.2–2%) and β-eudesmol (0.2–1%). Twenty sesquiterpene hydrocarbons and alcohols accounting for <2% of the oil remained unidentified. Apart from the reasonable amount of humulene in some samples there seems to be no commercial potential for this oil.

Leptospermum petersonii This species includes different chemotypes. One chemotype contains large amounts of citral and citronellal and has the characteristic lemon-scented smell (Penfold and Morrison 1950). The chemotype in this trial (S14555) is not that type but one rich in sesquiterpenes. It is characterised by the almost total lack of monoterpenes. The major compounds present were δ-cadinene (16%), germacrene-D (15%), an unidentified hydrocarbon $C_{15}H_{24}$ (15%), α-bergamotene (1–4%), β-elemene (2–4%), β-ylangene (1%), viridiflorene (3%), α-amorphene (4–6%), bicyclogermacrene (3%) and cadina-1,4-diene (1%). There were approximately eight other unidentified sesquiterpene hydrocarbons accounting for 8% of the oil.

The sesquiterpene alcohols were present in small amounts with at least 22 of them present. The major identified compounds were cubenol (0.7%), globulol (1.5%), viridiflorol (0.7%), spathulenol (1.5%), T-cadinol (1.7%), T-muurolol (1.7%), δ-cadinol (0.7%) and α-cadinol (3.6%). The other unidentified alcohols accounted for <5% of the oil. Small amounts of methyl eugenol (0.3%) and eugenol (0.4%) were also detected.

The oil, though complex and with a large number of the components still unidentified, has a quite pleasant aroma and may have potential in the perfumery field. This would, however, need confirmation from experienced perfumers.

Conclusion

Together with previous studies, this study has highlighted the importance of chemotypes within wild and cultivated populations of *Melaleuca* and *Leptospermum*. For example *M. leucadendra* can exist as different chemotypes and it is important to make sure that the correct chemotype is being collected. It has also been shown that the methylisoeugenol chemotype of *M. leucadendra* does not breed true to type (Brophy and Lassak 1988) and vegetative propagation is required to ensure an efficient means of propagating this chemotype.

Our survey has suggested the existence of regional chemotypes of certain *Melaleuca* species whereby particular chemotypes occur over a large geographic area. Both *M. cajuputi* and *M. leucadendra* appear to have chemotypes from the central (Northern Territory) and western (Western Australia) regions of Australia which differ (markedly in the case of *M. leucadendra*) from those in the eastern region.

As a general rule, the yields of oil obtained under the cohobation conditions used in these analyses are often at least one-third greater than those obtained under field conditions. It has been suggested that for a tree to have a commercial potential it should yield at least 1.5% oil on fresh foliage (equal to approximately 3% on dry weight). On this basis *M. bracteata*, *M. leucadendra*, *M. linariifolia*, *M. symphyocarpa* and *L. flavescens* may have some potential based on oil yield. Another general rule for *Melaleuca* seems to be that if yields are low then the oils are rich in sesquiterpenes and if high they are rich in monoterpenes.

In assessing the viability of commercial propagation of these species for oil production, it is worth noting that total oil yields are also influenced by the weight of leaf produced per tree over time. In addition, vegetative characteristics such as ability to root cuttings of superior chemotypes and oil yielders and coppicing ability of plants (after harvest) needs to be assessed. The ACIAR field trials at Gympie, Queensland, and in other countries, particularly in Thailand, will provide some of the required information to enable a more complete assessment of the commercial potential of these promising melaleucas and leptospermums.

Our survey of the tropical *Melaleuca* is, in its present stage, incomplete, though we have examined a representative of at least one population of most of the tree *Melaleuca* and *Leptospermum* which occur in tropical Australia (Table 1). For the sake of completeness, reference to other published work on *Melaleuca* species oils is included (Lassak 1979; Flynn et al. 1979; Brophy and Lassak 1985). Much more work is needed to complete this survey but at least a start has been made and a guide to future work has been indicated.

Acknowledgments

We would like to thank Mr P. Ryan and Mr D. Taylor, Department of Forestry, Gympie, Queensland, for collection of material in the ACIAR field trials. We also thank staff of the Australian Tree Seed Centre, Division of Forestry and Forest Products, CSIRO, for collection of wild material in Western Australia. Mr J. Clarkson, Queensland Department of Primary Industries, is thanked for the collection of *M. citrolens*. Dr B. Barlow, Division of Plant Industry, CSIRO, authenticated most of the *Melaleuca* botanical specimens and Mrs J. Thompson, National Herbarium and Botanic Gardens of NSW authenticated the *Leptospermum* botanical specimens. Dr Barlow and Mrs Thompson are especially thanked for permission to refer to unpublished taxa in the interest of completeness for this study.

Chapter 21

Leaf Essential Oil of *Eucalyptus bakeri*

J.J. Brophy and D.J. Boland

Abstract

The essential oil of fire-induced coppice leaves of *Eucalyptus bakeri* was steam-distilled and the oil composition analysed by gas chromatography and mass spectrometry. Oil yield on a fresh-weight basis was relatively high, ranging from 1.8 to 3%. The main oil component was 1,8-cineole (85–96%). High yields of cineole-rich leaf oils are rarely found in tropical/subtropical eucalypts. *Eucalyptus bakeri* has potential for oil production and field trials should be established to assess growth rates to appraise commercial potential.

Introduction

Eucalyptus bakeri varies in form from a multi-stemmed bush up to 5 m high to a small tree up to 12 m high and 0.5 m in diameter. It occurs in restricted locations over a wide range of central and southeastern Queensland and northern New South Wales (NSW) from about latitude 22°S to 30°S (Fig. 1). *Eucalyptus bakeri* occurs typically on gentle rises in country of low relief (Hall and Brooker 1974).

The essential oils of *E. bakeri* leaves were first investigated by Penfold (1927). He examined leaves collected near Inverell, NSW, and near Eidsvold in southern Queensland. The Inverell material yielded 1.12% oil and the Eidsvold material (3 lots) 1.8–2.2% on air-dried leaf basis. The cineole yield was high and ranged from 70 to 76%. Penfold remarked that the oils were bright reddish-yellow and resembled in all general physical characters the well-known commercial oils obtained from *E. polybractea* and *E. cneorifolia*. 1,8-cineole is a medicinal compound used in a wide range of pharmaceutical products and cineole-rich oil is currently the major eucalypt oil harvested in Australia (Small 1981).

The impetus for the present study came from an Australian Tree Seed Centre CSIRO seed collection team, led by Mr C. Gardiner, who noted the strong 'eucalyptus' smell of crushed leaves from *E. bakeri* coppice resulting from wild fires. The aim of the study was to assess the composition and oil yield from *E. bakeri* coppice.

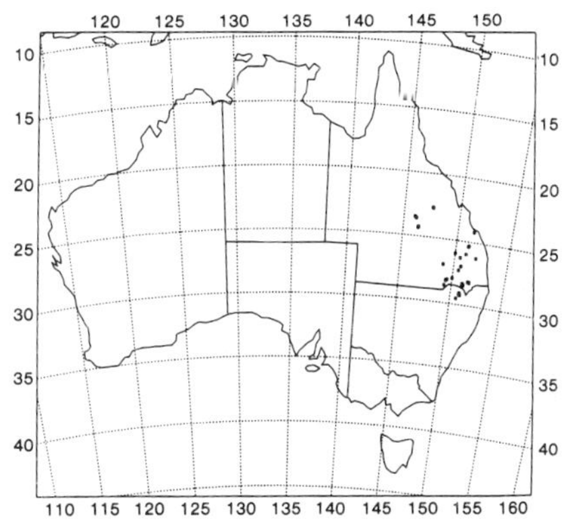

Fig. 1. Distribution of *E. bakeri* as derived from the original data collected for Eucalist (Chippendale and Wolf 1981).

Materials and Methods

Collection of Leaves and Isolation of Volatile Oils

Coppice leaves were collected from five trees in Durakai State Forest near Warwick, southern Queensland, and stored at about 5°C. About 50 g of leaves from two individual trees and one bulk sample from the three remaining trees were steam-distilled with cohobation as previously described (Lassak 1979) for 8 hours to yield colourless oils. Yields ranged from 1.8 to 3% on a fresh leaf weight basis.

Identification of Components

Analytical gas liquid chromatography (glc) was carried out on a Shimadzu GC6 AMP gas chromatograph. A SCOT column of SP 1000 (85 mm × 0.5 mm) which was programmed from 65°C to 225°C at 3°C/min was used with helium carrier gas. For combined gas liquid chromatography/mass spectrometry (glc/ms) the gas chromatograph was connected to an AEI MS12 mass spectrometer through an all-glass straight split interface. The mass spectrometer was operated at 70 eV ionising voltage and 8000 V accelerating voltage with the ion source at 200°C. Gas liquid chromatography conditions for combined glc/ms were the same as for the analytical glc. Spectra were acquired every 6 sec and processed by a VG Display Digispec data system. Gas liquid chromatography integrations were performed on a Milton Roy CI-10 electronic integrator.

Compounds were identified by matching their glc retention time to that of known compounds and by comparison of their mass spectra with either known compounds or published spectra (Stenhagen et al. 1974; Heller and Milne 1978, 1980, 1983).

Results

Thirty-one compounds were detected in the steam-volatile oil of *E. bakeri* of which 26 have been identified. The principal feature of the oil is the extraordinarily high proportion of 1,8-cineole (Table 1). From the two individual trees sampled, the cineole content was 94 and 96% respectively, while in the bulk sample from three trees it was 85%. Several monoterpene alcohols were also detected but in very small quantities. The bulk sample contained some sesquiterpene alcohols, principally globulol (2%), viridiflorol (0.24%) and spathulenol (0.16%). It also contained a larger amount of the monoterpene α-pinene (5.7%) and the related compounds pinocarvone and pinocarveol. In the two trees containing the largest cineole content, monoterpenes (mainly limonene and α-pinene) accounted for the greater part of the remaining oil.

Both the yield of oil and the 1,8-cineole content from these trees was higher than those reported previously for *E. bakeri* (Penfold 1927). The two individual trees with a high proportion of 1,8-cineole also had the highest oil yields, at 2.7 and 3.0%. No phloracetophenone dimethyl ether was detected in any of the samples which was the case in all but one of the samples in the first study (Penfold 1927).

Table 1. Compounds identified in the steam-volatile leaf oil of *Eucalyptus bakeri*.

Compound	%	Compound	%
α-pinene	0.8–5.7	$C_{15}H_{24}$	tr–0.02
camphene	tr[a]–0.01	δ-terpineol	tr–0.03
β-pinene	0.09–0.11	α-terpineol	0.18–0.67
sabinene	0.19–1.0	terpenyl acetate	tr–0.05
myrcene	0.05–0.22	viridiflorene	tr–0.02
limonene	0.09–2.7	carvone	0.01–0.06
1,8-cineole	85.2–96.0	*cis*-mentha-1(7),8-dien-2-ol	tr–0.01
γ-terpinene	0.24–0.34	*trans*-mentha-1(7),8-dien-2-ol	tr–0.01
p-cymene	0.31–0.64	$C_{15}H_{26}O$	0.01–0.17
terpinolene	tr–0.08	$C_{15}H_{26}O$	tr–0.01
linalool	tr–0.04	$C_{15}H_{26}O$	tr–0.01
pinocarvone	tr–0.07	globulol	0.01–2.0
terpinen-4-ol	tr–0.35	viridiflorol	0.01–0.24
aromadendrene	tr–0.36	$C_{15}H_{26}O$	tr–0.06
α-bulnesene	tr–0.01	spathulenol	0.01–0.16
pinocarveol	tr–0.40		

[a]tr = trace, <0.01%.
Compounds are listed in order of elution from a SP1000 column.

Discussion

The yield of oil (1.8–3%) on a fresh-weight basis (approximately 4–6% on a dry-weight basis) is high for a eucalypt and ranks with the commercial yields obtained from *E. polybractea* in the West Wyalong area of New South Wales. The yield of cineole found in the *E. bakeri* samples (85–96%) is much higher than the average reported by Penfold and Willis (1961) for *E. polybractea* (77–84%).

Eucalyptus bakeri belongs taxonomically to *Eucalyptus* section *Bisectaria* (Pryor and Johnson 1971). The majority of species belonging to this section occur in the southwest part of Western Australia. There are three east coast members of this group: *E. pachycalyx* (near Atherton), *E. squamosa* (near Sydney) and *E. bakeri.* The essential oils of some of the Western Australian members of section *Bisectaria* (e.g. *E. oleosa* and related species) are also high in cineole and are being studied by staff at Murdoch University and the Division of Forestry and Forest Products CSIRO in Perth.

There is a paucity of eucalypts in the tropical/subtropical parts of Australia with high yields of cineole-rich oils. Penfold (1927) drew attention to this fact when he first described the oils of *E. bakeri.* Penfold and Willis (1961) list the following oil yields and cineole contents for eucalypts having a Queensland distribution: *E. sideroxylon* (1.5–2.5%; 65–75%), *E. banksii* (0.28%; 69%), *E. microcorys* (0.71–0.73%; 43–46%), *E. punctata* (0.63–1.19%; 46–64%), *E. resinifera* (0.42%; 50%), *E. dealbata* (0.86%; 52%) and *E. seeana* (0.78%; 52%). More recently *E. punctata* has been reinvestigated by Southwell (1973) who found that it contained 0.2–2.3% volatile oil of which 70% was 1,8-cineole.

There has been strong interest from several developing tropical countries in obtaining seed of eucalypts with high yields of cineole-rich oil in order to establish small, cottage-type, oil industries. The results from the study on *E. bakeri* are encouraging although little is known of the growth rate of the species outside Australia. The species grew slowly in a research arboretum in Malawi (Poynton 1979) although trees of attractive stature were formed. Further growth trials are needed to assess productivity. The ability of the species to form vigorous coppice after intense wild fires suggests it would thrive under a coppice form of management.

A comparison of the results of the current study with those obtained earlier by Penfold (1927) suggests the possibility of provenance variation in oil yield and composition, and the need to survey yields from individuals in a wider range of provenances. *Eucalyptus bakeri* is a species of promise for cineole production in the tropics.

Acknowledgments

We wish to acknowledge the role of colleagues, particularly Mr C. Gardiner of the CSIRO Australian Tree Seed Centre, for bringing the species and the populations reported to our attention. We thank Ms D. Crawford, Mr Gardiner, Mr M. McDonald and Mr T. Vercoe for the leaf collections. The work was partly funded by the Australian Centre for International Agricultural Research.

Chapter 22

Managing Nitrogen Fixation in *Casuarina* Species to Increase Productivity

P. Reddell, P.A. Rosbrook and P.A. Ryan

Abstract

Progress of research aimed at increasing the productivity of *Casuarina* plantations by enhancing symbiotic nitrogen fixation is discussed. Research has concentrated on: (i) selecting strains of *Frankia* effective in promoting growth of *Casuarina*; (ii) developing a simple but effective inoculation technology suitable for forest nurseries; and (iii) identifying soil factors that influence tree responsiveness to inoculation. Field trials have demonstrated substantial benefits of inoculation of nursery stock of *Casuarina* with *Frankia*, wood production increases in excess of 200% being recorded in some situations. Factors affecting tree response to inoculation include tree provenance, strain of *Frankia* and phosphorus status of the planting site. Nursery studies have shown inoculum placement can be critical for rapid nodulation and fast seedling growth, while glasshouse studies have identified isolates of *Frankia* effective in promoting seedling growth. Advances in methods for cultivation of *Frankia* in the laboratory make commercial inoculum production a realistic possiblity, provided that a suitable carrier system can be found and that problems of quantifying the amount of infective *Frankia* in an inoculant can be overcome.

Introduction

Casuarinas occur naturally in Australia and the western Pacific where they are found in environments ranging from tropical forests to arid woodlands and saline sites. A number of species of *Casuarina* are noted for their rapid growth rates on infertile soils and are potentially useful for fuelwood plantings, agroforestry and reclamation of marginal and disturbed lands in the humid and semi-arid tropics (Midgley et al. 1983). To date, however, only one species *Casuarina equisetifolia* is widely planted for these purposes (U.S. National Research Council 1984).

The successful establishment and growth of *Casuarina* on nitrogen-deficient soils is dependent on the formation of a symbiotic nitrogen-fixing association between the plant roots and the soil actinomycete *Frankia*. Nodules formed by *Frankia* in roots of *Casuarina* are capable of nitrogen fixation at rates comparable to those found for effectively nodulated legumes (Torrey 1978). Two field studies have estimated fixation rates of 40–60 kg N/ha/year in plantings of *C. equisetifolia* on coastal sand dunes in Senegal (Dommergues 1963; Gauthier et al. 1985). Despite these results demonstrating the potential importance of symbiotic nitrogen fixation in increasing both tree growth and soil fertility, the possible management of the nitrogen-fixing symbiosis to increase productivity of *Casuarina* has received little study.

There are two approaches to increasing symbiotic nitrogen fixation in *Casuarina* plantings. One involves the selection both of fast-growing host-plant genotypes, and compatible strains of *Frankia* that are highly effective in fixing atmospheric nitrogen (i.e. improved genetic potential for nitrogen fixation — for example see Sougoufara et al. 1987). The other is the alleviation of soil constraints that limit potential nitrogen fixation by the particular plant-*Frankia* combination.

Management of the symbiosis by selection of infective and efficient nitrogen-fixing strains of *Frankia* and by routine inoculation of nursery stock has proved impractical until recently as pure cultures of *Frankia* for *Casuarina* have been available only for the last 5 years (Diem et al. 1983; Zhang et al. 1984). Even now, the use of *Frankia* isolates has been restricted because they have slow growth rates and inadequate information is available on their growth requirements (Shipton and Burggraaf 1983; Zhang et al. 1986). Both of these aspects limit production of inoculum for nursery experiments. Prior to the isolation and in vitro culture of *Frankia*, attempts at inoculation of nursery stock had relied on the use of crushed nodule suspensions or soil from beneath existing plantations as inoculants (McCluskey and Fisher 1983; Torrey 1982); both of these proved unsatisfactory due to the variability of infection obtained and to the danger of accidental introduction of pathogens into the nursery.

The second approach to increasing symbiotic nitrogen fixation in *Casuarina* is to alleviate soil conditions that may limit nitrogen fixation. This has been neglected also in the past due to an inadequate understanding of how environmental factors influence nitrogen fixation in *Casuarina*. Studies of legume-*Rhizobium* symbioses in tropical soils have shown that nutrient deficiencies, high soil temperatures and seasonal moisture deficits are the most critical limitations to legume growth, nodulation and nitrogen fixation (Sprent 1979). It is probable that these factors would also strongly influence growth and nitrogen fixation of *Casuarina*.

In 1984, ACIAR commissioned the CSIRO Division of Soils to investigate the management of nitrogen fixation by *Casuarina* for fuelwood and agroforestry. This chapter describes progress and problems during the first 3 years and considers the prospects for applying this technology to forestry practice.

Inoculum Production and Methods

Two objectives of this project have been to: (1) select broad host-range strains of *Frankia* that are highly effective in nitrogen fixation; and (2) develop methods for the production of pure inoculants of *Frankia* that can be applied simply and reliably in forest nurseries. Research has focused on:

(i) the assessment of the efficiency of strains of *Frankia* in promoting seedling growth;
(ii) determining the optimal culture conditions for growth of *Frankia*; and
(iii) the effects of inoculum placement and soil factors on infectivity in soils and nursery potting mixtures.

Selection of *Frankia* Strains

Glasshouse trials were used to compare the effectiveness of six isolates of *Frankia* in promoting growth of seedlings of five species of *Casuarina*: *C. equisetifolia*, *C. cunninghamiana*, *C. glauca*, *C. obesa* and *C. junghuhniana* (Rosbrook and Bowen 1987; Rosbrook and Francis 1987). All of the strains of *Frankia* were able to nodulate each of the species of *Casuarina* examined, suggesting the lack of any host specificity in nodulation. All strains also proved equally effective in increasing seedling growth in nitrogen-deficient soils. This contrasts with earlier studies with crushed nodule inocula, that found large differences between *Frankia* sources in their abilities to nodulate and fix N_2 with species of *Casuarina* (Coyne 1973; Reddell and Bowen 1985), and with some studies on other actinomycete-nodulated plants in which marked variation in effectiveness of N_2 fixation between isolates of *Frankia* has been demonstrated (Dawson and Sun 1981; Dillon and Baker 1982; Hooker and Wheeler 1987). However, we suspect that the six isolates of *Frankia* from *Casuarina* that we have screened represent only a very limited genetic base, with most being isolated from species of *Casuarina* growing outside their natural range. Further work is therefore necessary to establish whether differences in effectiveness in N_2 fixation do exist between isolates of *Frankia* from nodules of *Casuarina*. To this end we are increasing the diversity of strains of *Frankia* available for testing by isolating *Frankia* from nodules collected at more than 20 locations in tropical northern Australia.

Culturing of Inoculants

The production of *Frankia* inoculants for use in the field has been restricted by the limited availability of isolates of *Frankia* and by the slow growth rates of these isolates in liquid culture (doubling times are up to 24 hours — Zhang et al. 1986). Different media have been assessed to determine if *Frankia* growth rates in liquid culture can be enhanced. Fastest growth rates were obtained with two formulations — P media (Burggraaf and Shipton 1983) and BAP (Murry et al. 1984) — that are used widely already for *Frankia* cultivation. Moderate growth was shown on FMC (Benson 1982), defined propionate media (Baker and O'Keefe 1984) and on Qmod (Lalonde and Calvert 1979); *Frankia* grew poorly on all other media tried including yeast Czapek's (Higgins and Lechevalier 1969) and yeast extract/dextrose broth (Baker and Torrey 1979).

Another approach to increasing growth of *Frankia* has been to change the culture conditions. *Frankia* were assumed originally to be microaerophilic and consequently were often grown in stationary liquid culture. However, recent studies suggest that this assumption is incorrect and have demonstrated increased growth rates of *Frankia* in response to high oxygen levels in the culture medium (Murry et al. 1985). In the light of these findings we have commenced studies on the growth of *Frankia* in simple fermentors with constant aeration. Preliminary results suggest this to be a promising method for growing *Frankia*. Cultures grown in this way could readily be incorporated in peat or other suitable carriers in a similar way to *Rhizobium*.

Form of Inoculant

The inoculation of experimental plants in our nursery trials has relied on the use of liquid cultures of *Frankia*. Although these have worked successfully, there are obvious logistical problems in applying this method on a broader scale. With present methods, large volumes of liquid inoculant (at approximately 2–5 ml/seedling) are required to inoculate a reasonable size forest nursery, and transport of these liquid cultures from the laboratory to isolated nurseries is impractical. Methods developed for applying *Rhizobium* to agricultural plants (Roughley and Pulsford 1982) are being tested to see if they can be adapted for use with *Frankia*. These methods include the use of peat or synthetic carriers to produce an inoculant that is easy to transport and resistant to rough handling.

Method of Inoculum Placement

The placement of inoculant in relation to the seedling root system can markedly influence the pattern of nodulation of nursery-grown plants. In experiments with seedlings grown in dibbling tubes (Rosbrook 1988), placement of a crushed nodule inoculant had no effect on nodulation when relatively high levels of inoculum were applied (0.2 g of crushed nodule per seedling) — inoculum being equally infective whether mixed through the potting medium, syringed next to the base of each seedling or watered onto the base of the seedling. However, when lower levels of inoculum were applied, placement was important, with the most rapid nodulation and fastest growth responses to inoculation occurring where the inoculum was positioned close to the root system. When a liquid culture was used as inoculum, placement was again important at low levels of inoculum, with inoculum watered onto the base of the seedling resulting in greater shoot and nodule dry weights per plant than inoculum mixed through the soil. These results demonstrate the importance of placement and amount of inoculant in attaining rapid nodulation of seedlings of *Casuarina*.

Nursery Hygiene

A key aspect of the successful introduction of effective, growth-promoting *Frankia* into forest nurseries is the absence of any other potentially less effective strains of *Frankia*. However, in some nurseries the potting soil and the water supply may contain an 'indigenous' *Frankia* that competes with the inoculant strain. In these situations, positive responses to inoculation may not occur, particularly if the indigenous *Frankia* is a more competitive root coloniser than the introduced strain. This problem can be eliminated only by changing nursery production methods or by more rigorous nursery hygiene. In nurseries in which soil-less potting media (e.g. peat:vermiculite) are used or where potting soils are sterilised to control pathogens, this competition between inoculant and indigenous *Frankia* should not be a problem. There is a need to develop a simple quality control test (e.g. ELISA) to determine if nodules on nursery plants are formed by the inoculant strain.

Effect of Potting Media

The potting media used in the nursery can have a significant influence on the success of inoculants in nodulating the host seedling. Not only can the presence of indigenous *Frankia* compete with the inoculant strain for infection sites (and hence reduce the 'apparent' infectivity of the inoculant), but chemical and physical characteristics of the potting media may inhibit infection, nodule formation and nodule development. Soil factors shown to have a detrimental influence on nodulation of roots by an inoculant strain of *Frankia* include:

(i) low pH (pH range 5.5–7.0 is optimal for nodulation of *Casuarina* — Coyne 1973);
(ii) high levels of available nitrogen (Stewart 1963; Rodriguez-Barrueco et al. 1970);
(iii) low phosphorus status (Diem and Gauthier 1982; Reddell et al. 1986);
(iv) poor moisture-holding capacity (Kant and Narayana 1978); and
(v) temperatures above 25°C (Reddell et al. 1985).

These effects operate either by limiting host plant growth rate (and rate of root production) or directly on growth of *Frankia* in the rhizosphere and on nodulation processes. The effects of some of these factors can be alleviated by adopting appropriate nursery management practices (e.g. liming of the potting mix to increase soil pH or avoiding high levels of available nitrogen by not using animal manures in the mix). In other cases, further work is needed to identify the symbiotic stages affected by soil factors before practical strategies to overcome these effects can be developed.

Field Responses to Inoculation

Field studies assessing the long-term effects on growth of inoculation of seedlings of *Casuarina* with *Frankia* have been undertaken in cooperation with forestry organisations in Australia, Zimbabwe and Thailand. These studies, using either pure cultures or nodule suspensions as inoculants, aimed to identify the factors influencing the responsiveness of *Casuarina* to inoculation with *Frankia*. The effects of tree provenance, strain of *Frankia* and phosphorus nutrition on growth responses to inoculation have been examined.

Effect of Tree Provenance

Two trials conducted in Australia have demonstrated the importance of tree provenance in determining the magnitude of the growth response to inoculation with *Frankia*.

One trial in the Adelaide Hills in South Australia involved three provenances of *Casuarina cunninghamiana*, either inoculated with a crushed nodule preparation of *Frankia* or left uninoculated (Reddell et al. 1988). Twelve months after planting, inoculation had increased heights of trees from two provenances by more than 40%, but had no effect on the growth of trees from the third provenance. These trends continued over the course of the experiment and were still evident 44 months after planting, with trees from the two 'responsive' provenances producing 2.2–2.6 times more wood volume than did uninoculated trees of these same provenances.

Other than the differences between provenances in their response to inoculation, a notable feature of this experiment was that even at 44 months after planting nodules were not detected on uninoculated plants. The lack of nodules on these plants was surprising as these plots were randomised amongst the inoculated plots, and the soil at the planting site had also once supported native stands of *Allocasuarina verticillata*. We have also observed this apparent poor mobility of *Frankia* in other field trials in Australia. This suggests it is necessary to inoculate all nursery stock, even for areas where existing *Casuarina* plantings are nodulated.

The second indication of potential tree provenance-*Frankia* interactions is derived from one of the ACIAR provenance trials established by the Queensland Department of Forestry near Gympie, a subtropical area in southern Queensland. Three provenances of *C. cunninghamiana* were grown at this site, either inoculated with *Frankia* and largely dependent on symbiotic N_2 fixation or provided with a nitrogen fertiliser in split dressings (the equivalent of 233 kg N/ha over the first 19 months). All plots were also fertilised to provide phosphorus at a total of 100 kg P/ha.

The major effects of the inoculation treatment have been on basal area production, resulting in an overall wood volume increase of more than 70% compared to the nitrogen fertiliser treatment at 41 months (Table 1). Of particular note were the age differences between provenances when responses to inoculation first became apparent (Table 1). The response in the Gympie provenance to inoculation was established by age 22 months and has continued to increase with time. However, the response in the other two provenances has developed only after the cessation of nitrogen fertiliser to the uninoculated plots (at age 24 months). This is particularly the case for the Mareeba provenance which is the least productive of the three (Table 1).

Branchlets in the uninoculated plots became yellowish after nitrogen fertilisation ceased. This pattern has been repeated in other trials which have included uninoculated *Casuarina*. In contrast, the branchlets of trees in the inoculated plots of the Gympie and Mt Morgan provenances have maintained a deep green colour. The branchlets of the trees in the inoculated plots of the Mareeba provenance, while not as visually healthy as the

Table 1. The effect of inoculation with *Frankia* on estimated wood production[a] (m^3/ha) by three provenances of *Casuarina cunninghamiana* at three ages after planting out.

Tree provenance	Inoculation treatment	Age (months after planting) 22	29	41
Gympie	inoculated	38	59	111
	uninoculated +N	21	32	54
Mareeba	inoculated	15	26	69
	uninoculated +N	16	26	41
Mt Morgan	inoculated	21	42	84
	uninoculated +N	16	34	59

[a] Wood volume estimated assuming $V \approx 1/3\ d^2h$, where d = stem diameter at ground level, and h = tree height.

other two provenances, are still superior to trees in the uninoculated plots.

This differential inoculation response that is dependent on tree provenance is important as it shows that one criterion on which fast-growing provenances of *Casuarina* need to be selected is their ability to form effective associations with highly efficient N_2-fixing strains of *Frankia*.

Effect of *Frankia* Strain

Possible differences between isolates of *Frankia* in their abilities to increase the growth of trees from one provenance of *Casuarina cunninghamiana* were examined in a field trial established at Kadoma in Zimbabwe (Reddell et al. 1988). There were six treatments used in this study; four 'strains' of *Frankia* (three isolates and one crushed nodule suspension) and two uninoculated treatments, one of which involved application of nitrogen fertiliser at planting.

Fourteen months after planting, N fertiliser and all four *Frankia* treatments had increased tree growth in comparison to the uninoculated treatment. Three of the *Frankia* treatments and the N fertiliser treatment produced similar increases in tree growth, whereas one strain of *Frankia* (ORS 020607) was much more effective in stimulating tree growth, height of these plants being almost three times that of the uninoculated treatment.

Effect of Phosphorus Supply

In a number of preliminary field trials in Thailand and Zimbabwe there was no growth response by *Casuarina* to inoculation with *Frankia*. At some of these sites it was suspected that the absence of a positive growth response to inoculation with *Frankia* was caused by deficiencies of nutrients other than nitrogen acting to limit tree growth and/or nodulation and nitrogen fixation processes. This illustrates the need to identify likely nutritional constraints on tree growth that may occur at potential planting sites.

As phosphorus deficiency has been reported widely in tropical soils, two experiments examining the effects of phosphorus nutrition on response of *Casuarina* to inoculation with *Frankia* were planted at a site near Gympie in southeastern Queensland.

The first trial, established in early 1985, involved three nitrogen treatments (uninoculated, no nitrogen fertiliser; uninoculated, but provided with 160 kg N/ha as NH_4NO_3; inoculated with *Frankia*, no nitrogen fertiliser) and two phosphorus addition treatments (110 kg P/ha applied as double superphosphate; no P fertiliser applied). After 22 months, inoculation with *Frankia* or the addition of 160 kg N/ha as nitrogen fertiliser had no effect on the growth of *Casuarina cunninghamiana* unless P fertiliser had been applied also. P application increased wood production for all three nitrogen treatments, however, there was also a strong positive interaction between P application and both the nitrogen fertiliser and *Frankia* inoculation treatments. This interaction resulted in proportionally much larger increases in wood volume for these two treatments (262 and 249% for the N fertiliser and inoculation treatments respectively) than occurred for the uninoculated treatment without nitrogen fertiliser (a 138% increase).

The second trial, examining the response of *Casuarina cunninghamiana* over five rates of P supply, was planted in April 1987. A complete factorial combination of three nitrogen treatments (uninoculated, no nitrogen fertiliser; uninoculated, but provided with 200 kg N/ha as NH_4NO_3 in split applications; inoculated with *Frankia*, no nitrogen fertiliser) and five P application rates (equivalent to 0, 2.5, 5, 10 and 50 kg P/ha/year) were used. After 12 months, inoculation with *Frankia* had substantially increased wood volume in comparison to both uninoculated treatments (Table 2). Increasing P supply to inoculated trees also

Table 2. The effect of P application and inoculation with *Frankia* on estimated wood production[a] (m^3/ha) of *Casuarina cunninghamiana*, 12 months after planting.

N treatment	P applied (kg P/ha/year)				
	0	2.5	5	10	50
Uninoculated, no N	0.34	0.78	1.44	0.86	1.10
Uninoculated, +N	0.55	1.64	1.77	1.58	1.21
Inoculated	1.14	2.22	2.80	2.61	3.98

SE_d for comparing P rates in the same N treatment is 0.37, to compare different N treatments the SE_d is 0.38. ANOVA showed N treatment, P application and the interaction between these two factors to all be significant sources of variation ($P < 0.001$).

[a] Wood volume estimated assuming $V \approx 1/3\ d^2h$ and a stocking rate of 2667 trees/ha, where d = stem diameter at ground level, and h = tree height.

increased wood volume; three times more wood was produced at the highest P application rate (50 kg P/ha) than was produced when no P was applied (Table 2). As in the previous P nutrition trial, there was a positive interaction between P application and inoculation with *Frankia*.

The magnitude of these positive interactions between P supply and inoculation with *Frankia* demonstrates that application of P fertiliser to increase the growth of nitrogen-fixing plants may be a worthwhile forestry practice that could, in many situations, be justified on economic grounds by the appreciable increases in wood yield that result.

Prospects for Broad-Scale Application

Although the isolation and culture of *Frankia* from *Casuarina* has made commercial production of *Frankia* inoculants feasible, major limitations to its implementation still exist. These include the slow growth rate of *Frankia* in pure culture (as discussed above) and, perhaps more importantly, a lack of basic information on suitable carriers for *Frankia* and on inoculant quality control techniques. Peat has been used extensively as a carrier for rhizobial inoculants (Roughley and Pulsford 1982) and may prove suitable for *Frankia*. However, the relatively slow growth rate of *Frankia* and its colony-forming growth pattern (and consequently clumped distribution) may pose special problems here. Additionally, for any carrier system to be effective, it is essential that a quality control test be developed for accurately quantifying the amount of infective *Frankia* in the inoculant. This is especially difficult because in culture *Frankia* forms a number of potentially infective structures — hyphae, spores and vesicles — the relative infectivity of which is unknown. These aspects of inoculant production require urgent study before any larger-scale commercial production can be contemplated.

If the problems highlighted above can be overcome, the potential for routine inoculation of forest nurseries which use soil-less, fumigated or *Frankia*-free potting mixtures is high. In nurseries where the potting mix contains an indigenous *Frankia*, the situation is more complex (see earlier section), with further information on the relative competitiveness and infectivity of indigenous and inoculant strains required. There is also little known about the longevity of nodules formed in the nursery and the persistence of introduced strains once the seedlings are planted into the field. Both of these factors are likely to influence the success of the introduced strain in promoting long-term tree productivity.

The field trials described earlier demonstrate the importance of the interaction between P nutrition and N_2 fixation in *Casuarina*. The role of mycorrhizal associations in the P nutrition of species of *Casuarina* is being studied at present in our laboratory with the view to identifying superior P-scavenging fungi that enhance growth of *Casuarina*, are fast-growing in pure culture and could be made available readily as inoculants for forest nurseries. Dual inoculation with both infective *Frankia* and mycorrhizal fungi has potential to increase productivity of *Casuarina* significantly, while minimising the need for costly fertiliser application, particularly in the highly weathered soils commonly used for tree plantings in tropical regions.

Chapter 23

Susceptibility to Termite Attack of Various Tree Species Planted in Zimbabwe

M.R. Mitchell

Abstract

A trial containing 52 seedlots from 41 species of Australian, Central American and Zimbabwean trees at Kadoma, Zimbabwe, gave results showing major differences in overall survival and susceptibility to the fungus-growing termites *Ancistrotermes latinotus* and *Macrotermes michaelseni.* Species with better than 80% survival and less than 10% termite deaths were: *Acacia holosericea, A. albida, A. salicina, A. plectocarpa, A. leptocarpa, A. difficilis, Enterolobium cyclocarpum* and *Senna atomaria.* The survival of the standard species, *Eucalyptus camaldulensis,* was 34% with all deaths due to termite attack. The basal area per tree, at 30 cm, of one provenance of the Australian species *A. holosericea* was significantly better than that of *E. camaldulensis* at the 5% level.

Introduction

The communal lands of Zimbabwe comprise 41.9% of the land area but carry about 70% of the rural population. Most communal areas are situated in the poorer ecological regions, typified by low rainfall, low soil fertility and poor crop yields. Population pressure has, in many areas, resulted in widespread clearance of indigenous woodland for fuelwood, building timber and land for agricultural cropping, resulting in a severe deficit in timber products in these areas. To redress this situation, the establishment of community woodlots is being undertaken, with the major species planted being *Eucalyptus camaldulensis, E. grandis* and *E. tereticornis.* The major pests of these species in Zimbabwe are termites of the subfamily Macrotermitinae, the fungus-growing termites, a group which is confined to the Ethiopian and Indo-Malayan Zoogeographical Regions (Ruelle 1970).

At least 80 species of termites are known in Zimbabwe (Mitchell 1980); of these, the following members of the subfamily Macrotermitinae have been recorded attacking growing eucalypts: *Macrotermes falciger, M. michaelseni, Pseudacanthotermes militaris, Odontotermes* spp., *Ancistrotermes latinotus* and *Microtermes* spp. (Mitchell, M.R., unpublished data). Other species of *Macrotermes* are also likely to cause tree deaths. Termite damage to recently planted eucalypt seedlings is typified by partial or almost complete destruction of the root system, ringbarking of the root collar and ringbarking of the lower stem. The larger species *Macrotermes* spp. and *Pseudacanthotermes militaris,* attack the roots and the lower stem from the outside, completely girdling the tree. *Microtermes* spp. and *Ancistrotermes latinotus,* being smaller, may also enter the roots of larger trees from below and feed internally, hollowing out the roots and stem from within.

Mortalities due to termites in *Eucalyptus* spp. in Zimbabwe are commonly 30–50% but approach 100% in some areas, unless the pests are controlled with insecticides. At present the organochlorines aldrin and dieldrin are used, chemicals which are likely to become unavailable in the near future. A controlled-release formulation of the carbamate, carbosulfan, has been identified as a potential replacement for the organochlorines (Mitchell 1986), but the product is not yet registered in

A species trial near Kadoma, Zimbabwe, in which a range of Australian species are being tested to determine their susceptibility to termite attack. The taller trees are *Acacia holosericea* and this species has survived termite attack well. The trial is 7 months old. Photographed June 1988.

Zimbabwe and may not become freely available for some time. The planting of tree species which are either not attractive to termites, are repellent to termites or that can tolerate termite attack, would remove the need for chemical control.

There has been much debate as to whether termites are primary pests or secondary pests that will only attack trees that have been weakened by predisposing factors, such as drought or fungal infection of the root system. Nair and Varma (1985) describe three situations in which termites attack trees in India: (1) primary termite attack on healthy, vigorous saplings; (2) secondary termite attack on saplings dead due to other causes; and (3) complementary attack in which the death of saplings results from the combined effect of termite damage and other factors.

In trials to test insecticides for termite control in Zimbabwe, it is often the most vigorous untreated trees in the trial that die, even in wet weather, suggesting that predisposing factors are absent and that the observed root destruction by termites is the primary cause of death (Mitchell, M.R. unpublished data). Also, the well-known fact that treatment with the insecticides dieldrin and aldrin, which have no known fungicidal action, greatly reduces mortality in plantations (Sands 1962; Nair and Varma 1985; Mitchell 1986) implies that secondary termite attack after fungal infection is generally of little importance. All mortality reported as being due to termites in this paper is considered to be primary attack, where that attack is the sole cause of death of the tree.

Literature Review

Differences in the susceptibility of tree species to termites have been recognised for some years, but little has been published on the subject. Parry (1959) gives the following species as being resistant to termites in adverse conditions: *Cassia siamea*, *Albizia lebbek*, *Jacaranda* sp., *Casuarina* spp., *Gmelina* sp. and *Callitris* spp. Brown (1962) also gives *C. siamea* as a resistant species together with *Albizia procera*, *Tectona grandis* and *Pinus* spp. *Callitris* spp. are stated as showing early promise but are attacked at 3–4 years of age.

Rajagopal (1982) reports differences in mortality in *Eucalyptus* spp. due to attack by the termites *Odontotermes obesus* and *O. wallonensis* in India. He found *E. grandis* to be least heavily attacked at 5.7% followed by *E. tereticornis* (6.9%), *E. resinifera* (11.4%), *E. punctata* (17.1%), *E. propinqua* (24.3%), *E. saligna* (27.9%), *Eucalyptus* hybrid (29.3%) and *E. microcorys* (52.2%).

Midgley and Weerawardane (1986), working in Sri Lanka, give the following species, planted in species trials, as significantly more tolerant to termites than *Eucalyptus camaldulensis* (22.0% survival) and *E. tereticornis* (27.3%) at the 5% level, assuming all deaths to be due to termites: *Acacia mangium* (79.3%), *A. auriculiformis* (77.6%), *A. leptocarpa* (75.0%), *A. polystachya* (69.0%), *A. crassicarpa* (57.6%), *E. alba* (77.3%), *Azadirachta indica* (100.0%), *Terminalia arjuna* (97.6%), *Tamarindus indica* (88.6%), *Calliandra calothyrsus* (86.3%) and *Leucaena leucocephala* (77.6%). They found that all eucalypt species commonly grown in Sri Lanka's Community Forestry Project (*E. tereticornis*, *E. camaldulensis*, *E. grandis* and *E. torelliana*) were equally attacked by termites.

Mitchell et al. (1988) give mortalities in an introduction trial (MV05) of Australian tree species on the same site in Zimbabwe as the trial reported below. Inspection of this trial showed early differences in survival between species, and an assessment was undertaken, at 6 months after planting, to quantify these differences. Results for 47 seedlots from 35 species were analysed. Unfortunately, this trial was destroyed by fire before a planned further assessment at 18 months, so only a single assessment of mortality was made. It was therefore impossible to assess reliably mortality due to termites, and so total tree mortality only gives an indication of susceptibility to termites. *Acacia* spp. ranged from having the best survival (*A. melanoxylon*, 98% survival) to among the worst (*A. murrayana*, 7% survival). Overall, *Acacia* was the most promising genus tested, in terms of survival. The survival of the best 12 seedlots of *Acacia* spp. was significantly better than that of *E. camaldulensis*, the standard species. Six seedlots of *Eucalyptus* spp. had survivals ranging from 93 to 41%. The survival of the most resistant species, *E. gibsonensis*, was significantly better than that of all other *Eucalyptus* spp. *Eucalyptus camaldulensis*, with 75% survival, was the next best performer. Eight seedlots from six *Melaleuca* species were in the lower half of the survival rankings, with survivals ranging from 68 to 46%. The survival of *Casuarina equisetifolia* was similar to that of *Melaleuca* spp., but *C. glauca* and *C. cunninghamiana* were two of the poorest species with only 33 and 21% survival, respectively. The lowest survival recorded was for *Allocasuarina huegliana*, which had only 6% survival 6 months after planting. No clear relationship was found between survival and height between seedlots.

The nine seedlots that were regarded as successful in terms of survival and height were: *Acacia melanoxylon* (both seedlots); *A. leptocarpa* (one seedlot); *A. auriculiformis* (two seedlots); *A. cowleana* (one seedlot); *A. polystachya* (one seedlot); *A. crassicarpa* (one seedlot) and *Eucalyptus gibsonensis*.

There is much contradiction to be found between authors, where resistance or susceptibility to termites is concerned. For example, in Zimbabwe, *Casuarina* spp. are very susceptible to termites, whereas Parry (1959) found them to be resistant; in Zimbabwe, *E. grandis* is the most susceptible eucalypt species tested (Mitchell, unpublished data), contrary to the findings of Rajagopal (1982) in India. Thus, species susceptibility appears to vary with different conditions or termite species. Screening trials should therefore be carried out prior to the extensive planting of new species, preferably in the form of untreated provenance trials.

Sands (1960), Barrett and Mullin (1968), Browne (1968), Kudler (1970), Harris (1971), Lee (1971), Abdel Nour (1975), Parihar (1978, 1981), Roonwal (1979), U.S. National Academy of Sciences (1980, 1983), Selander and Nkunika (1981) and Webb et al. (1984) and others give tree species recorded as being susceptible to termites in various parts of the world.

Methods

This paper gives results for a trial containing Australian species, established for the prime purpose of studying differences in susceptibility to termites between species, which was planted at an altitude of 1180 m at Kadoma in Zimbabwe. The soil is a reddish brown clay loam which carries an indigenous vegetation of *Acacia*, *Terminalia* and *Combretum* scrub. Mean annual rainfall is 780 mm.

Fifty-two seedlots from 41 species were planted in a layout generated by a computer program designed for seed orchard plans, based on the Permutated Neighbourhood Design Concept, where single tree plots of each treatment are randomly distributed, but individuals of each treatment are isolated from each other. In this case, two individuals of the same treatment are separated by at least three individuals of other treatments, and no two treatments occur as diagonal, horizontal or vertical neighbours on more than three occasions, resulting in a good spread of the different treatments throughout the trial area. This type of design was used in an attempt to overcome the patchy distribution of termite attack encountered on many sites. Replication was uneven, with stock numbers having between 24 and 51 individuals in the trial with a mean of 46.4. This variation is in part due to the nature of the trial design, but mainly to poor germination in some species, such as *Alphitonia excelsa* (29 replications) and *Leptospermum longifolium* (24 replications). Plant spacing was 2 m × 2 m. Assessments of mortality and cause of death were carried out on 13 occasions between planting and 18 months after planting, to ensure accurate diagnosis of cause of death. Cause of death was assessed by the removal of any dead tree followed by careful examination of the root system. A dead tree was only recorded as having been killed by termites if the characteristic symptoms of primary termite attack were evident. A missing tree was recorded as having unknown cause of death unless there was clear evidence of termite foraging in the soil where the tree roots had been. Trees killed by people or excavated by animals were excluded from the data. The major termite species identified as attacking growing trees in this trial were *Macrotermes michaelseni* and *Ancistrotermes latinotus*.

Due to low rainfall, this trial was watered on four occasions in January-February 1987.

At 10 months after planting, mortality, cause of death, tip height, root collar diameter and number of stems were assessed. In species with a recumbent habit, such as *Acacia victoriae*, tip height was measured as the distance from the base of the stem to the tip of the longest stem. Root collar was measured at ground level below the origin of separate stems, which were taken as any branch originating within 10 cm of the soil surface.

At 18 months after planting, mortality, cause of death, tip height, number of stems and diameter at 30 cm from the base of the tree of all stems over 5 mm were assessed. Tip height was measured in the same way as at 10 months.

Survival figures were calculated as follows:

$$\% \text{ overall survival} = \frac{\%(\text{number planted–total deaths})}{\text{number planted}}$$

$$\% \text{ termite mortality} = \frac{\%(\text{deaths due to termites})}{\text{number planted}}.$$

Basal area was calculated as follows:

Basal area, $BA_t(\text{cm}^2) = BA_1 + \ldots BA_n$
where: BA_1 = Basal area of main stem
BA_2 = Basal area of second stem
BA_n = Basal area of last stem over 5 mm

The estimated total volume was calculated as:

Estimated total volume,

$$V_t(dm^3) = V_1 + V_2 + \ldots V_n$$

where:

$$V_1 = \frac{(BA_1 \times H_0)}{100} + (BA_1 \times F \times (H_1\text{-}H_0)).$$

The volumes of the subsidiary stems were based on the height to diameter ratio of the main stem, as follows:

$$V_2...V_n = BA_2...BA_n \times F \times \frac{(H_1 \times D_2...D_n)}{D_1} - H_0$$

where: H_0 = Reference height (3 *dm*),
H_1 = Tip height (*dm*),
F = Form factor (0.62),
D_1 = Diameter of main stem at 3 *dm*,
D_2 = Diameter of second stem,
D_n = Diameter of last stem.

In no case is the ratio D_n:D_1 less than 0.1, which would result in V_n being negative. If this had occurred, that volume should have been taken as zero.

The form factor value of 0.62 was derived from biomass studies, including branch wood, of mixed coppice species in Britain (Crockford 1987).

The two estimates of production per hectare given were derived as follows:

Untreated estimates

$= 1600 \times (1 - ((T+B)/P) \times BA_t$ or V_t

Treated estimates

$= 1600 \times (1 - (T/P)) \times BA_t$ or V_t

Where: P = number of trees planted,
T = number of trees killed by termites,
B = number of trees killed by other factors.

Spacing of 2.5m × 2.5m (1600 sph) is now considered more suitable for the species in this trial.

Eucalyptus camaldulensis is the most commonly planted species in Communal Lands, and the same seedlot of the species as used in trial MV05 on an adjacent site (stock number 10875) was included as a standard, against which the other seedlots could be compared.

Results and Discussion

Overall percentage survival and percentage termite mortality at both 10 and 18 months for the 52 seedlots in the trial are given in Table 1. The ranked means and Duncan's Multiple Range Test for tip height are shown in Table 2. Seedlots not sharing a common vertical bar are significantly different at the 5% level.

The ranked means and Duncan's Multiple Range Test for total basal area per tree as measured at 30 cm are given in Table 3, and the ranked means and Duncan's Multiple Range Test for total estimated volume per tree in Table 4.

Mean height, mean diameter of all stems over 5 mm, and mean number of stems over 5 mm are given in Table 5. Also shown are estimates of basal area per hectare and estimated volume per hectare when plants are untreated, and in the absence of deaths due to termites.

The Untreated Basal Area and Volume Estimate give measures of the basal area per hectare for each seedlot adjusted for the overall survival.

The Treated Basal Area and Volume Estimate give measures of the volume per hectare for each seedlot adjusted for the survival in the absence of any termite deaths, i.e. with 100% effective chemical control of termites.

Mortality Due to Termites

Due to the unequal replication of individuals within seedlots in the trial, no analysis of the survival and mortality data was carried out.

Overall survival for the 52 seedlots ranged from 100% to 0%, with termite mortality ranging from 0 to 95.7% of the trees planted. Those species with overall survival in excess of 80%, and termite survival of over 90%, may have potential in communal land afforestation.

Survival of *Acacia* spp. ranged from 100 to 16.7%. Eighteen of the 30 *Acacia* seedlots had termite mortalities of less than 10%, while only three species had mortalities due to termites over 30%. These were: the Central American species *A. pennatula* (42.9% termite mortality), the Brooklyn provenance of *A. flavescens* (31.6% termite mortality) and *A. crassicarpa* (30.4% termite mortality). This fast-growing species, which had 8.7% termite mortality at 10 months, appears more susceptible to *Macrotermes michaelseni*, which attacks mainly older and larger trees, than the smaller *Ancistrotermes latinotus*, which causes most of the damage in the early stages of growth on this site. It is not possible, at this time, to predict further mortality due to *M. michaelseni* in *A. crassicarpa* or other species. *Acacia crassicarpa* is also susceptible to an unidentified fungal root rot that caused many deaths which, together with the termite attack, resulted in only 26.1% survival. The two seedlots of *A. victoriae* are also beginning to succumb to *M. michaelseni* attack. In the Central American species *A. pennatula* and *A. farnesiana* (12.2.% termite mortality), almost all deaths due to termites occurred in the first few months after planting and were due to *A. latinotus*. The two *A. cowleana* provenances, which had few termite deaths, appear to be dying from drought and associated attack by the stem-boring bostrychid *Sinoxylon doliolum*.

It is interesting to note that *Acacia melanoxylon*, which performed well in the early stages of trial MV05, had only 56.8% survival, with 15.9% termite deaths 18 months after planting in a much drier year.

Those *Acacia* species with better than 80% survival and less than 10% termite deaths were both provenances of: *A. holosericea*, the indigenous *A. albida*, *A. salicina*, *A. plectocarpa* and *A.*

Table 1. Mean overall survival and termite mortality at 10 months and 18 months of 52 seedlots on a termite-infested site, at Kadoma, Zimbabwe.

Seedlot no.	Species	10 months Overall survival %	10 months Termite mort. %	18 months Overall survival %	18 months Termite mort. %
15365	*Acacia holosericea*	100.0	0.0	100.0	0.0
13270	*Acacia albida*	100.0	0.0	100.0	0.0
15464	*Acacia albida*	100.0	0.0	97.7	0.0
15367	*Acacia holosericea*	100.0	0.0	96.0	2.0
15402	*Acacia salicina*	97.9	0.0	95.8	0.0
15103	*Peltophorum africanum*	97.8	0.0	95.7	0.0
12538	*Enterolobium cyclocarpum*	93.8	0.0	93.8	0.0
12515	*Senna atomaria*	93.6	4.3	93.6	4.3
15401	*Acacia salicina*	97.9	2.1	91.5	4.3
15388	*Acacia difficilis*	95.7	4.3	91.5	8.5
15399	*Acacia plectocarpa*	95.6	0.0	90.7	2.3
15369	*Acacia leptocarpa*	93.9	4.1	89.8	4.1
12526	*Acacia farnesiana*	87.8	12.2	87.8	12.2
15368	*Acacia leptocarpa*	93.3	0.0	86.7	2.2
15419	*Cassia brewsteri*	90.9	0.0	86.4	0.0
15398	*Acacia plectocarpa*	91.8	4.1	83.3	6.2
12516	*Prosopis juliflora*	91.8	6.1	81.6	14.3
15418	*Atalaya hemiglauca*	86.4	11.6	81.4	11.6
15361	*Acacia brassii*	82.2	2.2	75.6	2.2
15400	*Acacia victoriae*	95.9	2.0	75.0	10.4
15390	*Acacia maconochieana*	100.0	0.0	73.5	0.0
12536	*Parkinsonia aculeata*	79.6	16.3	73.5	18.4
15381	*Acacia adsurgens*	86.0	4.0	67.4	8.2
15363	*Acacia cowleana*	85.1	2.1	66.7	4.4
15405	*Acacia simmsii*	77.1	2.1	62.5	9.3
15412	*Acacia victoriae*	89.8	4.1	61.2	12.2
12326	*Acacia melanoxylon*	62.8	15.9	56.8	15.9
15349	*Eucalyptus brassiana*	68.1	29.8	55.3	42.6
12525	*Acacia pennatula*	59.2	40.8	55.1	42.9
15362	*Acacia cowleana*	85.4	2.1	53.2	6.4
15348	*Eucalyptus brassiana*	58.7	37.0	53.2	42.6
15380	*Acacia aneura*	83.0	2.1	50.0	4.3
15376	*Alphitonia excelsa*	48.3	10.3	48.3	10.3
15406	*Acacia simmsii*	50.0	13.0	44.4	15.6
12519	*Leucaena leucocephala*	55.1	40.8	38.8	51.1
15397	*Acacia flavescens*	56.5	13.0	37.0	19.6
10875	*Eucalyptus camaldulensis*	44.7	55.3	34.0	66.0
15403	*Acacia shirleyi*	47.8	13.0	33.3	15.6
12257	*Parinari nonda*	36.7	38.8	31.3	39.6
12532	*Pithecellobium dulce*	42.9	57.1	27.1	68.7
15386	*Acacia crassicarpa*	71.7	8.7	26.1	30.4
15379	*Acacia aneura*	59.6	23.4	21.3	29.8
15389	*Acacia flavescens*	43.9	19.5	21.1	31.6
15382	*Acacia brachystachya*	84.0	6.0	16.7	16.7
15417	*Lophostemon suaveolens*	22.0	48.8	12.2	53.7
12521	*Leucaena shannonii*	39.6	52.1	8.5	63.8
12520	*Leucaena diversifolia*	50.0	40.0	8.0	68.0
15339	*Eucalyptus argophloia*	14.3	73.3	6.1	81.6
15422	*Brachychiton populneus*	34.8	32.6	4.4	32.6
8028	*Eucalyptus punctata*	8.5	80.9	4.3	85.1
15414	*Leptospermum longifolium*	8.3	29.2	4.2	29.2
15350	*Angophora costata*	4.3	91.5	0.0	95.7
Means:		70.1	87.6	56.7	77.0

Table 2. Ranked mean height and Duncan's multiple range test ($P = 0.05$) (1324 degrees of freedom; standard deviation 0.5992).

			Tip height (m)
15365	*Acacia holosericea*	1	3.07
10875	*Eucalyptus camaldulensis*	2	2.88
15367	*Acacia holosericea*	3	2.73
15349	*Eucalyptus brassiana*	4	2.40
12519	*Leucaena leucocephala*	5	2.26
12521	*Leucaena shannonii*	6	2.25
15386	*Acacia crassicarpa*	7	2.20
12526	*Acacia farnesiana*	8	2.15
12515	*Senna atomaria*	9	2.15
15405	*Acacia simsii*	10	2.08
15402	*Acacia salicina*	11	2.06
15369	*Acacia leptocarpa*	12	1.97
13270	*Acacia albida*	13	1.90
12520	*Leucaena diversifolia*	14	1.90
15361	*Acacia brassii*	15	1.90
15388	*Acacia difficilis*	16	1.89
15348	*Eucalyptus brassiana*	17	1.86
12538	*Enterolobium cyclocarpum*	18	1.84
15363	*Acacia cowleana*	19	1.81
15401	*Acacia salicina*	20	1.80
12536	*Parkinsonia aculeata*	21	1.74
15368	*Acacia leptocarpa*	22	1.68
15398	*Acacia plectocarpa*	23	1.67
15464	*Acacia albida*	24	1.66
12326	*Acacia melanoxylon*	25	1.61
15362	*Acacia cowleana*	26	1.58
15399	*Acacia plectocarpa*	27	1.57
15403	*Acacia shirleyi*	28	1.55
12525	*Acacia pennatula*	29	1.53
15389	*Acacia flavescens*	30	1.50
15414	*Leptospermum longifolium*	31	1.50
15381	*Acacia adsurgens*	32	1.42
15406	*Acacia simsii*	33	1.42
15103	*Peltophorum africanum*	34	1.31
15376	*Alphitonia excelsa*	35	1.27
12532	*Pithecellobium dulce*	36	1.25
8028	*Eucalyptus punctata*	37	1.15
15417	*Lophostemon suaveolens*	38	1.14
15400	*Acacia victoriae*	39	0.99
15339	*Eucalyptus argophloia*	40	0.97
15397	*Acacia flavescens*	41	0.94
15380	*Acacia aneura*	42	0.93
15379	*Acacia aneura*	43	0.90
15382	*Acacia brachystachya*	44	0.90
15418	*Atalaya hemiglauca*	45	0.81
12516	*Prosopis juliflora*	46	0.78
15390	*Acacia maconochieana*	47	0.69
15412	*Acacia victoriae*	48	0.58
15419	*Cassia brewsteri*	49	0.45
15422	*Brachychiton populneus*	50	0.35
12257	*Parinari nonda*	51	0.26
15350	*Angophora costata*	52	0.00

Table 3. Ranked mean total basal area per tree, and Duncan range test ($P = 0.05$) (1324 degrees of freedom; standard deviation 5.417960).

		Total basal area per tree (cm^2)	
15365	*Acacia holosericea*	1	34.31
15367	*Acacia holosericea*	2	16.47
10875	*Eucalyptus camaldulensis*	3	15.01
12515	*Senna atomaria*	4	12.32
12538	*Enterolobium cyclocarpum*	5	12.03
15401	*Acacia salicina*	6	11.31
15386	*Acacia crassicarpa*	7	10.88
15402	*Acacia salicina*	8	10.84
15349	*Eucalyptus brassiana*	9	9.53
12526	*Acacia farnesiana*	10	9.22
12519	*Leucaena leucocephala*	11	7.92
12536	*Parkinsonia aculeata*	12	6.96
13270	*Acacia albida*	13	6.87
15388	*Acacia difficilis*	14	6.76
12525	*Acacia pennatula*	15	6.38
12520	*Leucaena diversifolia*	16	5.72
15464	*Acacia albida*	17	5.66
15103	*Peltophorum africanum*	18	5.43
15348	*Eucalyptus brassiana*	19	4.63
15398	*Acacia plectocarpa*	20	4.43
15400	*Acacia victoriae*	21	4.23
15361	*Acacia brassii*	22	4.21
15405	*Acacia simsii*	23	4.09
15369	*Acacia leptocarpa*	24	4.03
15362	*Acacia cowleana*	25	3.80
15368	*Acacia leptocarpa*	26	3.78
15381	*Acacia adsurgens*	27	3.17
15389	*Acacia flavescens*	28	3.02
15363	*Acacia cowleana*	29	2.91
12326	*Acacia melanoxylon*	30	2.87
15417	*Lophostemon suaveolens*	31	2.84
12532	*Pithecellobium dulce*	32	2.64
15406	*Acacia simsii*	33	2.51
15399	*Acacia plectocarpa*	34	2.50
15403	*Acacia shirleyi*	35	2.45
8028	*Eucalyptus punctata*	36	2.18
15376	*Alphitonia excelsa*	37	2.16
12521	*Leucaena shannonii*	38	1.86
15339	*Eucalyptus argophloia*	39	1.48
15397	*Acacia flavescens*	40	1.13
15412	*Acacia victoriae*	41	1.08
15379	*Acacia aneura*	42	1.04
15414	*Leptospermum longifolium*	43	0.95
15380	*Acacia aneura*	44	0.88
12516	*Prosopis juliflora*	45	0.66
15382	*Acacia brachystachya*	46	0.55
15418	*Atalaya hemiglauca*	47	0.48
15422	*Brachychiton populneus*	48	0.43
15390	*Acacia maconochieana*	49	0.43
15419	*Cassia brewsteri*	50	0.26
12257	*Parinari nonda*	51	0.03
15350	*Angophora costata*	52	0.00

Table 4. Ranked mean total volume estimate per tree, and Duncan's multiple range test ($P = 0.05$) (1324 degrees of freedom; standard deviation 1.967224).

		Total volume estimate per tree (dm^3)	
15365	*Acacia holosericea*	1	5.754
10875	*Eucalyptus camaldulensis*	2	3.267
15367	*Acacia holosericea*	3	2.828
15349	*Eucalyptus brassiana*	4	1.708
15386	*Acacia crassicarpa*	5	1.701
12515	*Senna atomaria*	6	1.651
15402	*Acacia salicina*	7	1.573
12538	*Enterolobium cyclocarpum*	8	1.383
15401	*Acacia salicina*	9	1.311
12519	*Leucaena leucocephala*	10	1.202
12526	*Acacia farnesiana*	11	1.092
12520	*Leucaena diversifolia*	12	1.045
13270	*Acacia albida*	13	0.920
12536	*Parkinsonia aculeata*	14	0.891
15388	*Acacia difficilis*	15	0.867
15369	*Acacia leptocarpa*	16	0.788
15348	*Eucalyptus brassiana*	17	0.728
15405	*Acacia simsii*	18	0.701
15368	*Acacia leptocarpa*	19	0.643
15398	*Acacia plectocarpa*	20	0.642
15464	*Acacia albida*	21	0.630
15361	*Acacia brassii*	22	0.611
12525	*Acacia pennatula*	23	0.588
15103	*Peltophorum africanum*	24	0.485
15389	*Acacia flavescens*	25	0.458
12326	*Acacia melanoxylon*	26	0.430
15363	*Acacia cowleana*	27	0.424
15400	*Acacia victoriae*	28	0.423
15376	*Alphitonia excelsa*	29	0.401
15362	*Acacia cowleana*	30	0.371
15417	*Lophostemon suaveolens*	31	0.325
15403	*Acacia shirleyi*	32	0.305
15399	*Acacia plectocarpa*	33	0.303
12521	*Leucaena shannonii*	34	0.292
15406	*Acacia simsii*	35	0.250
15381	*Acacia adsurgens*	36	0.238
12532	*Pithecellobium dulce*	37	0.213
15397	*Acacia flavescens*	38	0.180
8028	*Eucalyptus punctata*	39	0.156
15339	*Eucalyptus argophloia*	40	0.116
15380	*Acacia aneura*	41	0.115
15414	*Leptospermum longifolium*	42	0.099
15379	*Acacia aneura*	43	0.090
15418	*Atalaya hemiglauca*	44	0.073
15382	*Acacia brachystachya*	45	0.061
15390	*Acacia maconochieana*	46	0.052
15412	*Acacia victoriae*	47	0.051
15419	*Cassia brewsteri*	48	0.044
12516	*Prosopis juliflora*	49	0.042
15422	*Brachychiton populneus*	50	0.024
12257	*Parinari nonda*	51	0.013
15350	*Angophora costata*	52	0.000

leptocarpa; and the single seedlot of *A. difficilis.*

In *Eucalyptus* species, almost all deaths were due to termite attack, with mortalities ranging from 42.6 to 85.1%, with the standard species *E. camaldulensis* at 66%. No trees of *Angophora costata*, which is closely related to the eucalypts, survived, with a termite mortality of 95.7%.

The three *Leucaena* species are also susceptible to termite attack, with termite mortalities in the range of 51.1-68% and with survival in the range 38.8-8%.

Of the other species indigenous to Australia, the only two that showed any promise were *Cassia brewsteri*, with 86.4% survival and no deaths due to termites, and *Atalaya hemiglauca* at 81.4% survival, but both are slow-growing.

Of the other Central American species, *Enterolobium cyclocarpum* and *Senna atomaria* both survived well (93.8 and 93.6% respectively) and proved resistant to termite attack, while *Parkinsonia aculeata* and *Prosopis juliflora* were moderately resistant with termite mortalities of 18.4 and 14.3%, and *Pithecellobium dulce* proved highly susceptible, at 68.7% termite mortality.

The indigenous species *Peltophorum africanum* survived well and proved resistant to termite attack.

Growth

Both seedlots of *Acacia holosericea* and *A. salicina*, the single seedlots of *E. camaldulensis*, *S. atomaria*, *E. cyclocarpum*, and *A. crassicarpa* had heights greater than 2 m, basal area per tree of more than 10 cm^2 and estimated volume exceeding 1 dm^3.

The height of the Hooker Creek provenance of *A. holosericea*, a multistemmed species, was significantly better than all seedlots other than *E. camaldulensis*, while both its basal area per tree and estimated volume per tree were significantly greater than all other seedlots. The Mount Molloy provenance, which has fewer stems, was also performing well.

Height growth in *Leucaena leucocephala* and *L. shannonii* was good, but their relatively small diameters result in lower basal area and volume. Another species with good height and estimated volume was *A. farnesiana*.

Growth Per Hectare

Basal area per hectare and estimated volume per hectare, shown in Table 5, give indices of the expected woody biomass production in plantations. For example, Hooker Creek *A. holosericea*, planted at 2.5 m × 2.5 m spacing, should have given a basal area, at 0.3 m, of 5.49 m^2/ha at 18 months on the Kadoma site, whether treated against termite attack or not. *Eucalyptus camaldulensis*, on the other hand, would have given 0.82 m^2 when untreated and 2.40 m^2 if a 100% effective termite control insecticide had been applied. Thus, the *A. holosericea* is 2.29 times as productive in terms of basal area per hectare and 1.76 times as productive in terms of estimated volume per hectare, than *E. camaldulensis*, the most commonly planted species in communal lands, even with the latter given effective termite control.

Those species with provenances showing untreated basal areas, or estimated volume per hectare greater than the mean of the respective treated values, are considered to merit formal provenance trials, conditional on their having attributes useful to the rural population. Those species are: *Acacia holosericea*, *A. salicina*, *A albida*, *A. difficilis*, *Senna atomaria*, *Enterolobium cyclocarpum*, *Eucalyptus camaldulensis*, *E. brassiana*, *Parkinsonia aculeata*, and *Peltophorum africanum*.

Conclusions

The subfamily Macrotermitinae, whose members are responsible for attack on the roots of living trees, are restricted to the Ethiopian and Indo-Malayan zoogeographical regions. In these areas, where tree species have evolved sympatrically with fungus-growing termites, those trees may be expected to be resistant to termite attack. This is generally true with *Acacia albida* and *Peltophorum africanum* above. Some tree genera originating in areas free from macrotermitinae termites — *Eucalyptus* and *Casuarina* from Australasia and *Leucaena* from Central America — are, on the other hand, generally susceptible to attack by the fungus-growing termites, presumably as there was no selective pressure for resistance to that particular group of termites. *Acacia*, a genus found in every continent except Europe and Antarctica, appears to retain a degree of resistance to termite attack.

Varma (1982), using extracts from *Eucalyptus* roots in choice experiments, has identified a phenolic acid which is attractive to *Odontotermes guptai.* Further work should be undertaken to ascertain whether the roots of African tree species contain toxic or repellent compounds or whether their observed resistance to termite attack is due to the lack of such attractants. This work should also be extended to cover Australian *Acacia* species.

In Zimbabwe, it is recommended that all *Eucalyptus* species be given prophylactic treatment, in the nursery, against termite attack. This recommendation will be extended to cover *Casuarina* spp. and *Leucaena* spp. Such treatment is both hazardous and expensive.

It appears, from the above results, that *Acacia holosericea* and several other species could be successfully established without any such treatment,

Table 5. Mean height, mean stem diameter and mean number of stems at 18 months after planting, with basal area per hectare and estimated volume per hectare when plants are untreated and when plants are successfully treated, resulting in no termite deaths, listed in order of decreasing estimated untreated volume per unit area.

Seedlot no.	Species	Mean height (m)	Mean stem diameter (mm)	Mean no. stems	Basal area per hectare Untreated (m^2)	Basal area per hectare Treated (m^2)	Volume estimate per hectare Untreated (m^3)	Volume estimate per hectare Treated (m^3)
15365	*Acacia holosericea*	3.07	31.46	4.02	5.49	5.49	9.21	9.21
15367	*Acacia holosericea*	2.73	33.51	1.77	2.53	2.58	4.34	4.44
12515	*Senna atomaria*	2.15	19.90	4.14	1.85	1.93	2.47	2.59
15402	*Acacia salicina*	2.06	23.24	2.41	1.66	1.66	2.41	2.41
12538	*Enterolobium cyclocarpum*	1.84	23.32	2.69	1.81	1.81	2.07	2.07
15401	*Acacia salicina*	1.80	22.08	3.26	1.66	1.73	1.92	2.01
10875	*Eucalyptus camaldulensis*	2.88	38.56	1.25	0.82	2.40	1.78	5.23
12526	*Acacia farnesiana*	2.15	14.09	5.19	1.29	1.48	1.53	1.75
15349	*Eucalyptus brassiana*	2.40	33.12	1.00	0.84	1.49	1.51	2.67
13270	*Acacia albida*	1.90	17.33	2.76	1.10	1.10	1.47	1.47
15388	*Acacia difficilis*	1.89	13.73	4.14	0.99	1.08	1.27	1.39
15369	*Acacia leptocarpa*	1.97	17.67	1.91	0.58	0.61	1.13	1.18
12536	*Parkinsonia aculeata*	1.74	20.04	2.22	0.82	1.02	1.05	1.31
15464	*Acacia albida*	1.66	13.74	3.47	0.89	0.89	0.97	0.99
15368	*Acacia leptocarpa*	1.68	17.90	1.15	0.52	0.54	0.89	0.91
15398	*Acacia plectocarpa*	1.67	13.80	2.58	0.59	0.64	0.86	0.92
15103	*Peltophorum africanum*	1.31	13.87	4.70	0.83	0.83	0.74	0.74
15361	*Acacia brassii*	1.90	13.16	2.56	0.51	0.52	0.74	0.76
15386	*Acacia crassicarpa*	2.20	20.15	2.58	0.45	0.98	0.71	1.54
15405	*Acacia simsii*	2.08	17.02	1.70	0.41	0.46	0.70	0.80
12519	*Leucaena leucocephala*	2.26	18.12	3.16	0.46	1.11	0.70	1.68
15348	*Eucalyptus brassiana*	1.86	22.04	1.08	0.39	0.71	0.62	1.12
12525	*Acacia pennatula*	1.53	13.91	4.30	0.56	1.00	0.52	0.92
15400	*Acacia victoriae*	0.99	7.91	5.00	0.51	0.58	0.51	0.58
15363	*Acacia cowleana*	1.81	13.26	1.73	0.31	0.33	0.45	0.48
15399	*Acacia plectocarpa*	1.57	12.78	1.69	0.36	0.37	0.44	0.45
12326	*Acacia melanoxylon*	1.61	14.13	1.88	0.26	0.33	0.39	0.50
15362	*Acacia cowleana*	1.58	11.36	3.16	0.32	0.36	0.32	0.35
15376	*Alphitonia excelsa*	1.27	11.11	2.29	0.17	0.20	0.31	0.38
15381	*Acacia adsurgens*	1.42	7.55	7.85	0.34	0.38	0.26	0.29
15406	*Acacia simsii*	1.42	9.62	2.90	0.18	0.24	0.18	0.24
15403	*Acacia shirleyi*	1.55	11.19	2.13	0.13	0.19	0.16	0.24
15389	*Acacia flavescens*	1.50	16.94	1.13	0.10	0.25	0.15	0.39
12520	*Leucaena diversifolia*	1.90	17.19	4.50	0.07	0.70	0.13	1.27
15397	*Acacia flavescens*	0.94	8.26	1.12	0.07	0.10	0.11	0.16
15418	*Atalaya hemiglauca*	0.81	5.83	1.20	0.06	0.07	0.09	0.11
12532	*Pithecellobium dulce*	1.25	7.92	4.15	0.11	0.41	0.09	0.33
15380	*Acacia aneura*	0.93	5.81	2.43	0.07	0.08	0.09	0.10
15417	*Lophostemon suaveolens*	1.14	15.70	1.20	0.06	0.30	0.06	0.34
15419	*Cassia brewsteri*	0.45	2.84	2.03	0.04	0.04	0.06	0.06
15390	*Acacia maconochieana*	0.69	3.52	3.25	0.05	0.05	0.06	0.06
12516	*Prosopis juliflora*	0.78	4.27	6.60	0.09	0.10	0.05	0.06
15412	*Acacia victoriae*	0.58	4.12	6.37	0.11	0.13	0.05	0.06
12521	*Leucaena shannonii*	2.25	15.00	1.00	0.03	0.22	0.04	0.34
15379	*Acacia aneura*	0.90	6.35	4.20	0.04	0.09	0.03	0.07
15382	*Acacia brachystachya*	0.90	3.91	1.88	0.02	0.03	0.02	0.03
8028	*Eucalyptus punctata*	1.15	13.17	2.00	0.02	0.31	0.01	0.22
15339	*Eucalyptus argophloia*	0.97	8.22	2.67	0.02	0.21	0.01	0.16
15414	*Leptospermum longifolium*	1.50	11.00	1.00	0.01	0.05	0.01	0.05
12257	*Parinari nonda*	0.26	0.31	2.87	0.00	0.00	0.01	0.01
15422	*Brachychiton populneus*	0.35	3.00	2.00	0.00	0.03	0.00	0.01
15350	*Anghophora costata*	0.00	0.00	0.00	0.00	0.00	0.00	0.00
Means:		1.52	13.82	2.77	0.58	0.73	1.64	2.08

resulting in reduced hazards and costs to the rural population.

Acknowledgments

We wish to thank Dr K. Crockford, Oxford Forestry Institute, for assistance in the 18-month measurements and data capture direct onto his portable computer, and advice on volume equations. Mr W.R. Mills gave advice on design and analysis of the experiment. Mr R. Dube assisted in the measurements.

Appendix 1

Provenance details of seedlots planted in trial B08K at Kadoma, Zimbabwe.

Seedlot no.	Species	Country	Provenance
15381	*Acacia adsurgens*	AUS	Milton Park, NT
15464	*Acacia albida*	ZW	Mana Pools
13270	*Acacia albida*	ZW	Mana Pools
15380	*Acacia aneura*	AUS	Charleville, QLD
15379	*Acacia aneura*	AUS	Vaughan Springs, NT
15382	*Acacia brachystachya*	AUS	Unknown
15361	*Acacia brassii*	AUS	Coen, QLD
15363	*Acacia cowleana*	AUS	Hooker Creek, NT
15362	*Acacia cowleana*	AUS	Werriaddo Well, WA
15386	*Acacia crassicarpa*	PNG	Mata
15388	*Acacia difficilis*	AUS	Borroloola, NT
12526	*Acacia farnesiana*	GCA	Llanos de la Fragua, W of Zacapa
15389	*Acacia flavescens*	AUS	Brooklyn, QLD
15397	*Acacia flavescens*	AUS	Cooktown, QLD
15365	*Acacia holosericea*	AUS	Hooker Creek, NT
15367	*Acacia holosericea*	AUS	Mt. Molloy — Mareeba, QLD
15369	*Acacia leptocarpa*	AUS	Musgrave, QLD
15368	*Acacia leptocarpa*	AUS	Starke Hld., QLD
15390	*Acacia maconochieana*	AUS	Lake Gregory, WA
12326	*Acacia melanoxylon*	AUS	Nambour, QLD
12525	*Acacia pennatula*	HON	Moracelli, Upper Cloluteca Valley
15398	*Acacia plectocarpa*	AUS	Kimberley area, WA
15399	*Acacia plectocarpa*	AUS	Middle Springs, WA
15401	*Acacia salicina*	AUS	Mitchell, QLD
15402	*Acacia salicina*	AUS	W. Banana, QLD
15403	*Acacia shirleyi*	AUS	Daly Waters, NT
15406	*Acacia simsii*	AUS	Mt. Molloy, QLD
15405	*Acacia simsii*	PNG	Rouku Province
15412	*Acacia victoriae*	AUS	Alice Springs, NT
15400	*Acacia victoriae*	AUS	Blackall, QLD
15376	*Alphitonia excelsa*	AUS	Dingo, QLD
15350	*Angophora costata*	AUS	Stockton peninsula, NSW
15418	*Atalaya hemiglauca*	AUS	34km W of Georgetown, QLD
15422	*Brachychiton populneus*	AUS	Dalby, QLD
15119	*Cassia brewsteri*	AUS	Blackwater, QLD
12538	*Enterolobium cyclocarpum*	HON	Rio Otoro region
15339	*Eucalyptus argophloia*	AUS	Ballon, QLD
15349	*Eucalyptus brassiana*	AUS	Bamaga, QLD
15348	*Eucalyptus brassiana*	PNG	Woroi
10875	*Eucalyptus camaldulensis*	AUS	Irvine — Petford Rd., QLD
8028	*Eucalyptus punctata*	ZW	Nyangui F.R.
15414	*Leptospermum longifolium*	AUS	Weipa, QLD
12520	*Leucaena diversifolia*	GCA	Puerto del Golpe, Montagua Valle
12519	*Leucaena leucocephala*	HON	Finca san Felipe, Nr. Duyure
12521	*Leucaena shannonii*	HON	Valle Comayagua
15417	*Lophostemon suaveolens*	AUS	Ravenshoe, QLD
12257	*Parinari nonda*	AUS	Weipa, QLD
12536	*Parkinsonia aculeata*	NIC	NE of Cuidad Dario
15103	*Peltophorum africanum*	ZW	Kadoma — Chegutu road
12532	*Pithecellobium dulce*	NIC	Nr. Lago de Managua
12516	*Prosopis juliflora*	HON	Valle Comayagua
12515	*Senna atomaria*	HON	Valle Comayagua

Future Perspectives

Top — A glasshouse trial investigating geographic variation in seedling morphology of *Acacia auriculiformis* provenances (photograph J. Oros, December 1988). *Bottom* — Research trials growing Shiitaki mushrooms on composite samples of *Acacia mearnsii* wood. Photograph taken at Subtropical Crops Institute, Wenzhou City, Zhejiang Province, People's Republic of China. New by-products such as mushrooms increase the final value of the end-product in multipurpose trees (photographed April 1988).

Chapter 24

Realising the Potential of Australia's Lesser-Known Trees and Shrubs: A Summary and Future Perspectives

D.J. Boland

Australia has a large number of lesser-known tree and shrub species suitable for use in other countries with similar environments. Some scientists have referred to these species as 'uncut diamonds' because their potential is great but their utility unrealised. Many species in northern Australia are poorly known as the region is sparsely populated. In order to exploit these tree resources we must first identify potentially useful species and then examine their growth in scientifically conducted field trials. Each species can be assessed for useful biological characteristics such as fodder value, coppicing ability, useful essential leaf oils, nitrogen fixing capacity and ability to resist termite damage in the field. The development of the ACIAR forestry program, which has become an important stimulus to the exploration of these species, has coincided with heightened international interest in the use of trees in sustainable farming systems.

Australia is not alone in having an exceptional forest resource to exploit for new international needs, but many trees from this country have attributes of value elsewhere. In the Australian tropics most of the tree and shrub flora has evolved from a rich array of rainforest progenitors that adapted to seasonally dry environments and to soils of low fertility. Some genera (e.g. eucalypts) have developed specialised leaves to avoid drought, while others such as acacias have evolved expanded associations with microorganisms (e.g. rhizobia) to enhance the supply of nutrients. Coupled with these attributes are other properties such as fast early growth, an ability to tolerate wildfires through the possession of dormant vegetative bud traces under the bark, and a great capacity to coppice and thus survive and regenerate after harvesting. These attributes make many Australian tropical species ideally suited to the harsh environments usually available for tree planting in developing countries.

The aim of this Chapter is to summarise the main findings reported in this Monograph for those readers who want a quick overview of the results of the ACIAR work, and to indicate productive research directions for the future. The Monograph has been divided into three main sections:

(1) ACIAR forestry program development;
(2) Field trials; and
(3) Resource evaluation.

Program Development

This section includes three chapters describing the development of the initial ACIAR forestry project, the utilisation of a remarkable Australian multipurpose tree *Acacia mearnsii*, and the first ACIAR seed collection program in Australia. In the first chapter the philosophy of the ACIAR forestry program is given together with the reasons why particular countries and organisations were chosen for field trials. This chapter describes the development of a network of trials in Thailand, China, Kenya and Zimbabwe and assessment procedures, which we sought to standardise. In order to further strengthen the network, a workshop was held in 1986 in Gympie, Queensland, at which participants from all countries were invited to compare experiences in growing acacias. Three research foresters (one each from Thailand, China and Zimbabwe) are undertaking postgraduate studies in Australia on aspects related to the field trials. These training programs were an additional means of strengthening the program and the forest research capability of collaborating countries.

The second chapter explains the international importance of *Acacia mearnsii* to the tannin industry, and more recent developments in tannin-formaldehyde adhesives. These adhesives are becoming increasingly valuable for use in wood-composite products such as particle boards, plywood, laminated beams, and more recently Scrimber. Silvicultural practices developed in plantations of *A. mearnsii* in the Republic of South Africa have significantly affected the silviculture of modern-day forest plantations involving other species. This species was arguably the first native Australian tree species on which tree improvement was practiced intensively. The chapter concludes with details of the ACIAR program with *A. mearnsii* in China.

The ACIAR forestry program is still expanding and the challenge is to develop useful new lines of inquiry and practical research that will benefit both Australia and collaborating countries. To date the program has concentrated on Australia's genetic resources for fuelwood and agroforestry. Australian scientists have much to offer in the field of tree nutrition and soil science. In future, attention may shift towards industrial projects such as pulping research, high-value furniture timbers, etc., as perceived needs change in Australia and partner countries. Currently, ACIAR is exploring opportunities to extend partnership arrangements, especially to evaluate forest genetic resources for tolerance to soil salinity and soil acidity. In one species, *Acacia mearnsii*, a program is under way to examine genetic variation in frost resistance of seedlings under laboratory conditions, because intense cold is a major constraint to the use of this tree in China.

Field Trials

This section includes 11 chapters and contains early data from the ACIAR field trials in Australia, China, Thailand, Zimbabwe and Kenya. The section begins with a summary of the climatic conditions at 19 ACIAR trial sites, then compares the climate of each of 17 sites outside Australia with areas of Australia having approximately similar climates. The section concludes with a comparison of all sites to indicate the climatic similarities and differences among them. This approach enables the reader to quickly assess the range of climatic conditions covered by trials in the ACIAR network. This climatic technique could be further developed in the future to include more specific site factors such as soil chemical and physical properties. The matching climatic profile could also be used to determine areas of Australia suitable for the introduction of exotic tree species.

Chapter 5 provides detailed results of field tests on 148 lesser-known species at two Australian trial sites. These trials were conducted by the Queensland Department of Forestry and were a cornerstone in the overall ACIAR program. Their establishment reflected the view that Australia should establish trials to complement those in other countries in order to identify difficulties and problems associated with lesser-known species and to help solve these problems. The chapter gives data on tree growth, flowering and coppicing abilities. It also provides preliminary information on the effect of moving species 'off-site' and on variation within species. Promising species are listed in four categories according to the mean annual rainfall of the origin of the seed source. The trials showed that most myrtaceous species coppice well whereas acacias have variable coppicing ability; some shoot well, others do not. Flowering data reflect to some extent life cycles of species whereby short-lived species often flower early and profusely, whereas long-lived species flower much later in their life cycles. Some species may have the potential to become weeds.

Four chapters cover ACIAR studies in southern China, ranging from trials of temperate eucalypts at high altitudes in Yunnan Province to tropical eucalypts, acacias and casuarinas in coastal lowlands at Fujian, Guangdong and Hainan provinces. Details of early growth of *Acacia mearnsii* provenance trials are also given. In Yunnan the most commonly planted eucalypt today is *E. globulus* subsp. *globulus*; early trial data confirm that this is a suitable species. The trials also suggest that new introductions such as *E. globulus* subsp. *bicostata*, *E. nitens*, *E. viminalis* and *E. camphora* are well-adapted and grow vigorously, while *E. smithii*, *E. badjensis* and *E. scoparia* are less well-known species worthy of closer attention. In tropical China, early results suggest that *E. camaldulensis* from northern Western Australia, *E. tereticornis* from North Queensland and *E. urophylla* from Indonesia grow faster than the traditionally widely grown species *E. exserta* and *E. citriodora*.

Tropical acacias have exciting potential with *A. crassicarpa*, *A. mangium*, *A. auriculiformis*, *A. cincinnata* and *A. aulacocarpa* growing rapidly on infertile soils. The results of provenance trials of *A. mearnsii* have justified the decision to introduce new genetic material, as many of the newly introduced provenances grow faster than local sources. A surprising finding was the early flowering of local provenances compared with all the new introductions. The reason for this is not known but it may reflect past hybridisation with an early-flowering species or the development of a land race created because of past seed collection practices favouring small, early-flowering trees. The fast initial growth of *Casuarina junghuhniana* is noteworthy and leads to the conclusion that range-wide seed collections in Indonesia and the establishment of international provenance trials are required urgently.

The ACIAR trials in Thailand have been well managed by the Royal Forest Department and the results should have wide relevance in tropical Asia. Seven trials are reported; particularly promising species are *Acacia crassicarpa*, *A. auriculiformis*, *A. torulosa*, *A. holosericea* and *A. julifera*. Other species such as *Acacia oraria*, *A. polystachya* and *Albizia procera* survive well but grow slowly. The rapid growth of *Grevillea pteridifolia* in Thailand provides a good example of the untapped potential that exists within this genus. Chapter 14 documents the processes required for site/genotype interaction studies and illustrates the highly variable height growth of *Acacia crassicarpa* across sites in Thailand compared with *Eucalyptus camaldulensis*, which is fairly stable across sites. *Acacia aulacocarpa* falls into two separate groups, the Papua New Guinea provenances being better than the Queensland provenances. Chapter 13 indicates the need for more tree nutrition studies in Thailand; this is clearly one area worthy of additional research.

ACIAR field trials in Africa have been limited to Zimbabwe and Kenya. Six trials were established in Zimbabwe and early data suggest that *Casuarina cunninghamiana*, *C. glauca*, *Grevillea glauca*, *Acacia holosericea*, *Acacia auriculiformis*, *A. crassicarpa*, *A. cowleana*, *A. torulosa*, *A. podalyriifolia* and *A. leptocarpa* are promising. ACIAR is now supporting species trials in the drier (<500 mm) regions of Zimbabwe and Kenya, and the results should have wide applicability in Africa. In Kenya, very early data have confirmed that *Eucalyptus saligna* and *E. grandis* are fast-growing species and that *E. urophylla* is promising in the wet–humid zones.

Provenance field trials are warranted for some acacias such as *Acacia auriculiformis* and *A. crassicarpa*. For the former species ACIAR/CSIRO is now cooperating with F/FRED (USAID) to assist establishment of a series of 10 provenance trials across a range of countries in Asia during 1989. In developing this species further, CSIRO/F/FRED commenced work on a bibliography and a glasshouse trial is being conducted in Canberra to determine geographic variation in seedling morphology. Eventually, this information will be compared with the results of the field trials in order to assist in interpreting the results. Other species worthy of provenance studies are *Grevillea robusta*, an important, established agroforestry species, and *Grevillea pteridifolia* which exhibits at least two tree forms (bushy and columnar) related to seed source in our early trials. Such work requires extensive seed collections with well-organised follow-up field provenance trials, activities which must be adequately funded to be successful.

The ACIAR forestry program has concentrated on evaluating the growth potential of new species in standardised field species trials rather than attempting at this stage to conduct strict agroforestry (tree/crop mixture) experiments. Any new Australian tree species will first have to be proved successful before performance in tree/crop mixtures is studied. We are hopeful that regional nurseries may be a vehicle by which better-performing species will be released for on-farm field testing. Despite this low-key approach one enterprising agroforester at Si Sa Ket, Thailand, has commenced agroforestry research experiments with *Acacia leptocarpa*. This species captured his attention for agroforestry use because of its light, open crown and its propensity to produce a single stem. Further experiments are also warranted on productivity of eucalypt/acacia species mixtures, and on the use of dense-canopied, multistemmed species as rapid-growing cover crops for slow-growing but high-value indigenous tree species.

The previous section on field trials detailed the environmental range of ACIAR trial sites and indicated that a wide range of lesser-known species has been field-tested. Despite this, it is still possible that some potentially important species have escaped our attention. To date, the ACIAR program has concentrated on fast-growing trees suitable for fuelwood and agroforestry for use on infertile soils in the seasonally dry tropics. This approach has ignored many high-value tropical rainforest species that grow on better-watered, more fertile soils, such as *Flindersia* spp. and *Agathis* spp., and has also overlooked tree species suitable for horticultural development (e.g.*Macadamia* spp. and *Davidsonia* spp.). Suitable field testing programs for such species could be developed in the future.

Resource Evaluation

In this third section there are nine chapters exploring a range of utilisation or biological attributes in order to improve the potential utility of particular species. Subjects covered include values for fuelwood and fodder, essential leaf oils, susceptibility to termite attack, propagation and management of nitrogen fixation in *Casuarina* species.

The first chapter of the section describes vegetative propagation in *Casuarina* and *Acacia*, and contains a strong recommendation for continued investigation of traditional means of vegetative propagation (hardwood cuttings, etc.), rather than embarking exclusively on more resource-dependent tissue culture techniques. Casuarinas and acacias appear to be easier than many eucalypts to propagate from cuttings and may be easier to use in clonal forestry. It is highly likely that ease of propagation will be dependent on certain individual plants having a greater propensity than others to strike from cuttings, as well as being species-dependent.

Chapters 16–18 are devoted to the evaluation of wood of several Australian species for fuel. Chapter 16 gives a brief review of terms used in fuelwood testing and indicates values to consider in identifying good fuelwoods. Chapter 17 gives a description of the development of a specialised crib to compare, under standardised conditions, the burning properties of wood of various species. In Chapter 18 critical attention is given to drying rates of timbers and how these vary amongst species. An attempt was made to compare the fuelwood properties of different species using roundwood (as it was felt that this was closer to a real-life situation) but the authors consider now that this method has limited usefulness.

Chapter 19 reports the fodder value of selected species in ACIAR trials at Gympie. It is surprising that scientific study of this subject has been so neglected in Australia given the importance of 'top-feed' in times of drought to the Australian pastoral industry. Protein levels were generally low but some species worthy of further study are documented. Details of the nutrient content of tree foliage are given because, as one reviewer of the Chapter commented, this kind of basic data is seriously lacking for much of our Australian tree flora. In addition, it is conceivable that certain trees will be used in the future in agroforestry systems as nutrient 'pumps' to capture nutrients deep in the profile, and there is evidence that some species recycle certain elements better than others.

Draught animals have a significant requirement for fodder at the end of the dry season to regain strength and condition quickly before the first rains and ploughing commences. Tree species that produce tender, new, nutritious shoots prior to the first heavy rains are valuable fodder trees, and a search for such species amongst the Australian tree flora would be worthwhile. There is also a need to conduct fodder tests using livestock in which preferences for species and animal weight gains or losses after feeding are monitored closely.

Chapters 20 and 21 detail the essential oils in species belonging to three genera of the family Myrtaceae (*Eucalyptus*, *Leptospermum* and *Melaleuca*). Past work in Australia has concentrated on temperate eucalypts while most tropical species have been somewhat neglected. Previously it was thought that tropical species contained little cineole but high yields have been found in *Eucalyptus bakeri* and several *Melaleuca* species. The work reported on tropical melaleucas is of a pioneering nature; several interesting and useful compounds were found. The value of this work for cottage industries is apparent when one considers that only simple steam-distillation techniques are required to extract oils (as is already happening with *Eucalyptus globulus* in temperate areas of India and the People's Republic of China). The extracted oils can be used for low-cost medicines, disinfectants, flavouring and antiseptics.

Much work still has to be done to evaluate Australia's lesser-known tree flora. Fast-growing tropical acacias are likely to be valuable for pulpwood and bark-derived tannin. Miscellaneous additional studies could be conducted on perfumes from acacia flowers and on the potential as cut-flowers and foliage for the horticultural industry. Work on shiitaki mushroom cultivation using acacia wood with various additives is being conducted in China. There has been surprisingly little research in essential oils of tropical eucalypts; current work suggests species belonging to the box, red gum and ironbark groups produce useful oils. Work on eucalypt leaf extracts in China has identified compounds that increase crop yields when sprayed on certain vegetables.

The final two chapters deal with the management of nitrogen fixation to increase productivity in casuarinas, and to determine the resistance to termite attack of Australian tree species planted in Zimbabwe.

Chapter 22 examines the issues of selecting effective *Frankia* strains for casuarinas, developing simple but effective inoculation technology and identifying soil factors that influence tree responsiveness to inoculation. So far, commercial quantities of these organisms are not available for routine use.

The role of bacteria (especially *Rhizobium*) in stimulating the growth of agricultural legumes is well known, and inoculum is commonly applied (e.g. seeds of legumes are coated with inoculum prior to sowing). Such techniques are not used in forestry but bacteria could be easily applied to seedlings in the nursery. Thus rapid early growth could be achieved in the nursery and in at least the critical first few months after field planting.

Research work in understanding and using microorganisms to aid tree health and productivity are important lines of inquiry. It is possible that particular strains of mycorrhiza and rhizobia are specific to particular taxonomic groups of acacias, and basic research is needed to determine if such relationships exist. In addition, mulching experiments examining rates of litter turnover and nutrient release for different tree species could be rewarding.

The work reported in Chapter 23 on the susceptibility of species to root damage by termites in Zimbabwe is relevant to common field problems. Many poor farmers in Africa and especially those on communal lands of Zimbabwe have no access to termiticides to prevent attack. Termite-resistant species are therefore essential. Over 41 species were tested for survival and susceptibility to the fungus-growing termites *Ancistrotermes latinotus* and *Macrotermes michaelseni*. Promising Australian species displaying early resistance were *Acacia holosericea*, *A. salicina*, *A. plectocarpa*, *A. leptocarpa* and *A. difficilis*. This work may also be significant in understanding species' ecology and their distribution patterns in natural ecosystems.

Conclusion

In conclusion, the papers presented cover a wide range of activities in the ACIAR forestry program. All reflect the central theme of identifying and evaluating lesser-known Australian trees. The characterisation of some well-known, widely grown Australian species has been a haphazard process spread over 100 years or more, and even in that time the work has not always been very thorough. In this program, through a collaborative, planned effort, it has been possible to condense that time scale by a factor of 10, and to do a thorough, systematic job. The work is by no means finished, but the accounts in this book show that it is proceeding rapidly. In addition to the completion of research itself, the collation and dissemination of resulting information is a key step towards the ultimate objective of enhancing living standards. This task is being approached in a number of ways: the development of data bases, workshops, publications, training and demonstrations. In addition, basic genetic resources are being provided. Initially these have come from indigenous forests in Australia, but increasingly planted seed sources will become more significant. Ultimately, the choice of useful trees available for planting in many countries will be significantly expanded, and that choice will be backed up by supplies of high-quality seed.

Chapter 25

References

Abo el-Nil, M.M. 1987. Micropropagation of *Casuarina*. In: Bonga, J.M., and Durzan, D.J. ed., Cell and tissue culture in forestry. (Martinus Vijihoff: The Netherlands) 400–410.

Abdel Nour, H.O. 1975. Drought and termites: a threat to *Cupressus lusitanica* plantations in Jebeb Marra-Sudan. Sudan Silva 3(20), 12–17. Cited in: Ernst, E., Araujo, R.L., and Broughton, P. 1986. A bibliography of termite literature, 1966–1978. Chichester, U.K., Wiley.

Anon. 1963. Annual Report of Wattle Research Institute. Pietermaritzburg, University of Natal.

1967. Annual Report of Wattle Research Institute. Pietermaritzburg, University of Natal.

1969. Annual Report of Wattle Research Institute. Pietermaritzburg, University of Natal.

1988. Wattle be the result for ACIAR? Action China No.10 (July), 23.

Applegate, G.B., and Nicholson, D.I. 1987. Growth rate of selected *Acacia* species in north and southeast Queensland, Australia. In: Turnbull, J.W., ed., Australian acacias in developing countries: proceedings of an international workshop held at the Forestry Training Centre, Gympie, Queensland, Australia, 4–7 August 1986. ACIAR Proceedings No.16, 196p.

Arola, R.A. 1976. Woods fuels — how do they stack up? In: Energy and wood products industry proceedings. Madison, Wisconsin, USA, Forest Products Research Society, 76–14.

Baker, D., and O'Keefe, D. 1984. A modified sucrose fractionation procedure for the isolation of frankiae from actinorhizal root nodules and soil samples. Plant and Soil, 78, 23–28.

Baker D., and Torrey, J.G. 1979. The isolation and cultivation of actinomycete root nodule endophytes. In: Gordon, J.C., Wheeler, C.T. and Perry, D.A., ed., Symbiotic nitrogen fixation in the management of temperate forests. Corvallis, Oregon State University Press, 38–56.

Barlow, B.A. 1986. Contributions to a revision of *Melaleuca* (Myrtaceae): 1–3 Brunonia, 9, 163–177.

Barrett, R.L., and Mullin, L.J. 1968. A review of introductions of forest trees in Rhodesia. Salisbury, Rhodesia Forestry Commission, Bulletin of Forest Research No.1.

Bayfield, R.F., Humphreys, F.R., and Martin, P.J. 1952. Survey of the tannin content of bark of plantation grown *Pinus radiata*. Forestry Commission of New South Wales, Wood Technology Division, Report No.1, 5p.

Benson, D. 1982. Isolation of *Frankia* strains from alder actinorhizal root nodules. Applied and Environmental Microbiology, 44, 461–465.

Bhimaya, C.P., and Kaul, R.N. 1966. Levels of macro and microelements in *Eucalyptus camaldulensis*. Nature 212, 319–320.

Bleakley, S., and Matheson, C. 1988. Patterns of morphological variation in seedlings of *A. mearnsii* De Wilde. Paper delivered to ACIAR/CAF Workshop in Guangzhou, People's Republic of China, 2–11 Oct.

Boland, D.J. 1986. Selection of species and provenances for tree introduction. In: Turnbull, J.W., ed., Multipurpose Australian trees and shrubs: lesser-known species for fuelwood and agroforestry. ACIAR Monograph No.1, 316p.

1987. Genetic resources and utilisation of Australian bipinnate acacias (Botrycephalae). In: Turnbull, J.W., ed., Australian acacias in developing countries: proceedings of an international workshop held at the Forestry Training Centre, Gympie, Queensland, Australia, 4-7 August 1986. ACIAR Proceedings No.16, 29–37.

Boland, D.J., and Turnbull, J.W. 1981. Selection of Australian trees other than eucalyptus for trials as fuelwood species in developing countries. Australian Journal of Forestry, 44, 235–246.

Bolza, E., and Keating, W.G. 1972. African timbers — the properties, uses and characteristics of 700 species. Melbourne, CSIRO Australia Division of Building Research.

Booth, T.H. 1985. A new method for assisting species selection. Commonwealth Forestry Review, 64, 241–250.

1988. Which wattle where?: selecting Australian acacias for fuelwood plantations. Plants Today, 1, 86-90.

Booth, T.H., Nix, H.A., Hutchinson, M.F., and Busby, J.R. 1987. Grid matching: a new method for homoclime analysis. Agricultural and Forest Meteorology, 39, 241-255.

Booth, T.H., Nix, H.A., Hutchinson, M.F., and Jovanovic, T. 1988. Niche analysis and tree species introduction. Forest Ecology and Management, 23, 47-59.

Bootle, K.R. 1971. Commercial timbers of New South Wales and their use. Halstead Press, Sydney.

1983. Wood in Australia: types, properties and uses. Sydney, McGraw-Hill, 443p.

Bowyer, R.C., and Jefferies, P.R. 1962. New sources and synthesis of torquatone. Australian Journal of Chemistry, 15, 145-150.

1963. Structure of spathulenol. Chemistry and Industry, 1245-1246.

Brockway, G.E., and Hillis, W.E. 1955. Tan bark eucalypts of the semi-arid regions of southwestern Australia. Empire Forestry Review, 34, 31-41.

Brophy, J.J., and Lassak, E.V. 1983. The volatile leaf oils of *Melaleuca armillaris*, *M. dissitiflora* and *M. trichostachya*. Journal and Proceedings of the Royal Society of New South Wales, 116, 7-10.

1985. The volatile leaf oils of two cultivars of *Callistemon viminalis*. Journal and Proceedings of the Royal Society of New South Wales, 118, 101-104.

1988. The leaf oil of *Melaleuca leucadendra* L. Flavour and Fragrance Journal, in press.

Brophy, J.J., Lassak, E.V., and Boland, D.J. 1987. Volatile leaf oils of the two subspecies of *Melaleuca acadioides*. Journal and Proceedings of the Royal Society of New South Wales, in press.

1988a. Volatile leaf oils of six northern Australian *Melaleucas*. Journal and Proceedings of the Royal Society of New South Wales, 121, 29-33.

1988b. Volatile leaf oils of *Melaleuca symphyocarpa* and *Melaleuca uncinata*. Flavour and Fragrance Journal, submitted.

Brown, H.P., Panshin, A.J., and Forsaith, C.C. 1952. Textbook of wood technology. New York, McGraw-Hill, Volume 2, 783p.

Brown, K.W. 1962. Termite control research in Uganda. Paper presented at 8th British Commonwealth Forestry Conference, East Africa 1962.

Browne, F.G. 1968. Pests and disease of forest plantation trees. Oxford, Clarendon Press.

Browning, B.L. 1967. Methods of wood chemistry. New York, Interscience, Volume 1, 384p.

Burggraaf, A.J.P., and Shipton, W.A. 1983. Studies on the growth of *Frankia* isolates in relation to infectivity and nitrogen fixation (acetylene reduction). Canadian Journal of Botany, 61, 2774-2782.

Burley, J., and Wood, P.J. 1976. A manual on species and provenance research with particular reference to the tropics. University of Oxford, Department of Forestry, Tropical Forestry Papers No.10.

Busgen, M., and Munch, E. 1929. Structure and life of forest trees (translated by T. Thomson). New York, John Wiley and Sons, 436p.

Cameron, D.M., Rance, S.J., and Williams, E.R. 1986. Nutrition studies with irrigated *Eucalyptus camaldulensis* and *Anthocephalus chinensis* on a cracking clay in northern Australia. Tropical Agriculture (Trinidad), 63, 11-16.

Chijioke, E.O. 1980. Impact on soils of fast-growing species in lowland humid tropics. Rome, FAO Forestry Paper 21, 111p.

Chippendale, G.M., and Wolf, L. 1981. The natural distribution of *Eucalyptus* in Australia. Canberra, Australian National Parks and Wildlife Service, Special Publication No.6, 192p.

Chittachumnonk, P. 1981. Silviculture of *Casuarina junghuhniana* in Thailand. In: Midgley, S.J., Turnbull, J.W., and Johnston, R.D., ed., *Casuarina* ecology, management and utilisation: proceedings of an international workshop, Canberra, Australia, 17-21 August 1981. Melbourne, CSIRO, 286p.

Clarke, B. 1975. Establishment of eucalypt plantations. Australian Forest Industries Journal, 41, 45-51.

Cochran, W.G., and Cox, G.M. 1957. Experimental Designs (2nd edition). New York, Wiley.

Coyne, P.D. 1973. Some aspects of the autoecology of *Casuarinas*, with particular reference to nitrogen fixation. Australian National University, Ph.D. thesis.

Craib, I.J. 1933. The place of thinning in wattle silviculture and its bearing on the management of exotic conifers. Zeitschrift für Weltforstwirtschaft, 1 (2/3), 77-108.

Crocker, W.N. 1916. Mechanics of dormancy in seeds. American Journal of Botany, 99-120.

Crockford, K. 1987. The evaluation of British woodlands for fuelwood and timber production. University of Oxford, unpublished D. Phil thesis.

Cronquist, A.J. 1981. An integrated system of classification of flowering plants. New York, Columbia University Press.

Dann, P., and Low, S. 1988. Assessing the value of browse plants as alternative fodder sources. Agricultural Science, February-March 1988, 20-27.

Dawson, J.O., and Sun, S-H. 1981. The effect of

Frankia isolates from *Comptonia peregrina* and *Alnus crispa* on the growth of *Alnus glutinosa*, *A. cordata* and *A. incana* clones. Canadian Journal of Forest Research, 11, 758-762.

Diem, H.G., and Gauthier, D. 1982. Effet de l'infection endomycorrihizienne (*Glomus mosseae*) sur la nodulation et la croissance de *Casuarina equisetifolia.* Comptes Rendus de l'Academie des Sciences (Paris) 294, ser.3, 215-218.

Diem, H.H., Gauthier, D., and Dommergues, Y. 1983. An effective strain of *Frankia* from *Casuarina* sp. Canadian Journal of Botany, 61, 2815-2821.

Digby, P. 1979. Modified joint regression analysis for incomplete variety x environment data. Journal of Agricultural Science, 93, 81-86.

Dillon, J.T., and Baker, D. 1982. Variation in nitrogenase activity among pure-cultured *Frankia* strains tested on actinorhizal plants as an indication of symbiotic compatibility. New Phytologist, 92, 215-219.

Dommergues, Y.R. 1963. Evaluation du taux de fixation de l'azote dans un sol dupaire reboise en filao (*Casuarina equisetifolia*). Agrochimica, 105, 179-187.

Duhoux, E., Sougoufara, B. and Dommergues, Y. 1986. Propagation of *Casuarina equisetifolia* through auxiliary buds of immature female inflorescences cultured in vitro. Plant Cell Reports 3, 161-164.

Duncan, D.B. 1955. Multiple range and multiple F tests. Biometrics, 11, 1-42.

Duryea, M.L., and Landis, T.D. 1984. Forestry Nursery Manual. Production of bareroot seedlings, 385p.

Evans, J. 1982. Plantation forestry in the tropics. Oxford, Clarendon Press, 472p.

Evans, J., Khemnark, C., and Stewart, H. 1987. Australian hardwoods for fuelwood and agroforestry. Report to ACIAR of the triennial review of ACIAR Projects 8320 and 8331. International Institute for Environment and Development.

Everist, S.L. 1969. Use of fodder trees and shrubs. Brisbane, Australia, Queensland Department of Primary Industries, Division of Plant Industry, Advisory leaflet No.1024.

FAO 1979. Eucalypts for planting. Rome, Food and Agriculture Organization of the United Nations, Forestry Series No.11. 677p.

1981. Eucalypts for planting (2nd edition). Rome, Food and Agriculture Organization of the United Nations, Forestry Series No.11. 677p.

Feeny, D. 1984. Agricultural expansion and forest depletion in Thailand, 1900-1975. Paper prepared for Economic Growth Centre, Yale University. Discussion Paper No. 458. 65p.

Finlay, K.W., and Wilkinson, G.N. 1963. The analysis of adaptation in a plant breeding program. Australian Journal of Agricultural Research, 14, 742-754.

Flynn, T., Lassak, E.V., and Smyth, M.P. 1979. The volatile leaf oils of three species of *Leptospermum.* Phytochemistry, 18, 2030-2031.

Fung, P.Y.H. 1984. Factors influencing the choice of wood energy systems for timber. Presented at timber drying seminar, Forest Research Institute. March 1984, Rotorua, New Zealand.

Gauthier, D., Diem, H.G., Dommergues, Y.R., and Ganry, R. 1985. Assessment of N2 fixation by *Casuarina equisetifolia* inoculated with *Frankia* ORS 021001 using 15N methods. Soil Biology and Biochemistry, 17, 375-379.

Gordon-Gray, C.G. 1957. A comparison of the results of estimating black wattle tannin by the official hide power method and a proposed ultra-violet spectrophotometric method. Journal of the Society of Leather Trades Chemists, 41, 269-275.

Granhof, J.J. 1974. Early results of a combined species/soil cultivation/fertiliser trial in Northern Thailand. In: Thai-Danish pine project, 1969-1974. Bangkok, Allied Printers, 114p.

Hall, N., and Brooker, I. 1974. Bakers mallee, *Eucalyptus bakeri* Maiden. Leaflet, Forest Tree Series No.145, Forestry and Timber Bureau, Department of Primary Industry.

Halos, S.D. 1981. Production practices of *Casuarina equisetifolia.* In: Midgley, S.J., Turnbull, J.W., and Johnston, R.D., ed., *Casuarina* ecology, management and utilisation: proceedings of an international workshop, Canberra, Australia, 17-21 August 1981. Melbourne, CSIRO, 286p.

Hannah, B.C., Fergus, B.J., and Jones, R.N. 1977. Kraft pulping and bleaching studies on young hardwood species. Appita, 30, 483-487.

Harker, A.P., Sandels, A., and Burley, J. 1982. Calorific values for wood and bark and a bibliography for fuelwood. London, Tropical Products Institute, Report G162, 23p.

Harris, W.V. 1971. Termites, their recognition and control (2nd edition). Longmans, London.

Heffernan, B. 1985. A handbook of methods of inorganic chemical analysis for forest soils, foliage and water, CSIRO Division of Forest Research, Canberra. 281p.

Heller, S.R., and Milne, G.W.A. 1978, 1980, 1983. EPA/NIH mass spectral data base, Washington, D.C., U.S. Government Printing Office.

Hellyer, R.O. 1968. The occurrence of β-triketones in the steam volatile oils of some myrtaceous Australian plants. Australian Journal of Chemistry, 21, 2825-2828.

Hellyer, R.O., and Lassak, E.V. 1968. The steam volatile constituents of *Melaleuca viridiflora* Sol.

ex Gaertn. Australian Journal of Chemistry, 21, 2585–2587.

Higgins, M.L., and Lechevalier, M.P. 1969. Poorly lytic bacteriophage from *Dactylosporangium thailandensis*. Journal of Virology, 3, 210–216.

Hillis, W.E. 1956. The production of mangrove extract in the delta region of Papua. Empire Forestry Review, 35, 420–436.

1962. Wood extractives. New York, Academic Press, 62p.

Hingston, F.J., Malajczuk, N., and Grove, T.S. 1982. Acetylene reduction (N2-Fixation) by Jarrah forest legumes following fire and phosphate application. Journal of Applied Ecology, 19, 631–645.

Hooker, J.E., and Wheeler, C.T. 1987. The effectivity of *Frankia* for nodulation and nitrogen fixation in *Alnus rubra* and *A. glutinosa*. Physiologia Plantarum, 70, 333–341.

Howard, E.T. 1973. Heat of combustion of various southern pine materials. Wood Science, 5, 194–197.

Humphreys, F.R., and Johnstone, R.S. 1957. Analysis of tannin barks from experimental plots of *Acacia mollissima*: Forestry Commission Project C1 Final Report, Sydney, Forestry Commission of New South Wales, 16p.

Hutchinson, M.F. 1988. A new objective method for spatial interpolation of meteorological variables. In: E.A. Fitzpatrick, ed., Proceedings of United Nations University workshop on the need for climatic and hydrologic data in agriculture of South-East Asia. Tokyo, United Nations University.

Ince, P.J. 1977. Estimating effective heating value of wood or bark fuels at various moisture contents. USDA Forest Service, General Technical Report FPL-13, 9p.

Janick, J.W., Schery, W.R., Woods, F.W., and Ruttan, V.W. 1981. Plant science. An introduction to world crops. San Francisco, W.H. Freeman and Co., 868p.

Jeffers, J.N.R. 1978. An introduction to systems analysis: with ecological applications. London, Edward Arnold, 198p.

Johnston, A.D., Simmons, J.G., Hansen, R.W., and Daniel, R.A. 1985. Chemical procedures for the analysis of plant material: multielement, oil, sugars, and gum. St Lucia, Queensland, Australia, CSIRO Division of Tropical Agronomy, Tropical agronomy technical memorandum Number 40.

Jorgensen, J.R., and Wells, C.G. 1986. Tree nutrition and fast-growing plantations in developing countries. International Tree Crops Journal, 3, 225–244.

Kant, S., and Narayana, H.S. 1978. Effect of water stress on growth, nodulation and nitrogen fixation in *Casuarina equisetifolia*. Annals of the Arid Zone, 17, 216–222.

Kehl, J.C. 1980. The biology and population of *Rattus tunneyi* culmorum in hoop pine plantations in southeast Queensland. University of Queensland, St Lucia, Australia, MSc thesis.

Kingston, R.J.T., and Risdon, C.J.E. 1961. Shrinkage and density of Australian and other southwest Pacific woods. Technological Paper No.13, Division of Forest Products CSIRO, 65p.

Kondas, S. 1981. Mist tent vegetative propagation of *Casuarina junghuhniana*. In: Midgley, S.J., Turnbull, J.W., and Johnston, R.D., ed., *Casuarina* ecology, management and utilisation: proceedings of an international workshop, Canberra, Australia, 17–21 August 1981. Melbourne, CSIRO, 286p.

Kramer, P.J., and Koslowski, T.T. 1979. Physiology of woody plants. New York, Academic Press, 811p.

Krilov, A., Gardner, W.D., and Meredith, S.J. 1986. ACIAR fuelwood project, Report No. 11490/1, 28 February 1986. Sydney, Wood Technology and Forest Research Division, Forestry Commission of New South Wales, 28p.

Kudler, J. 1970. The attack and protection of *Eucalyptus* plantations in Ghana. Ghana Journal of Agricultural Science, 3, 39–44.

Lalonde, M., and Calvert, H.E. 1979. Production of *Frankia* hyphae and spores as an infective inoculant for *Alnus* species. In: Gordon, J.C., Wheeler, C.T., and Perry, D.A., ed., Symbiotic nitrogen fixation in the management of temperate forests. Corvallis, Oregon State University Press, 95–110.

Lassak, E.V. 1979. The volatile leaf oils of three species of *Melaleuca*. Journal and Proceedings of the Royal Society of New South Wales, 112, 143–145.

Lassak, E.V., and McCarthy, T.M. 1983. Australian medicinal plants. Sydney, Methuen Australia, 65p.

Lassak, E.V., and Southwell, I.A. 1977. Essential oil isolates from the Australian flora. International Flavour and Food Additives, 4, 126–132.

Le Houerou, H.N. 1980. Browse in Africa: the current state of knowledge. Papers presented at the international symposium on browse in Africa, Addis Ababa, 8–12 April 1980, and other submissions. International Livestock Centre for Africa, 499p.

Lee, R.F. 1971. A preliminary annotated list of Malawi forest insects. Malawi Forest Research Institute, Research Record No.40.

Logan, A.F. 1987. Australian acacias for pulpwood. In: Turnbull, J.W. ed., Australian acacias in developing countries: proceedings of an international workshop held at the Forestry Training Centre, Gympie, Queensland, Australia, 4-7 August 1986. ACIAR Proceedings No.16, 89-94.

Lunquist, R., and Torrey, J.G. 1984. The propagation of *Casuarina* species from rooted stem cuttings. Bot. Gaz. 145, 378-384.

Mahadevan, P. 1981. Pastures and animal production. In: Hacker, J.B., ed., Nutritional limits to animal production from pastures. Proceedings of an international symposium held at St. Lucia, Queensland, Australia, August 24-28. Farnham Royal, U.K., Commonwealth Agricultural Bureaux, 1-17.

Mandel, J. 1971. A new analysis of variance model for non-additive data. Technometrics, 13, 1-18.

McCluskey, D.N., and Fisher, R.F. 1983. The effect of inoculum source on nodulation in *Casuarina glauca*. Commonwealth Forestry Review, 62, 117-124.

McDonald, W.J.F., and Ternouth, J.H. 1979. Laboratory analysis of the nutritional value of western Queensland browse feeds. Australian Journal of Experimental Agriculture and Animal Husbandry, 19, 344-349.

McLeod, M.N. 1973. The digestibility and the nitrogen, phosphorus and ash contents of the leaves of some Australian trees and shrubs. Australian Journal of Experimental Agriculture and Animal Husbandry, 13, 245-250.

Metz, 1963. Fire protection of wood. Zeitschrift für Ver. deut. Ingr. 80, 660-667 (cited in Shelton 1983).

Midgley, S.J., Turnbull, J.W., and Johnston, R.D. 1983. *Casuarina* ecology, management and utilisation: proceedings of an international workshop, Canberra, Australia, 17-21 August 1981. Melbourne, CSIRO, 286p.

Midgley, S.J., and Weerawardane, N.D.R. 1986. Termites and their control in eucalypt plantations of the Community Forestry Project. Sri Lankan Forester 16(3,4), in press.

Milford, R., and Haydock, K.P. 1965. Nutritive value of protein in subtropical pasture species grown in southeast Queensland. Australian Journal of Experimental Agriculture and Animal Husbandry, 5, 13.

Minson, D.J., and McLeod, M.N. 1978. The accuracy of the pepson-cellulase technique for estimating the dry matter digestibility in vivo of grasses and legumes. Animal Feed Science Technology, 3, 277-287.

Mitchell, B.L. 1980. Report on a survey of the termites of Zimbabwe. Occasional paper, National Museums of Zimbabwe, Series B, Natural Sciences, 6(5), 187-323.

Mitchell, M.R. 1986. The use of controlled release carbosulfan for termite control in *Eucalyptus* plantations. Paper presented at XVIII IUFRO World Congress, Group 2.07.07 Insects threatening forest plantations, 7-21 September 1986, Ljubljana, Yugoslavia.

Mitchell, M.R., Gwaze, D.P., and Stewart, H.T.L. 1988. Survival and early growth of Australian tree species planted at a termite-infested site in Zimbabwe. East African Agricultural and Forestry Journal, 52(4), in press.

Murry, M.A., Fontaine, M.S., and Torrey, J.G. 1984. Growth kinetics and growth induction in *Frankia* sp.HFPAr13 grown in batch culture. Plant and Soil, 78, 61-78.

Murry, M.A., Zhang, Z., and Torrey, J.G. 1985. Effect of O_2 on vesicle formation, acetylene reduction and O_2 uptake kinetics in *Frankia* HFPCc13 isolated from *Casuarina cunninghamiana*. Canadian Journal of Microbiology, 31, 804-809.

Nair, K.S.S., and Varma, R.V. 1985. Some ecological aspects of the termite problem in young eucalypt plantations in Kerala, India. Forest Ecology and Management, 12, 287-303.

Palmberg, C. 1981. A vital fuelwood gene pool is in danger. Unasylva, 33, No.133, 22-30.

Parihar, D.R. 1978. Field observations on the nature and extent of damage by Indian desert termites and their control. Annals of the Arid Zone, 17, 192-199.

1981. Termites affecting *Eucalyptus* plantations and their control in the arid region of India. Zeitschrift für Angewandte Entomologie, 92, 106-111.

Parry, M.S., 1959. Control of termites in *Eucalyptus* plantations. Empire Forestry Review, 38, 288-292.

Patterson, H.D., and Silvey, V. 1980. Statutory and recommended list trials of crop varieties in the United Kingdom (with discussion). Journal of the Royal Statistical Society, A143, 219-252.

Patterson, H.D., Williams, E.R., and Hunter, E.A. 1978. Block designs for variety trials. Journal of Agricultural Science, 90, 395-400.

Penfold, A.R. 1927. The essential oil of *Eucalyptus bakeri* Maiden. Journal and Proceedings of the Royal Society of New South Wales, 61, 179-189.

Penfold, A.R., and Morrison, R.F. 1950. In: Guenther, E., ed., The essential oils, Vol.IV. New Jersey, Van Nostrand & Co., 526-548.

Penfold, A.R., and Willis, J.L. 1961. The eucalypts, botany, cultivation, chemistry and utilisation. New York, Leonard Hill.

Pinyopusarerk, K., and Boland, D.J. 1987. Assessment of early growth of some Australian *Acacia* and *Casuarina* tree species in field trials in Thailand. Paper presented to Symposium on The Contribution of Biological Nitrogen Fixation to Plant Production, held 3-7 August 1987, Cisarua, Indonesia.

Poynton, R.S. 1979. Tree planting in southern Africa. Volume 2 The eucalypts. Pretoria, Department of Forestry, Republic of South Africa.

Pryor, L.D. 1976. Biology of eucalypts. Studies in Biology No.61. Edward Arnold. 80p.

Pryor, L.D., and Johnson, L.A.S. 1971. A classification of the eucalyptus. Canberra, Australian National University Press.

Pushparajah, E. 1966. Studies on the effects of rock phosphate on growth and yield of *Hevea brasiliensis*. Kuala Lumpur, Malaysia, University of Malaya, M.Ag.Sc. thesis, 92p.

1983. Nutrient availability in acid soils of the tropics following clearing and cultivation with plantation crops. In: Proceedings of the International Workshop on Soils, Townsville, Queensland, Australia, 12-16 September 1983. ACIAR/IBSRAM. 47-51.

Rajagopal, D. 1982. Relative incidence of termites on exotic species of *Eucalyptus* in India and their control. Zeitschrift für Angewandte Entomologie, 98, 225-230.

Raymond, C. 1987. *Acacia mearnsii*: a breeding plan for China. Canberra, CSIRO Australia Division of Forestry and Forest Products, ACIAR Project 8418, unpublished report.

1988. *Eucalyptus globulus*: a breeding plan for China. Unpublished report, ACIAR Project 8457, CSIRO Canberra.

Reddell, P., and Bowen, G.D. 1985. *Frankia* source affects growth, nodulation and nitrogen fixation in *Casuarina* species. New Phytologist, 100, 115-122.

Reddell, P., Bowen, G.D., and Robson, A.D. 1985. The effect of soil temperature on plant growth, nodulation and nitrogen fixation in *Casuarina cunninghamiana* Miq. New Phytologist, 101, 441-450.

1986. Nodulation of Casuarinaceae in relation to host species and soil properties. Australian Journal of Botany, 34, 435-444.

Reddell, P., Rosbrook, P.A., Bowen, G.D., and Gwaze, D. 1988. Growth responses in *Casuarina cunninghamiana* plantings to inoculation with *Frankia*. Plant and Soil, 108, 79-86.

Richardson, S.D. 1986. The Cotchell Report. Published by Cotchell Pacific Ltd. p.42.

Rodriguez-Barrueco, C., MacIntosh, A.H., and Bond, G. 1970. Some effects of combined nitrogen on the nodule symbiosis of *Casuarina* and *Ceanothus*. Plant and Soil, 33, 129-139.

Roonwal, M.L. 1979. Termite life and termite control in tropical South East Asia. Jodhpur, Jodhpur Scientific Publishers.

Rosbrook, P.A. 1988. Effect of inoculum type and placement on nodulation and growth of *Casuarina cunninghamiana* seedlings. Australian Forest Research, in press.

Rosbrook, P.A., and Bowen, G.D. 1987. The ability of three *Frankia* isolates to nodulate and fix nitrogen with four species of *Casuarina*. Physiologia Plantarum, 70, 373-377.

Rosbrook, P.A., and Francis, I.W. 1987. Screening of *Frankia* isolates to assess their nodulation and nitrogen fixation ability when associated with five species of *Casuarina* sens.strict. CSIRO Australia Division of Soils Technical Memorandum 74/87, 5p.

Roughley, R.J., and Pulsford, D.J. 1982. Production and control of legume inoculants. In: Vincent, J.M., ed., Nitrogen fixation in legumes. Sydney, Academic Press, 193-209.

Roux, D.G. 1951. Photometric methods of tannin analysis for black wattle tannin. Journal of the Society of Leather Trades Chemists, 35, 322-337.

1957a. The estimation of tannins in black wattle barks and commercial 'mimosa' extracts, Part II. Journal of the society of Leather Trades Chemists, 41, 275-286.

1957b. The estimation of tannins in black wattle barks and commercial 'mimosa' extracts, Part III. Journal of the Society of Leather Trades Chemists, 41, 287-300.

1957c. Ultraviolet photometric methods of tannin extracts in relation to wattle extract utilisation. Journal of the Leather Chemists Association, 52, 319-329.

Ruelle, J.E. 1970. A revision of the termites of the genus *Macrotermes* from the Ethiopian Region (Isoptera: Termitidae). Bulletin of the British Museum (Natural History) Entomology, 24, 365-444.

Ryan, P.A., Podberscek, M., Raddatz, C.G., and Taylor, D.W. 1987. *Acacia* species trials in southeast Queensland, Australia. In: Turnbull, J.W., ed., Australian acacias in developing countries: proceedings of an international workshop held at the Forestry Training Centre, Gympie, Queensland, Australia, 4-7 August 1986. ACIAR Proceedings No.16, 81-85.

Sands, W.A. 1960. Observations on termites destructive to trees and crops. In: Termite research in West Africa. Centre for Overseas Pest Research, London, Report of the Termite Research Unit 1955-1956.

1962. The evaluation of insecticides as soil and mound poisons against termites in agriculture and forestry in West Africa. Bulletin of Entomological Research, 58, 179-192.

Schönau, A.P.G. 1975. Ensure satisfactory eucalypt coppice regeneration by applying correct felling and bark stripping practices. Pietermaritzburg,

South African Timber Growers Association, Newsletter No.52.

1983. The effects of fertilising on the foliar nutrient concentration in *Eucalyptus grandis*. Fertilizer Research, 2, 73–87.

Selander, J., and Nkunika, P.O.Y. 1981. Termites as pests of *Eucalyptus* in Zambia. Paper presented at XVII IUFRO World Congress, Group 2.4 Entomology, insects and diseases of exotic eucalypts, 6–17 September 1981, Kyoto, Japan, 10p.

Sharma, E.R. 1985. Harvesting and transport systems: clearcutting and replanting, coppicing, pollarding, lopping. In: Burley, J., and Stewart, J.L. ed. Increasing productivity of multipurpose species planning workshop for Asia on forest research and technology transfer, Kandy, Sri Lanka, 16–28 July 1984. IUFRO, Vienna, 401–411.

Shelton, J. 1983, Jay Shelton's solid fuels encyclopedia. Charlotte, Vermont, USA, Gordon Way Publishing, 268p.

Shepherd, K.R. 1979. Energy from the forests: exercise in community forestry for developing countries. Development Studies Monograph No.19, Australian National University, Canberra.

Sherry, S.P. 1971. The black wattle (*Acacia mearnsii* De Wild.). Pietermaritzburg, University of Natal Press.

Shipton, W.A., and Burggraaf, A.J.P. 1983. Aspects of the cultural behaviour of *Frankia* and possible ecological implications. Canadian Journal of Botany, 61, 2783–2792.

Simsiri, A. 1988. Vegetative propagation of *Acacia auriculiformis*. Bangkok, Thailand, Division of Silviculture, Royal Forest Department, unpublished research report.

Small, B.E.J. 1981. The Australian eucalyptus oil industry — an overview. Australian Forestry, 170–177.

Smelstorius, J.A. 1971. Chemical composition of wood of Australian grown *Pinus radiata* D. Don. Holzforschung, 25 (2), 33–39.

Somasundaram, T.R., Jagadees, S.S. 1977. Propagation of *Casuarina equisetifolia* forest by planting shoots. Indian Forestry 103, 735–737.

Sougoufara, B., Duhoux, E., Corbasson, M., and Dommergues, Y. 1987. Improvement of nitrogen fixation by *Casuarina equisetifolia* through clonal selection: a research note. Arid Soil Research and Rehabilitation, 1, 129–132.

Southwell, I.A. 1973. Variation in the leaf oil of *Eucalyptus punctata*. Phytochemistry, 12, 1341–1343.

Specht, R.L., Roe, E.M., and Boughton, V.H. 1974. Conservation of major plant communities in Australia and Papua New Guinea. Australian Journal of Botany, Supplementary Series No.7, 667p.

Sprent, J.I. 1979. The biology of nitrogen-fixing organisms. London, McGraw-Hill, 196p.

Stenhagen, E., Abrahamsson, S., and McLafferty, F.W. 1974. Registry of mass spectral data. New York, Wiley.

Stewart, W.D.P. 1963. The effect of combined nitrogen on growth and nodule development of *Myrica* and *Casuarina*. Zeitschrift für Allgemeine Mikrobiologie, 3, 152–156.

Streets, R.J. 1962. Exotic forest trees in the British Commonwealth. Oxford, Clarendon Press, 750p.

Swain, E.H.F. 1928. The timbers and forest products of Queensland. Brisbane, Australia, Queensland Forest Service, 500p.

Teel, W. 1985. A pocket directory of trees and seeds in Kenya. Nairobi, Kenya Energy Non-governmental Organizations, 151p.

Thailand Royal Forest Department. 1986. Forestry statistics of Thailand 1986. Bangkok, Forest Statistics Sub-Division, Planning Division, Royal Forest Department.

1987. Australian hardwoods for fuelwood and agroforestry. Project summary report October 1987. Bangkok, Division of Silviculture, Royal Forest Department, 63p.

Torrey, J.G. 1978. Nitrogen fixation by actinomycete nodulated angiosperms. Bioscience, 28, 586–592.

1982. *Casuarina:* actinorhizal nitrogen-fixing tree of the tropics. In: Graham, P.H., and Harris, S.C., ed., Biological nitrogen fixation technology for tropical agriculture. Cali, CIAT, 427–439.

Turnbull, J.W. 1986. Multipurpose Australian trees and shrubs: lesser-known species for fuelwood and agroforestry. ACIAR Monograph No.1, 316p.

Turnbull, J.W., and Pryor, L.D. 1978. Choice of species and seed sources. In: Hillis, W.E., and Brown, A.G., ed., Eucalypts for wood production. Melbourne, CSIRO/Academic Press, 6–65.

U.S. National Academy of Sciences. 1983. Firewood crops, Vol.2. Washington, D.C., National Academy Press.

U.S. National Research Council. 1980. Firewood crops. Washington D.C., National Academy Press, 237p.

1984. Casuarinas: nitrogen-fixing trees for adverse sites. Washington, D.C., National Academy Press, 118p.

Underwood, E.J. 1981. The mineral nutrition of livestock (2nd edition). Farnham Royal, U.K., Commonwealth Agricultural Bureaux.

Varma, R.V. 1982. Investigations on the possibility of noninsecticidal control of termites. Kerala

Forest Research Institute Research Report No.11, 28p.

Vercoe, T.K. 1987. Fodder potential of selected Australian tree species. In: Turnbull, J.W., ed., Australian acacias in developing countries: proceedings of an international symposium held at the Forestry Training Centre, Gympie, Queensland, Australia, 4-7 August 1986. ACIAR Proceedings No.16, 95-100.

VITA (Volunteers in Technical Assistance). 1982. Testing the efficiency of wood burning stoves, provisional international standards. 1815 North Lynn Street, Suite 200, Arlington, Virginia, U.S.A.

Walter, H. 1970. Vegetationszonen und Klima. Stuttgart, Ulmer, 372p.

Wang, S., and Huffman, J.B. 1982. Effect of extractives on heat content of *Melaleuca* and *Eucalyptus*. Wood Science, 15, 33-38.

Ward, S.C., Pickersgill, G.E., Michaelsen, D.V., and Bell, D.T. 1985. Responses to factorial combinations of nitrogen, phosphorus and potassium fertilisers by saplings of *Eucalyptus saligna* Sm., and the prediction of the responses by DRIS indices. Australian Forestry Research, 15, 27-32.

Watson, G.A. 1973. Soil and plant nutrient studies in rubber cultivation. FAO/IUFRO international symposium on forest fertilisation. Paris, 3-7 December 1973.

Webb, D.B., Wood, P.J., Smith, J.P., and Henman, G.S. 1984. A guide to species selection for tropical and sub-tropical plantations. Oxford, Commonwealth Forestry Institute, Tropical Forestry Paper No.15, 256p.

White, T. 1956. The scope of vegetable tannin chemistry. In: Chemistry of vegetable tannins — a symposium. Croydon (U.K.), Society of Leather Trades Chemists, p.7-29.

Williams, W.T. 1976. Pattern analysis in agricultural science. Amsterdam, Elsevier Scientific Publishing Company.

Wilson, A.D., and Harrington, G.N. 1980. In: Le Houerou, H.N., ed., Browse in Africa: the current state of knowledge. Papers presented at the international symposium on browse in Africa, Addis Ababa, 8-12 April 1980, and other submissions. International Livestock Centre for Africa. 491p.

Wissing, A. 1955. The utilisation of bark, Part II. Svensk Papperstidning, 50, 745-750.

Yates, F., and Cochran, W.G. 1938. The analysis of groups of experiments. Journal of Agricultural Science, 28, 556-580.

Yost, R.S., DeBell, D.S., Whitesell, C.D., and Miyasaka, S.C. 1987. Early growth and nutrient status of *Eucalyptus saligna* as affected by nitrogen and phosphorus fertilisation. Australian Forest Research, 17, 203-214.

Zhang, Z., Lopez, M.F., and Torrey, J.G. 1984. A comparison of the cultural characteristics and infectivity of *Frankia* isolates from root nodules of *Casuarina* species. Plant and Soil, 78, 79-90.

Zhang, Z., Murry, M.A., and Torrey, J.G. 1986. Culture conditions influencing growth and nitrogen fixation in *Frankia* sp. HFPCc13 isolated from *Casuarina*. Plant and Soil, 91, 3-15.

Zheng, G., and Yazaki, Y. 1988. Extractives yields and polyflavinoid contents of *Acacia mearnsii* barks in Australia. Australian Forest Research, in press.

Zummer-Linder, M. 1983. Re-assessment of the Turkana forestry project. A report prepared for Kenya Forestry Department, Forest Research Organization, K.A.R.I., and Norwegian Agency for International Development, 60p.

Chapter 26

List of Publications for ACIAR Forestry Projects

(Covering journals, workshop proceedings, newsletter items, etc.)

Anon 1986. Proceedings of a workshop on seed handling and Eucalypt taxonomy. Held 8–12 July 1985, Harare, Zimbabwe. Produced by IDRC/ACIAR/ZFC.

1988. Black wattle cultivation and use of tannin extracts. *Forest Research* Vol. 1 No. 1 (CAF). (In Chinese language.)

Applegate, G.B. and Nicholson, D.I. 1987. Growth rate of selected Acacia species in north and southeast Queensland, Australia. p86–88. In: Turnbull, J.W. ed. Australian Acacias in developing countries, proceedings of an international workshop held at Gympie, Qld, Australia, 4–7 August 1986. ACIAR Proceedings No. 16, 196p.

Aswathappa, N. 1987. *Sesbania formosa* for waterlogged-saline soils. ACIAR Forestry Newsletter, 4, 3.

Aswathappa, N. and Marcar, N.E. 1988. Evaluation of 60 Australian tropical/subtropical tree species for salt tolerance. (In press). *Aust. J. For. Res.*

Aswathappa, N., Marcar, N.E. and Thomson, L.A.J. 1986. Screening Australian trees for salt tolerance. ACIAR Forestry Newsletter No. 2 p.3.

1987. Salt tolerance of Australian tropical and subtropical acacias. Paper, Acacia workshop 4–7 August 1986, Gympie, Australia.

1987. Screening of tropical and subtropical Australian tree species for salt tolerance. Paper, International Conference on afforestation of salt-affected soils, 16–20 February, 1987, Karnal, India. Vol. 2, 1-15. Plant Resources.

Boland, D.J. 1986. Selection of lesser-known Australian tree species for fuelwood and agroforestry — the ACIAR experience. Paper, ICRAF/ICAR workshop on agroforestry, 16 Oct.-1 Nov., Hyderabad, India.

1986. Testing Australian trees for fuelwood and agroforestry. ACIAR Forestry Newsletter No. 1, 1-2.

1986. Selection of species and provenances for tree introduction. 45-57 Chap 3. In: Multipurpose Australian trees and shrubs — lesser-known species for fuelwood and agroforestry. ACIAR Monograph No. 1.

1987. Genetic resources and utilization of Australian bipinnate acacias — section Botrycephalae. Paper, Acacia workshop, 4–7 August 1986, Gympie, Queensland.

1987. Essential leaf oils — good oil from trees. ACIAR Forestry Newsletter No. 4, p.2.

1989. Realising the potential of Australia's lesser-known trees and shrubs: a summary and future perspectives. (In this Monograph)

Boland, D.J. and Pinyopusarerk, K. 1987. Early growth and survival of some eucalypts and Australian tree species planted at Tung Kula Ronghai T.K.R. development project in northeastern Thailand. (Paper presented to ACIAR/ADAB/DLD/TKR seminar on Salinity in northeast Thailand 21–23 September 1987, Roi Et, Thailand.). *Thai J. For.* 6, 250–267.

Boland, D.J. and Turnbull, J.W. 1989. Australian tree species for fuelwood and agroforestry in China, Kenya, Thailand and Zimbabwe. (In this Monograph).

Boontawee, B. and Kuwalairat, P. 1987. Introduction of *Acacia mangium* to Thailand. Paper, Acacia workshop, 4–7 August 1986, Gympie, Australia.

Booth, T.H. 1985. A new method for assisting species selection. *Commonw. For. Rev.* 64, 241-250.

1987. Selecting *Acacia* species for testing outside Australia. Paper, Acacia workshop, 4-7 August 1986, Gympie, Australia.

1987. Recent developments in homoclime analyses to assist species selection. ICRAF/WMO Conference on Application of meteorology to agroforestry systems, planning and management. Held 9-13 February 1987, Nairobi, Kenya. (In press).

1988. Climatology of *Acacia mearnsii*. 2. Homoclime analysis of potential trial sites in China. *New Forests* 2, 31-40.

1988. Which wattle where? Selecting Australian acacias for fuelwood plantations. *Plants Today* 1, 86-90.

1989. Climatic conditions at trial sites in China, Kenya, Thailand and Zimbabwe compared to similar regions in Australia. (In this Monograph).

Booth, T.H. and Jovanovic T. 1988. Climatology of *Acacia mearnsii*. 1. Characteristics of natural sites and exotic plantations. *New Forests* 2, 17–30.

1988. Assaying natural climatic variability in some Australian species with fuelwood and agroforestry potential. *Common. For. Rev.* 67, 27-34.

Booth, T.H., Nix, H.A., Hutchinson, M.F. and Busby, H.J.R. 1987. Grid matching: a new method for homoclime analysis. *Agric. For. Meteorol.* 39, 241-255.

Booth, T.H., Nix, H.A. Hutchinson, M.F., and Jovanovic, J. 1988. Niche analysis and tree species introduction. *Forest Ecology & Management.* 23, 47-59.

Booth, T.H. Searle, S.D., and Boland, D.J. 1988. Selecting provenances of tree species for trials: a case study using *Acacia mearnsii* de Wild. Submitted to *New Forests.*

Booth, T.H., Stein, J.A., Nix, H.A. and Hutchinson, M.F. 1988. Mapping regions in Africa climatically suitable for particular species. *For. Ecol. Management* (In press).

Brophy, J.J. and Boland, D.J. 1989. Leaf essential oil of *Eucalyptus bakeri.* (In this Monograph.)

Brophy, J.J., Boland, D.J. and Lassak, E.V. 1989. Leaf essential oils of *Melaleuca* and *Leptospermum* species from tropical Australia. (In this Monograph.)

1988. *Melaleuca leucadendra* L. leaf oil: two phenylpropanoid chemotypes. *Flavour and Fragrance* 3, 43-46.

Brophy, J.J., Lassak, E.V. and Boland, D.J. 1987. Volatile leaf oils of the two subspecies of *Melaleuca acacioides* F. Muell, *J. Proc, Roy. Soc. N.S.W.* 120, 135-39.

1988. Steam volatile leaf oils of six northern Australian broadleaf *Melaleuca. J. Proc. Roy. Soc. N.S.W.*121, 29-33.

1989. Steam volatile leaf oils of *Melaleuca symphyocarpa* and *M. uncinnata. Flavour and Fragrance* (submitted).

Brown, A.G., 1988. Database applications in forestry research — an Australian experience. Paper, workshop on Database management applications in forestry research; CATIE, Turrialba, Costa Rica 20-21 June 1988. (Proceedings in press).

1988. Data analysis, management and sharing. Paper to workshop. Use of Australian trees in China, 2-11 October 1988, Guangzhou, China.

Brown, A.G. and Turnbull, J.W. 1986. The Australian environment. 11-27. Chap. 1. In: Multipurpose Australian trees and shrubs: lesser-known species for fuelwood and agroforestry. ACIAR Monograph No. 1.

Brown, A.G. Wolf, L., and Ryan, P. 1986. New tree crop database. ACIAR Newsletter No. 2, p.2.

Brown, A.G., Wolf, L., Ryan, P.A. and Voller, P. 1988. TREDAT: a tree crop database. (Submitted to Australian Forestry).

Cromer, R. 1989. Response of Australian tree species to nitrogen and phosphorus in Thailand. (In this Monograph.)

Gao Chuanbi. 1989. *Acacia mearnsii* provenance trials in the People's Republic of China. (In this Monograph.)

Gardner, W.D. 1989. Fuelwood evaluation using a simple crib test. (In this Monograph.)

Gough, D.K., Bell, R.E., Ryan, P.A. and Bragg, C.T. 1989. Drying and burning properties of the wood of some Australian tree species. (In this Monograph.)

Groves, K. and Chivuya, A.M. 1989. Fuelwood evaluation of four Australian-grown tree species. (In this Monograph.)

Gwaze, D. 1986. Status of Australian acacias in Zimbabwe. Paper, Acacia workshop, 4-7 August, Gympie, Australia.

1989. Growth and survival of Australian tree species in field trials in Zimbabwe. (In this Monograph.)

Gwaze, D. and Stewart, H. 1987. Use of poles in communal areas in Zimbabwe. ACIAR Forestry Newsletter No. 3, p.3.

Hillis, W.E. 1989. *Acacia mearnsii*: its past and potential use with reference to the development of plantations in the People's Republic of China. (In this Monograph.)

Kaumi, S.Y.S. 1987. Australian acacias in Kenya. Paper, Acacia workshop, 4-7 August 1986, Gympie, Australia.

Luyt, I.E., Mullin, L.J. and Gwaze, D. 1987. Black wattle (*Acacia mearnsii*) in Zimbabwe. Paper, Acacia workshop, 4-7 August 1986, Gympie, Australia.

Matheson, C. 1987. Tree experiments on computer in China. ACIAR Forestry Newsletter No. 3, 1 and 4.

Matheson, A.C. and Mullin, L.J. 1987. Variation among neighbouring and distant provenances of *Eucalyptus grandis* and *E. tereticornis* in Zimbabwean field trials. *Aust. For. Res.* 17, 233-50.

Milimo, P. 1989. Growth and survival of Australian tree species in field trials in Kenya. (In this Monograph.)

Mitchell, M. 1989. Susceptibility to termite attack of various tree species planted in Zimbabwe. (In this Monograph.)

Mitchell, M., Gwaze, D. and Stewart, H. 1986. Termite susceptibility of Australian trees in Zimbabwe. ACIAR Forestry Newsletter No. 2, p.2.

Pan Zhigang, and Yang Minquan. 1987. Australian acacias in the People's Republic of China. Paper, Acacia Workshop, 4-7 August 1986, Gympie, Australia.

Pinyopusarerk, K. 1987. Progress and prospects for improving *Acacia auriculiformis* through selection and breeding in Thailand. Paper, Acacia workshop, 4-7 August 1986, Gympie, Australia.

1989. Growth and survival of Australian tree species in field trials in Thailand. (In this Monograph.)

Pinyopusarerk, K. and Boland, D.J. 1987. Assessment of early growth of some Australian *Acacia* and *Casuarina* tree species in field trials in Thailand. Paper presented to symposium The contribution of biological nitrogen fixation to plant production. Held 3-7 August 1987, Cisarua, Indonesia.

Pinyopusarerk, K., and Puriyakorn, B. 1987. *Acacia* species and provenance trials in Thailand. Paper, Acacia workshop, 4-7 August 1986, Gympie, Australia.

Pryor, L. 1989. Vegetative propagation of casuarinas and acacias: potential for success. (In this Monograph.)

Queensland Department of Forestry 1987. Australian trees for fuelwood and agroforestry. Research Report 1985, 14-16.

Raymond, C.A. 1987. *Acacia mearnsii*: a breeding plan for China. Unpublished ACIAR report, Canberra, Australia.

1988. Developing a breeding strategy for *Acacia mearnsii.* Meeting, Research Working Group Number 1, Gympie, Qld, 31 Oct-4 Nov 1988.

1988. *Eucalyptus globulus*: a breeding plan for China. Unpublished ACIAR report, Hobart, Australia.

Reddell, P.L., Rosbrook, P.A. and Ryan, P. 1989. Managing nitrogen fixation in *Casuarina* species to increase productivity. (In this Monograph.)

Ryan, P.A. 1986. Lesser-known Australian trees tested in Queensland. ACIAR Forestry Newsletter No. 1, January 1986.

Ryan, P.A., and Bell, R.E. 1988. Acacias in the ACIAR species/provenance trials in southeast Queensland — a brief outline. Paper presented to a meeting of Research Working Group 1 (Forest Tree Breeding) at Gympie 31 October to 4 November 1988. 7p.

1989. Growth, coppicing and flowering of Australian tree species in trials in southeast Queensland, Australia. (In this Monograph.)

Ryan, P.A., Podberscek, M., Raddatz, C.G. and Taylor, D.W. 1987. Acacia species trials in southeast Queensland. 81–85 In: Turnbull, J.W. ed. Australian Acacias in developing countries, proceedings of an international workshop held at Gympie, Qld., Australia, 4–7 August 1986. ACIAR Proceedings No. 16, 196p.

Searle, S. 1987. Seed collection in northern Australia. ACIAR Forestry Newsletter No. 3, p.4.

1989. Seed collections of lesser-known trees and shrubs in Queensland, Australia. (In this Monograph.)

Turnbull, J.W. 1986. Multipurpose Australian trees and shrubs: lesser-known species for fuelwood and agroforestry. ACIAR Monograph No. 1, 316p.

Vercoe, T.K. 1987. Fodder potential of selected Australian tree species. Paper, Acacia workshop, 4–7 August 1986, Gympie, Australia.

1989. Fodder value of selected Australian tree and shrub species. (In this Monograph.)

Wang Huoran 1987. Chinese translation of 'A classification of the eucalypts' by Pryor, L.D. and Johnson, L.A.S.

1988. Transfer and cultivation of Australian hardwoods. *Forest Research* Vol. 1, No. 1 (CAF). (In Chinese language.)

Wang Huoran, Yan Hong and Zhang Rongqui 1989. Temperate eucalypt trials in southwest People's Republic of China. (In this Monograph.)

Williams, E.R. 1987. Genstat 4.03 E in China. Genstat Newsletter No. 19, 4.

Williams, E.R., and Luangviriyasaeng, V. 1989. Statistical analysis of tree species trials and seedlot: site interaction in Thailand. (In this Monograph.)

Yang Minquan, Bai Jiayu and Zeng Yutian. 1989. Tropical Australian acacia trials on Hainan Island, People's Republic of China. (In this Monograph.)

Zheng, G., Lin, Y., and Yazaki, Y. 1988. Molecular size distribution of tannin extracts from *Acacia mearnsii* bark comparing from different countries. *Holzforschung* (In press).

Zhou Wenlong and Bai Jiayu. 1989. Tropical eucalypt trials on Hainan Island, People's Republic of China. (In this Monograph.)